DATE DUE

CLIMATOLOGY

This book is produced in full compliance with the government's regulations for conserving paper and other essential materials.

CLIMATOLOGY

BY

BERNHARD HAURWITZ, Ph. D.

AND

JAMES M. AUSTIN, Sc. D.

Associate Professors of Meteorology, Massachusetts Institute of Technology

First Edition

McGRAW-HILL BOOK COMPANY, INC.

NEW YORK AND LONDON

1944

CLIMATOLOGY

PRINTED IN THE UNITED STATES OF AMERICA

THE MAPLE PRESS COMPANY, YORK, PA.

PREFACE

The aim of this book is to provide an introduction to climatology for the student of meteorology. Hence, the authors have stressed the physical causes of the climates and of the variations of the climatic elements in space and time. Since the book is written primarily for meteorologists, the description of the various climates is given in meteorological language, emphasis being placed on the interpretation in terms of dynamics of the atmosphere, of air mass types, and of frontal activity.

It is assumed that the reader is familiar with the basic concepts of meteorology, such as cooling by adiabatic lifting and the relation between the fields of pressure and wind in balanced motion. Wherever it seemed desirable to discuss a meteorological principle in its climatic effects, this has been done in connection with the phenomenon to be explained rather than in a separate section.

In view of the scope of the book a discussion of the research methods by which the climatologist prepares, analyzes, and presents his data has been omitted. For the same reason no mention, except incidentally, is made of the many applications of climatology in agriculture, industry, merchandising, transportation, medicine, and other fields.

Much of the material has been presented to students taking the meteorological course for Aviation Cadets during a thirty-hour course on climatology at the Massachusetts Institute of Technology. The present book, however, is intended rather for a two-term course in climatology which would naturally be divided into general and regional climatology. It is hoped that in addition to serving as a textbook, the book may also prove useful for private study by readers who want a general survey of the field of climatology or who desire information about the climate of a definite region.

The data for the tables, maps, and diagrams have been taken from a variety of sources, especially from the "Handbuch der Klimatologie" by W. Köppen and R. Geiger (Verlags. Gebr.

Bornträger, Berlin). The authors are keenly aware of the fact that a large amount of climatological material is now being accumulated and brought into a suitable form so that climatological information will be greatly enlarged and doubtless modified in many instances after the war when these data become generally accessible. The increase of observations is particularly conspicuous in the field of upper-air data, and the climatology of the upper atmosphere will expand considerably in the future.

Wherever practicable the authors have stated the authority on which the various charts are based. In all cases, however, revisions arising out of more recent data have been made on the existing charts. The Mercator projection has been used for the maps in spite of its disadvantages. This projection permits one to represent the whole globe without interruption on one map, except for the polar regions. While the Mercator projection shares this property with some others, such as Mollweide's homolographic projection, the latter distorts shapes too much near the margins for the purposes of this book. The reader is advised to have a good atlas at hand while studying this book. This is not only desirable in order to correct any wrong ideas of sizes and distances, but also, and perhaps mainly, in order to give a clear picture of the geographical conditions of the places and regions under discussion.

Considerable space has been given to Köppen's classification of climates because such an arrangement is of great assistance for an introduction into the variety of terrestrial climates. Moreover, it provides a means of expressing quickly some of the most important features of a climate.

The authors wish to express their appreciation to Dr. S. Petterssen for permission to reproduce Figs. 13 and 15; to Dr. H. U. Sverdrup and collaborators and Prentice-Hall, Inc., New York, for permission to reproduce, in modified form, Plates VI–VIII; to Dr. C. W. Thornthwaite for permission to reproduce Fig. 25; to Messrs. W. E. K. Middleton and G. F. Millar, Toronto, for permission to reproduce Fig. 27; and to Dr. H. R. Byers and Dr. V. P. Starr to use Fig. 12. In a number of cases it has been impossible, owing to the war, to obtain the author's permission for reproduction of a figure. In all these instances the diagram appeared first in a research paper, and proper reference to this paper has always been made.

The authors' thanks are due to Miss Dorothy Green and Miss Margaret Whitcomb who assisted in the preparation of the plates and to Mrs. Haurwitz and Mrs. Austin for their great aid in preparing the manuscript for the printer and in proofreading. To Professor H. G. Houghton, Executive Officer of the Department of Meteorology, the authors are indebted for his interest in the progress of this work.

B. HAURWITZ,
J. M. AUSTIN.

CAMBRIDGE, MASSACHUSETTS,
August, 1944.

CONTENTS

PART A

GENERAL CLIMATOLOGY

PART B

REGIONAL CLIMATOLOGY

PLATES

(Inside Back Cover)

CLIMATOLOGY

INTRODUCTION

The weather at a given place is the sum total of all the meteorological variables, such as temperature, pressure, wind, humidity, precipitation, and the state of the sky at a given time. A study of the simultaneous weather conditions over a large area and their development leads to *synoptic meteorology*. Even though the changes of the weather are proverbial, it is nevertheless possible at every place to arrive at a generalization and a composite of these variations. One speaks then of the *climate of the region*.

The science that discusses the climates found on the earth is called "climatology." It is obviously closely related to meteorology, and also to geography, since the atmosphere is strongly influenced by the physical nature of the earth's surface. Moreover, the simplest method of arranging a description of the various terrestrial climates is to enumerate the climatic conditions according to geographical units, that is, to describe the climates of the different continents and oceans. Owing to the intermediate position of climatology between meteorology and geography, climatological discussions stress either the meteorological, physical, and dynamical aspects or the descriptive, geographical viewpoint. Here the meteorological viewpoint will be emphasized.

It should be pointed out at the start that the climate of a place is not determined only by the annual means of the meteorological elements. For example, the temperatures of Boston and Edinburgh may be considered. Both cities have approximately the same annual mean temperature, 48.8°F and 47.6°F, respectively. But the highest and lowest monthly mean temperatures are at Boston 72.0°F and 27.8°F, at Edinburgh 58.8°F and 39.0°F. The annual range of temperature is much greater at Boston than at Edinburgh. The example shows clearly that the seasonal variation of the temperature is an important climatological quantity.

Another point is illustrated by Table 1. At Cairo, Egypt, and at Galveston, Tex., not only the annual mean temperatures but also the temperatures of the different months are very

TABLE 1.—TEMPERATURE AND PRECIPITATION AT TWO SELECTED STATIONS

Place	Temperature, deg F						Annual precipitation, in.
	Jan.	Apr.	July	Oct.	Year	Range	
Cairo, Egypt	52.7	67.6	81.0	71.8	68.2	28.3	1.3
Galveston, Tex	53.6	68.5	82.9	72.2	69.4	29.3	46.1

much alike. Nevertheless, the climate of the two cities is, of course, very different as shown by the annual precipitation. This example demonstrates that in comparing different climates all the main climatic elements have to be considered.

In order to prove that the climates of Cairo and Galveston, Tex., are entirely different another element could have been chosen instead of the annual precipitation. However, temperature and precipitation are the two most important climatological variables owing to their practical significance.

A complete description of the climate of a region should comprise not only the means, and the seasonal and diurnal variations of the climatic elements, but also information about their extremes and about the frequencies with which the various values occur. However, it is not possible to give the data in such completeness in this book since they are only rarely available in this form and since it would obviously be impossible to reproduce and discuss all this information in a volume of reasonable size.

The object of the first part of the book is to present a general survey of the various meteorological elements, their distribution over the globe, and their seasonal and diurnal variations. At the same time the factors that determine the distribution and variation of these elements will be analyzed. Finally, in order to arrive at a synthesis of the meteorological elements giving a picture of the different climates of the earth, the climatic types as introduced by Köppen will be discussed. This part is called General Climatology.

In the second part the special characteristics of each continent and of the oceans will be examined in greater detail. Besides the conventional explanation of the climates by means of the geographical features, an attempt has been made to describe the distribution and variation of the meteorological elements in terms of frontal and air-mass concepts. This part is called Regional Climatology.

PART A

General Climatology

CHAPTER I

SOLAR RADIATION AND TERRESTRIAL HEAT BALANCE

Practically all the energy that produces and maintains motions and variations in the earth's atmosphere comes from the sun in the form of radiation. The amounts of energy received from the stars and from the interior of the earth are entirely negligible. In view of this importance of the incoming solar radiation, its intensity and distribution over the earth and its variation with the seasons will be discussed first.

1. The Incoming Solar Radiation in the Absence of the Atmosphere. To obtain an orientation about the distribution of the solar radiation over the earth's surface it is convenient to disregard at first the effect of the atmosphere. In the absence of the atmosphere the intensity of the solar radiation received on a surface perpendicular to the solar beam is 1.94 cal/cm^2/min provided that the earth is at its mean distance of 93 million miles from the sun. This figure is called the solar "constant" although the energy output of the sun is not strictly invariable as indicated by the observed variations on the sun and by radiation measurements on the earth.

There is also a slight variation of the solar energy received at the earth owing to the variation of the distance between the earth and the sun. Shortly after the winter solstice of the Northern Hemisphere the distance is smallest, and the energy received at the outer limit of the atmosphere is 2.01 cal/cm^2/min. About the summer solstice of the Northern Hemisphere, when the distance between the sun and the earth is largest, the amount received at the outer limit of the atmosphere falls to 1.88 cal/cm^2/min. But these variations have very little effect on the seasonal changes as is indicated by the fact that the Northern Hemisphere has its summer when the earth is farthest away from the sun and its winter when the earth is closest to the sun.

The total solar radiation received in the absence of the atmosphere at the earth's surface is the product of the solar constant and the cross section of the earth. To find the mean value per cm^2 of the solar radiation over the whole earth this product has to be divided by the area of the earth. Since the area of a sphere is four times larger than its cross section, $1.94/4 = 0.485$ cal/cm^2/min represents the average amount of solar radiation that would be received on a horizontal plane at the earth's surface in the absence of the atmosphere.

The computation of the actual amount of solar radiation received at a given point of the earth's surface at a given time in the absence of the atmosphere is an astronomical rather than a climatological problem. Extensive calculations bearing on this problem have been carried out by Milankovitch,[1] and some of his results are represented in the following figures. Figure 1 shows the amount of solar radiation received at different latitudes during the winter half of the year, during the summer half of the year, and during the whole year. It is based on Milankovitch's computations, but the value for the solar constant has been reduced from 2 cal/cm^2/min, which was used by Milankovitch, to 1.94 cal/cm^2/min. The curves of Fig. 1 hold for the Northern Hemisphere as well as for the Southern Hemisphere because the greater length of the northern summer as compared with the southern summer is compensated by the greater distance between the earth and the sun. The summer half of the year of the Northern Hemisphere is reckoned from the spring to the fall equinox; the other seasons are limited correspondingly. The yearly total of the solar radiation is largest at the equator and smallest at the pole. For the summer half of the year the maximum is displaced from the equator about 23° latitude toward the pole of the summer hemisphere.[2]

[1] Milankovitch, M., Mathematische Klimalehre, in Köppen-Geiger, "Handbuch der Klimatologie," Vol. 1, A, Verlagsbuchhandlung Gebrüder Borntraeger, Berlin, 1930.

[2] By "summer hemisphere" is meant the hemisphere that has summer. Thus the Northern Hemisphere would be the summer hemisphere during July, and the Southern Hemisphere would be the summer hemisphere during January.

During the winter half of the year the total amount of incoming radiation decreases from the equator to the pole. At the pole itself the radiation is zero. It should be noted that the variation of the incoming solar radiation with the geographical latitude is

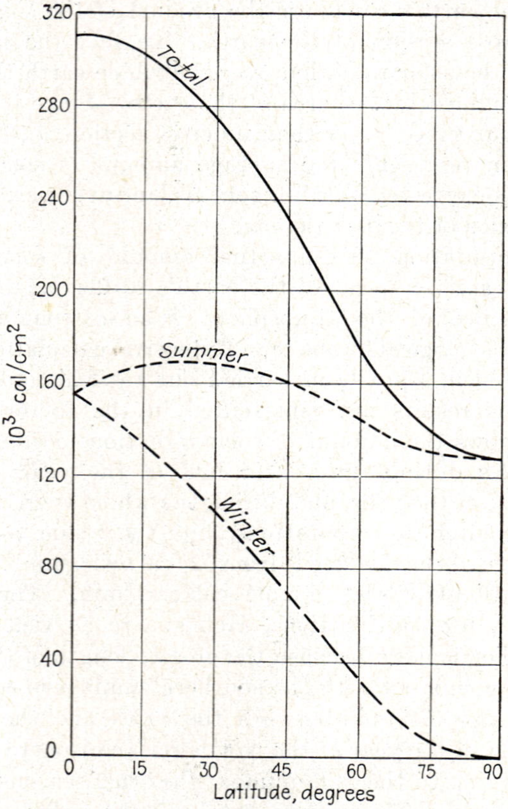

FIG. 1.—Radiation received at the ground per cm² in the absence of the atmosphere during the summer and winter half year and during the whole year. (*After Milankovitch.*)

much greater in the winter hemisphere than in the summer hemisphere.

The three curves for the solar radiation received during summer, winter, and the whole year show clearly its dependence on the latitude, which in its turn is, of course, due to the variation of the zenith distance of the sun with the latitude. As a result, the temperature in particular, but also all the other climatic elements show a dependence on the latitude. The latitude is the most important factor that determines the climate of a locality.

A more detailed representation of the geographical and seasonal distribution of the incoming solar radiation in the absence of the earth's atmosphere is given in Fig. 2, which shows the amount of solar radiation received at the earth's surface in cal per cm² and per day on a horizontal plane at different latitudes and at different times of the year. The amount of solar radiation is shown by lines of equal insolation.

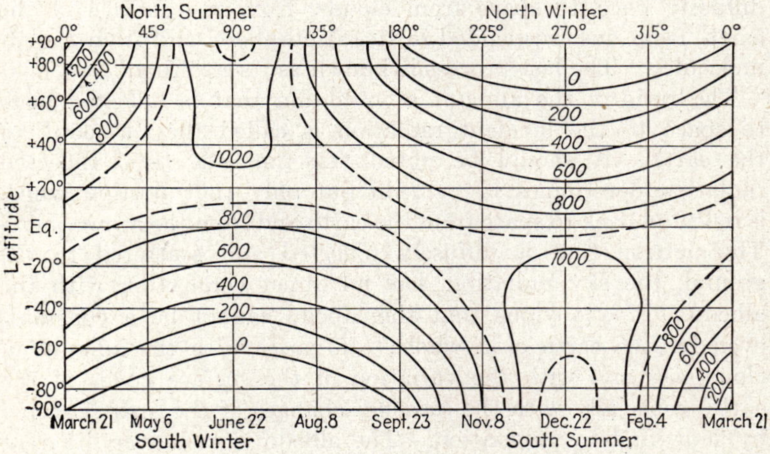

FIG. 2.—Daily insolation in cal per cm² received at the ground in the absence of the atmosphere. (*After Milankovitch.*)

Strong maxima of insolation are found in the polar regions about the time of the summer solstice in the hemisphere that has summer. In the winter hemisphere, on the other hand, the sun does not rise at all, so that there is a region without insolation. The maximum of insolation around the December solstice in the Southern Hemisphere is larger than the maximum around the June solstice in the Northern Hemisphere, owing to the smaller distance of the earth from the sun during December. At the summer solstice a secondary insolation maximum is found in the zone between 40° and 50° latitude in addition to the main maximum at the pole. The secondary maximum is indicated in Fig. 2 by the slight bulges of the 1000 cal isopleths (see also Table 3, part *A*).

In most parts of the earth the annual variation of insolation has one maximum. Only between the tropics of Cancer and Capricorn do two maxima occur throughout the year. They are reached when the sun is directly overhead, at the equator, for example, at the times of the equinoxes.

2. Depletion of the Incoming Solar Radiation by the Earth's Atmosphere. Owing to the presence of the earth's atmosphere the amount of solar radiation actually received at the ground differs considerably from the values presented in the preceding section. A part of the incoming solar radiation is reflected diffusely back to space from clouds, from the surface of the earth, from dust suspended in the atmosphere, and from the air molecules. The last process is known as "scattering."

The ratio of the amount of radiation that is reflected back to space to the incident radiation is called the "albedo" of the earth. It should be noted that not the total reflected radiation determines the albedo but only that portion which is reflected back to space from the earth and from the atmosphere- The portion that is diffusely reflected and scattered to the ground, the sky radiation, has no direct connection with the albedo. It is obvious that the albedo cannot be a constant, even for the earth as a whole. It varies with the amount of cloudiness and with the variation of the surface cover of the earth, especially with the seasonal changes of the areas covered by snow and by vegetation. The albedo is moreover different for different spectral regions. A figure of 42 per cent may here be adopted as a representative mean value for the earth as a whole. However, this figure cannot be regarded as final, especially since the differences in albedo of different types of clouds have not yet been sufficiently well studied to take them into account with the necessary accuracy.

The amount of solar radiation arriving at the surface is further reduced by absorption in the atmosphere, especially by water vapor, by carbon dioxide, and, at higher levels, by ozone. But the absorbed amount of solar radiation is not lost to the terrestrial heat balance, since it serves to raise the temperature of the absorbing layers and hence to increase their radiation.

If I_0 denotes the solar constant, the solar radiation I_S received by the unit area perpendicular to the sun's rays at the ground with the sun at the zenith is a certain fraction q of the solar constant,

$$I_S = qI_0$$

The quantity q is called the "transmission coefficient." If the zenith distance of the sun is, for example, 60°, the path of the light through the atmosphere is twice as long as with vertical incidence and q^2 has to be substituted for q in the preceding equation. For an arbitrary zenith distance ζ the path through the atmosphere increases as the secant of ζ so that, in general,

$$I_S = q^{\sec\zeta} I_0$$

except when the zenith distance is very nearly 90°. Then a more complicated expression has to be substituted for sec ζ.

The transmission coefficient is a function of the wave length. But for an approximation in climatological problems a uniform transmission coefficient is often assumed for the incoming solar radiation. The transmission coefficient represents a measure of the atmospheric turbidity. Even a dry and pure atmosphere does not transmit all the incoming solar radiation to the ground owing to the scattering by air molecules to which reference was made previously. Instead of using the transmission coefficient as an expression for the turbidity, it is more customary to give the number of pure dry atmospheres that would by their scattering produce the same depletion of the solar radiation as the actual atmosphere containing water vapor and dust. This measure of atmospheric turbidity, which was introduced by Linke, is called the "turbidity factor."[1] Since it depends mainly on the water-vapor content and on the dust content of the air it shows large unperiodic changes, which are determined by the type of air mass and by the stability of the air. There are also seasonal changes as shown in Table 2, which represents an excerpt from a larger compilation given by Steinhauser.[2] It might be expected *a priori* that the yearly variation of the turbidity factor reflects

[1] The above definition is not strictly correct but entirely adequate for the purpose of this book. For details see, for example, H. Wexler, *Trans. Am. Geophys. Union*, 14*th Ann. Meeting*, 1933. A theoretically more satisfactory measure of atmospheric turbidity than Linke's factor is Ångström's turbidity coefficient (*Geog. Annaler* **11,** 156, 1929), but this quantity seems too elaborate for general use in climatological studies.

[2] STEINHAUSER, F., *Gerl. Beitr. Geophys.*, **42,** 110, 1934.

the yearly variation of the atmospheric water-vapor content, which is higher in summer than in winter. As Table 2 shows, this supposition is correct for all the stations except for Batavia, which has no warm and cold seasons, owing to its tropical position. Here the maximum occurs in spring (September–November) at the end of the dry season. The smaller water-vapor content during this season may be more than compensated by the greater dust content at the same time so that the turbidity is actually greater than during the rainy season. Similar conditions prevail at many other locations. It is therefore difficult to anticipate for any station the annual range of the turbidity factor without actual measurements of the solar radiation. The

TABLE 2.—SEASONAL* VARIATION OF THE TURBIDITY FACTOR
(After Steinhauser)

Station	Height, m	Lat.	Long.	Winter	Spring	Summer	Autumn	Year
Treurenburg...	9	80°N	17°E			2.2		
Irkutsk........	474	52°N	104°E	2.3	2.6	3.3	2.4	2.65
Washington....	127	39°N	77°W	2.8	3.6	4.2	3.2	3.43
Batavia........	8	6°S	161°E	4.0	5.2	3.6	3.7	4.15
Rio de Janeiro..	18	23°S	43°W	3.3	3.6	3.8	3.4	3.52

* Winter in this table comprises in the Northern Hemisphere the months December, January, February; in the Southern Hemisphere, June, July, and August.

magnitude of the turbidity factor depends furthermore largely on the position of the observing station with respect to industrial centers in whose vicinity the amount of solar radiation reaching the ground is considerably reduced by smoke.

The variability of the geographical distribution of the turbidity factor shown in Table 2 reflects the variability of the solar radiation reaching the earth's surface. In order to plot the geographical distribution of the average solar radiation arriving at the ground during a given period of time, say a month, a season, or a year, it is necessary to have observations from a sufficient number of points well distributed over the globe. H. H. Kimball[1] has made an attempt to give such maps of the actual distribution of the sun's radiation, at least for the oceans where the conditions should be fairly uniform. Over land the

[1] KIMBALL, H. H., *Monthly Weather Rev.*, **56**, 393, 1928.

radiation seems to increase toward the interior of the continents, other things being equal, owing to the decrease of the atmospheric water-vapor content inland. But these observations are still too scanty and too unevenly distributed to give a reliable picture of the actual geographic distribution.

Baur and Philipps[1] have computed mean values of the zonal distribution of the amount of solar radiation arriving at the surface in the Northern Hemisphere with the aid of the relatively few available mean values of the turbidity factor, taking into account also the cloudiness. The results of their computations are summarized in Table 3, part *B*. The first part (*A*) of the

TABLE 3.—AVERAGE AMOUNT OF DIRECT SOLAR RADIATION, CAL/CM²/MIN
(After Baur and Philipps)

Date	0°–10°	10°–20°	20°–30°	30°–40°	40°–50°	50°–60°	60°–90°N
A. Received at the upper limit of the atmosphere							
Dec. 21.....	0.549	0.465	0.373	0.274	0.173	0.079	0.006
Mar. 21.....	0.619	0.601	0.563	0.509	0.441	0.358	0.211
June 21.....	0.579	0.629	0.664	0.684	0.689	0.683	0.703
Sept. 23.....	0.610	0.592	0.556	0.503	0.435	0.353	0.208
B. Received at the earth's surface if cloudiness and turbidity are considered							
Dec. 21.....	0.164	0.161	0.134	0.082	0.036	0.013	0.001
Mar. 21.....	0.191	0.224	0.206	0.161	0.116	0.096	0.055
June 21.....	0.144	0.170	0.216	0.233	0.183	0.159	0.133
Sept. 23.....	0.170	0.162	0.201	0.183	0.131	0.079	0.028

table gives in the same units the amount of solar radiation arriving at the top of the atmosphere in order to show clearly the effect of the atmosphere in depleting the solar radiation. Cloudiness, turbidity, and diffuse reflection in the atmosphere diminish considerably the amount of solar radiation received at the ground as is seen by a comparison of corresponding figures in the first and second part of the table. The polar maximum, which appeared at the summer solstice, has now disappeared, and the highest amount of radiation is found between 30° and 40° latitude. One might be inclined to attribute the absence

[1] BAUR, F., and PHILIPPS, H., *Gerl. Beitr. Geophys.*, **42**, 160, 1934.

of the polar maximum to the effect of atmospheric depletion of radiation, which should *a priori* be strongest at the pole where the path through the atmosphere is longest. In fact, if a uniform transmission coefficient is assumed for all latitudes and when the cloudiness is disregarded, the maximum at the pole disappears. But Baur and Philipps have found that owing to the smaller turbidity at higher latitudes (see small turbidity factor for Treurenburg, page 10) the polar regions would actually receive a larger amount of radiation than any other latitude at the summer solstice. Only the distribution of the cloudiness explains why the maximum radiation reaching the earth's surface is at 30° to 40° latitude. This latitude zone is in summer the region of the subtropical high-pressure belt where the cloudiness is small owing to the atmospheric subsidence (page 64). The effect of the low cloudiness in the subtropical high-pressure belt also makes itself felt at the equinoxes because even then the solar radiation arriving at the ground is largest not at the equator but at 10° to 20° north latitude during the spring equinox and at 20° to 30° during the autumn equinox. Similar conditions prevail presumably in the Southern Hemisphere, although the distribution of the radiation reaching the ground cannot be symmetrical with respect to the equator, owing to the unequal distribution of water and land, which produces differences in the distribution of turbidity and cloudiness.

In concluding the discussion of the incoming direct solar radiation it may be mentioned that there is a definite increase of the solar radiation with altitude since the amount of air through which the sun's rays have to pass decreases with elevation.

3. Diffuse Sky Radiation. The solar radiation that is scattered and diffusely reflected in the atmosphere is not completely lost in space. A considerable fraction arrives at the earth as diffuse sky radiation. It is by no means negligible compared with the direct solar radiation. When the altitude of the sun is sufficiently low, the diffuse radiation may even be stronger than the direct solar radiation. The relative importance of the diffuse sky radiation becomes still greater when the direct solar radiation is shielded by clouds. Table 4 after Ångström[1] shows the monthly sums of the solar and sky radiation measured at Stocksund, Sweden, at 59.4° north latitude. During 5 months of

[1] Ångström, A., *Meddel. Stat. Met.-Hydr. Anst.*, **4**, No. 3, 1928.

the year the diffuse sky radiation is here greater than the direct solar radiation.

For climatological investigations of the incoming radiation and its distribution over the earth it is helpful that the diffuse sky radiation can be related to the sunshine duration and the cloudiness by empirical formulas. While these relations cannot be applied to obtain data for any fixed instant, they permit a

TABLE 4.—SOLAR AND SKY RADIATION ON A HORIZONTAL SURFACE, CAL/CM², AT STOCKSUND, SWEDEN
(After Ångström)

Item	Jan.	Feb.	Mar.	Apr.	May	June	July	Aug.	Sept.	Oct.	Nov.	Dec.	Year
Sol. rad.	322	975	3,140	6,350	9,530	10,050	9,825	6,900	4,265	1,495	335	85	53,272
Sky rad.	543	1,225	2,335	2,340	2,920	3,225	2,375	2,480	2,175	1,650	1,095	550	22,913
$\frac{\text{Sol. rad.}}{\text{Sky rad.}}$	0.59	0.71	1.34	2.72	3.27	3.26	4.14	2.78	1.96	0.91	0.28	0.15	2.33

reasonably reliable determination of monthly or similar means of diffuse sky radiation when only cloud or sunshine data are available. Thus Kimball[1] in his discussion of the *monthly means* of the total radiation made use of the expression

$$R_C = R_0\,[0.29 + 0.71(1 - C)]$$

where R_C is the total radiation at cloudiness C, counted from 0.0 to 1.0, and R_0 the total radiation in the absence of clouds. It is, of course, necessary to keep in mind that this equation gives only approximate values and that the monthly sums of the radiation during any specific month may deviate greatly from the values obtained by the preceding expression.

Computations of the diffuse sky radiation received at the earth's surface were made by Baur and Philipps. Some of their results are compiled in Table 5. A comparison with Table 3 for the direct solar radiation shows again the importance of the diffuse radiation, especially at higher latitudes and in winter. But even in summer the diffuse sky radiation is always, at least, a very considerable fraction of the direct solar radiation. The seasonal variation is similar to that of the direct solar radiation arriving at the ground. Particularly striking is the maximum in the zone 60° to 90° north on June 21. In this zone the diffuse

[1] KIMBALL, *op. cit.*, **56**, 393, 1928.

sky radiation is greater than the direct solar radiation even at the summer solstice owing to the great cloudiness. The direct solar radiation is here greater than the diffuse sky radiation only in spring, as shown by the figures for Mar. 21.

TABLE 5.—AVERAGE AMOUNT OF THE DIFFUSE SKY RADIATION, CAL/CM^2/MIN
(After Baur and Philipps)

Date	0°–10°	10°–20°	20°–30°	30°–40°	40°–50°	50°–60°	60°–90°N
Dec. 21.....	0.091	0.079	0.066	0.052	0.034	0.016	0.001
Mar. 21.....	0.108	0.105	0.099	0.093	0.083	0.066	0.047
June 21.....	0.105	0.114	0.124	0.125	0.126	0.122	0.153
Sept. 21.....	0.107	0.104	0.097	0.091	0.081	0.065	0.048

4. General Survey of the Terrestrial Heat Balance. The earth and its atmosphere do not become progressively warmer or colder. Hence, the total amount of radiation received by the earth and by the atmosphere must be equal to the total amount of radiation emitted back into space, at least when the yearly totals for the whole earth are considered. For climatology, as well as for meteorology, the seasonal and geographical distribution of the incoming and the outgoing radiation is of fundamental importance. But for a first orientation it is useful to see how the different items of the terrestrial heat balance are distributed on the average for the whole year and for the whole earth. In the present state of our knowledge, the values given for the various items in the heat balance are not more than estimates, and these estimates vary considerably from author to author. Nevertheless, the following survey will at least list the different processes that affect the incoming and outgoing radiation and their relative importance.

It can only be claimed that the figures represent estimates for the Northern Hemisphere. The differences between the Northern and the Southern Hemisphere are, presumably, not large. But they must nevertheless be noticeable, even if the mean values for a whole hemisphere are considered, in view of the great differences between the physical features of both hemispheres. The numerical values quoted are largely those given by Baur and Philipps.[1]

[1] BAUR, F., and PHILIPPS, H., *Gerl. Beitr. Geophys.*, **42**, 160, 1934; **45**, 82, 1935.

According to page 5 the total amount of solar radiation received per square centimeter and per day on a horizontal surface is $1.94 \frac{1440}{4} = 700$ cal at the outer limit of the atmosphere. Of this amount, 27 per cent penetrates directly to the earth's surface, and 16 per cent arrives as diffuse sky radiation, so that 43 per cent reaches the ground altogether. Fifteen per cent is absorbed by the atmosphere, including clouds. The remaining 42 per cent is reflected back into space and is therefore lost for the terrestrial heat balance. This figure, 42 per cent, represents the albedo of the earth, which was discussed on

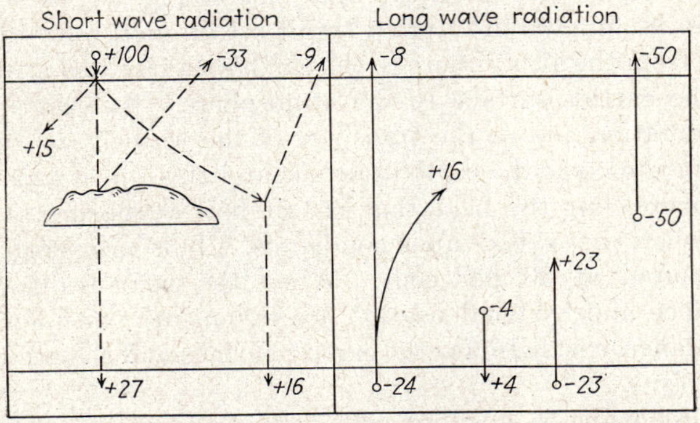

FIG. 3.—Mean heat balance in the Northern Hemisphere. The figures are percentages of the incoming solar radiation. (*After Baur and Philipps.*)

page 8. It is composed of the reflection on clouds and on the ground, which accounts for 33 per cent, and the diffuse reflection, which makes up the remaining 9 per cent. It will be noticed that the diffuse radiation toward the ground is considerably greater than that returned to space, the values being 16 per cent and 9 per cent, respectively. This difference is due to the fact that the larger dust particles scatter more radiation in the direction away from the sun than in the direction toward the sun. The distribution of the solar radiation over the various items is shown on the left side of Fig. 3. The radiation received by the earth and the atmosphere is counted positive; the radiation emitted or reflected and scattered to space, negative.

While 42 per cent of the incoming solar radiation is returned directly back to space, the remaining 58 per cent is absorbed by the ground and the atmosphere. This 58 per cent must be radiated back to space, since the yearly mean temperature of the earth as a whole remains the same. The radiation from the ground to space is called the "effective radiation." It amounts to 24 per cent of the 700 cal received at the top of the atmosphere. This 24 per cent represents the difference between the actual radiation from the ground and the radiation from the atmosphere to the ground. Of the 24 per cent that comprises the effective radiation from the ground 16 per cent is reabsorbed in the atmosphere, while 8 per cent returns directly to space. The other 50 per cent is radiated back to space by the atmosphere. These items are shown on the right side of Fig. 3.

There is, finally, an internal transport of energy by turbulent mixing from the atmosphere to the earth's surface and a transport from the earth's surface to the atmosphere. The latter is due to evaporation and to the transport of the water vapor upward. Evaporation cools the earth's surface and the lowest layer of the atmosphere, but the heat that is lost here reappears at higher levels when the water vapor condenses. This transport of heat is estimated at 23 per cent. When the heat transports by turbulence and by condensation are taken into consideration, a heat balance exists also for the earth's surface and the atmosphere separately.

The following is a summary of the separate heat balances of the earth and the atmosphere:

The surface of the earth

receives		loses	
by direct radiation	27%	by radiation	24%
by diffuse radiation	16%	by condensation (evaporation)	23%
by turbulent transfer	4%		
	47%		47%

The atmosphere

receives		loses	
by absorption of solar radiation	15%	by radiation	50%
by absorption of ground radiation	16%	by turbulent transfer	4%
by condensation	23%		
	54%		54%

To understand the effect of the atmosphere on the terrestrial heat balance, it is necessary to bear in mind the following facts about the terrestrial and solar radiation: Both the sun and the earth's surface can be regarded as perfect radiators (black bodies). Owing to the high temperature of the sun, 99 per cent of its radiative energy is contained between the wave lengths 0.17μ and 4μ, with a maximum at 0.475μ, in the visible part of the spectrum. On the other hand, 99 per cent of the radiative energy of a black body at a temperature of 300° abs is emitted between 3μ and 90μ, with a maximum at 10μ; 99 per cent of the radiative energy of a black body at 200° abs is in the wave-length region between 4μ and 120μ, with a maximum at 15μ. Thus, one may say that the radiation from the sun is of a much shorter wave length than the terrestrial radiation, since the latter is emitted mainly at temperatures between 200° and 300° abs. This difference between the wave lengths of the incoming solar radiation and of the outgoing terrestrial radiation is of very great practical importance. The absorptive power of the atmosphere, *i.e.*, principally of the water vapor and carbon dioxide, is small in the short-wave regions of the solar radiation but large in the wave lengths emitted by the earth and the atmosphere. Consequently, a large part of the terrestrial radiation is absorbed in the atmosphere instead of being radiated directly back to space. As a result of this selective absorption, the temperature of the atmosphere is higher than it would be otherwise (Greenhouse effect).

5. Geographical and Seasonal Distribution of the Outgoing Radiation. For the earth as a whole the yearly average of the total outgoing radiation from the earth and the atmosphere must equal the total incoming solar radiation. This radiation balance does not hold, however, for any given location and for any season or for a smaller time interval. At lower latitudes the incoming radiation is greater than the outgoing radiation; at higher latitudes the outgoing radiation is greater, even if the yearly averages are considered. The surplus of heat at lower latitudes is transported by the atmosphere and to a smaller extent by ocean currents toward higher latitudes and compensates these regions for their heat deficit. Thus the yearly mean temperature of each latitude remains the same.

To study this transport of heat the incoming and outgoing radiation has to be known for each latitude. Exact calculations of the outgoing radiation are complicated, and the physical basis for such calculations, the absorption spectra of water vapor and carbon dioxide, are only now becoming sufficiently well known for exact computations. The values for the outgoing radiation that have thus far been computed by various authors differ consequently to a certain extent. In the following discussion, values computed by Simpson[1] and Baur and Philipps[2] will be used in order to show that the results obtained by different methods give about the same picture in spite of the unavoidable differences due to the lack of sufficient information.

Table 6 contains the yearly averages of the outgoing radiation at different latitude zones according to both papers. For the

TABLE 6.—ANNUAL MEANS OF THE INCOMING AND OUTGOING RADIATION IN THE NORTHERN HEMISPHERE, CAL/CM²/MIN

A. After Simpson

Item	0°	10°	20°	30°	40°	50°	60°	70°	80°	90°N
Inc. rad.....	0.339	0.334	0.320	0.297	0.267	0.232	0.193	0.160	0.144	0.140
Outg. rad....	0.271	0.282	0.284	0.284	0.282	0.277	0.272	0.260	0.252	0.252
Difference...	0.068	0.052	0.036	0.013	−0.015	−0.045	−0.079	−0.100	−0.108	−0.112

B. After Baur and Philipps

Item	0°–10°	10°–20°	20°–30°	30°–40°	40°–50°	50°–60°	60°–90°N
Inc. rad......	0.354	0.346	0.336	0.297	0.236	0.185	0.145
Outg. rad....	0.296	0.299	0.298	0.291	0.269	0.253	0.240
Difference....	0.058	0.047	0.038	0.006	−0.033	−0.068	−0.105

sake of comparison the values of the incoming solar radiation reaching the ground are also included. The outgoing radiation in Table 6 is the sum of the radiation from the earth's surface and from the atmosphere back to space. The disagreement between the figures given by Simpson and by Baur and Philipps for the incoming solar radiation is due to the different assumptions about the distribution of clouds and to the different methods of determining the effect of atmospheric absorption and scattering of solar radiation.

The outgoing radiation changes much less with the latitude than the incoming radiation. Simpson's figures decrease from

[1] SIMPSON, G. C., *Mem. Roy. Met. Soc.*, **3,** No. 21, 1928; No. 23, 1929.
[2] BAUR, F., and PHILIPPS, H., *Gerl. Beitr. Geophys.*, **45,** 82, 1935.

the equator to the pole by less than 10 per cent; Baur and Philipps's figures by about 20 per cent. The smallness of the latitudinal variation is largely due to the presence of clouds. Where the sky is overcast, the radiation from the ground is replaced by the radiation from the top of the clouds. Simpson assumes at all latitudes a uniform temperature of $-12°C$ for the cloud tops, arguing that the cloud heights must be lower at higher latitudes with their cooler temperatures. Hence the values of the outgoing radiation become more uniform than they would be if all the outgoing radiation came from the ground and from the atmosphere whose temperatures change with the latitude. The figures of Baur and Philipps for the outgoing radiation show greater dependence on the latitude since in their calculations it was assumed that the temperature of the cloud tops is a function of the latitude. Nevertheless, even with this assumption the outgoing radiation changes considerably less with the latitude than the incoming radiation.

On the most important point both sets of figures agree: Between 30° and 40° from the equator the outgoing radiation becomes larger than the incoming radiation. At lower latitudes there is a surplus of incoming radiation. It follows that there must be a transport of heat to higher latitudes because the lower latitudes do not become progressively warmer. The surplus of heat in the zone from 0° to 10° latitude must flow across the tenth parallel poleward. Across the twentieth parallel the heat flow must be more intense since not only the heat surplus of the zone from 0° to 10° latitude but also the surplus of the second zone must flow across this parallel. Obviously, the heat flow reaches a maximum somewhere between 30° and 40° latitude. Farther toward the pole it diminishes again because the higher latitudes have a radiational deficit, which is compensated by this meridional heat flow from the surplus regions at lower latitudes. As already explained, this heat transport from the equator to the pole is effected mainly by the motion of the atmosphere, which will be discussed in Chap. III. A much smaller part of the heat transport is taken care of by the oceanic circulation.

The preceding discussion refers only to the annual means. The values of the incoming and outgoing radiation change considerably with the seasons. Figure 4 shows the difference between the incoming and outgoing radiation, the "net" radiation, in January and July. The curves are based on Simpson's computations; the dots indicate the values calculated by Baur and Philipps. For the Polar Zone from 60° to 90° north latitude Baur and Philipps determined only one value, which has been entered at 75° latitude. For the Southern Hemisphere only the figures computed by Simpson are available. The abscissa in Fig. 4 is the sine of the latitude in order to allow for the decrease of the area of the latitude zones toward the poles. On the whole, the values given in these two papers agree remarkably well; only the July figures for the Polar Zone differ considerably.

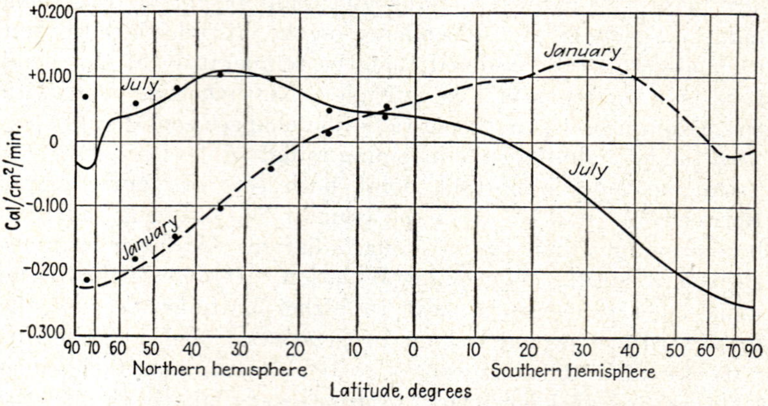

FIG. 4.—Difference between incoming and outgoing radiation in January and July. Curves: Simpon's values; dots: Baur and Philipps's values.

This difference is due to the fact that Simpson's value for the incoming radiation reaching the earth's surface is considerably smaller than that of Baur and Philipps. Wherever the curve for the net radiation is above the zero line, there is a net gain of heat; wherever it is below zero, there is a net loss. It should be kept in mind that these curves refer, not to the yearly average as do the values in Table 6, but to monthly mean values. When the yearly averages were discussed, it could be assumed that a net gain or loss of heat has to be compensated by transport of heat to or from other latitudes, respectively, since the annual mean temperature remains constant from year to year. But this is not correct for any month separately. At least a part of the surplus of energy is now used to increase the temperature.

Similarly, when the net radiation is negative, the temperature falls. The two curves of Fig. 4 show clearly how the regions of heat surplus and deficit shift with the seasons. During July the surplus region extends from 15° south to 70° north, or even to the North Pole if the value of Baur and Philipps for the outgoing radiation is accepted. In January, on the other hand, the Northern Hemisphere down to about 20° north has a heat deficit. The region of heat surplus is largely in the Southern Hemisphere although it extends according to Simpson only to 60° south latitude.

The values for the incoming and the outgoing radiation given in this section are mean values for the parallels. Owing to the unequal air temperatures over water and land and owing to the unequal cloudiness, the actual values for different longitudes are, of course, different even on one and the same parallel. Maps showing the distribution of the incoming and the outgoing radiation over the whole globe during January and July have been drawn by Simpson on the basis of his calculations. Such maps are naturally at present only rough approximations to actual conditions owing to the lack of reliable data about cloud cover and temperatures at the surface and aloft over large parts of the globe. It is therefore not possible at present to use these maps as a basis for a detailed discussion of the temperature distribution over the globe.

But, in a more general manner, we may briefly summarize the geographical distribution and the seasonal variation of the radiation as follows, bearing in mind its effects on the distribution of temperature, pressure, and wind. The intensity of the solar radiation received at the ground is primarily a function of the latitude since its intensity depends mainly on the obliqueness of the sun's rays. The outgoing radiation is also dependent on the latitude, but its variation from the equator to the poles is much less pronounced than that of the incoming radiation. Thus, there is a net gain of radiative energy at low latitudes, a net loss at high latitudes. If no air circulation were taking place between high and low latitudes, the temperature would rise at low latitudes and fall at high latitudes until the outgoing radiation—which depends on the temperature of the emitting surface—became equal to the incoming radiation. Actually, however, the temperature differences produce pressure gradients and air motions so that heat is transported from lower to higher latitudes. Thus, the temperature difference between the pole and the equator is less extreme than it would be if radiation alone were determining the temperature field without the regulating influence of the air circulation.

As was discussed earlier in this section the latitudinal distribution of the net radiative energy gain and loss undergoes seasonal variations, which are reflected in the seasonal variations of the temperature field, the pressure distribution, and the atmospheric circulation. These elements and their changes throughout the year will be considered in the next chapters.

CHAPTER II

TEMPERATURE

The atmospheric temperature is controlled principally by the incoming solar radiation. Hence its distribution depends largely on the latitude. Furthermore, it is affected by the nature of the surface of the earth, especially by the differences between water and land, by the altitude, and by the prevailing winds.

Except where conditions at higher elevations are considered the temperature referred to in this chapter is that measured regularly at meteorological stations, about 2 m above the ground. Monthly mean values have mainly been used in this chapter to describe temperature conditions. In many instances information about extreme temperatures and about frequencies with which various temperatures occur would have been desirable. But it was necessary to restrict the material to be presented for the sake of conciseness.

6. The Temperature at the Earth's Surface as a Function of the Latitude. For a first orientation it is useful to consider the mean temperature of each latitude. In this manner the dominance of the latitudinal control of the temperature is more evident. The mean temperatures are given in Table 7.[1] The yearly averages show that the warmest latitude is not at the equator, but at 10° north. Only during January is the equator the hottest circle of latitude. In July, on the other hand, the warmest parallel of latitude is somewhat north of 20° north.

The Northern Hemisphere as a whole has a higher annual mean temperature than the Southern Hemisphere. On the yearly average each latitude of the Northern Hemisphere is warmer than the corresponding latitude in the Southern Hemisphere. The temperature decrease with latitude is very small in tropical and subtropical latitudes. The temperature difference between the equator and 30° south is only 17.3°F. At higher latitudes it is considerably greater. Between 30° and 60° south it is 36.0°F;

[1] HANN, J. VON, and SÜRING, R., "Lehrbuch der Meteorologie," 5th ed., p. 180, Verlag W. Keller, Leipzig, 1937.

between 60° south and the South Pole, 53.5°F. In the Northern Hemisphere the temperature decrease with the latitude is on the whole somewhat smaller.

During January when the Northern Hemisphere has its winter the temperature is here, of course, lower than in the Southern Hemisphere, while in July the Southern Hemisphere is colder than the Northern Hemisphere. The temperature decrease from the equator toward either pole is considerably greater in

TABLE 7.—MEAN TEMPERATURE, DEG F, OF THE LATITUDE CIRCLES

Latitude	Year	January	July	Range
90°N	− 8.9	−42.0	30.0	72.0
80°N	− 1.0	−26.0	35.6	61.6
70°N	12.7	−15.3	45.1	60.4
60°N	30.0	3.0	57.4	54.4
50°N	42.5	19.2	64.6	45.4
40°N	57.4	41.0	75.2	34.2
30°N	68.7	58.1	81.1	23.0
20°N	77.5	71.2	82.4	11.2
10°N	80.1	78.4	80.4	2.5
Equator	79.2	79.5	78.1	1.8
10°S	77.5	79.3	75.0	4.3
20°S	73.2	77.7	68.0	9.7
30°S	61.9	71.4	58.5	12.9
40°S	53.4	60.1	48.2	11.9
50°S	42.4	46.6	38.1	8.5
60°S	25.9	35.8	15.6	20.2
70°S	7.5	25.7	− 9.4	35.1
80°S	−16.6	12.6	−39.1	51.7
90°S	−27.6	7.7	−54.0	62.0

winter than in summer. The greater magnitude of the meridional temperature gradient during the winter than during the summer is due to the seasonal variation of the radiation. As Fig. 4 (page 20) shows, the difference between incoming and outgoing radiation varies much less with the latitude on the hemisphere that has summer than on the other hemisphere. Consequently, the temperature, which depends largely on the heat balance, shows a similar variation. Of course, the temperature does not depend on the radiative balance alone, and therefore the temperature distribution is not entirely parallel to the distribution of the radiation. Thus, the greatest surplus of radiation during January occurs at about 30° south, while the temperature maximum is found at the equator. The meridional temperature gradient acts as a regulator of the intensity of the general circulation. Hence the general circulation is considerably more intense during the winter season than during the summer season.

The last column of Table 7 shows the annual range of the temperature, *i.e.*, the difference between the mean temperature of the warmest and of the coldest month. It will be noticed that at the equator and at 10° north latitude the range is not equal to the difference between the mean temperatures of July and January. The reason for this deviation is that at these latitudes the monthly maxima and minima are not reached in July and in January. The annual temperature range is very small in tropical latitudes where the daily amount of the solar radiation changes only slightly throughout the year. The annual range increases toward both poles since at higher latitudes the yearly variation of the incoming solar radiation becomes larger. In the Northern Hemisphere the annual temperature range at every latitude, except at 10°, is greater than in the Southern Hemisphere owing to the greater area covered by land.[1] Only at 10° south latitude is the temperature range greater than at 10° north. Here the land area comprises about the same percentage of the total surface north and south of the equator so that the continental effect is the same in both hemispheres. South of 30° south latitude the annual temperature range decreases to a secondary minimum at 50° south. This drop is due to the sharp decrease of the continental area beyond 30° south latitude. At this latitude about 20 per cent of the earth's circumference is covered by land, at 40° south only 4 per cent. Thus, the effect of the oceans with their smaller annual temperature range makes itself felt more strongly beyond 30° south latitude than the influence of the increasing latitude, which would tend to give a greater annual temperature range. The result of both these counteracting factors is the secondary minimum of the annual range at 50° south latitude.

Owing to the difference in the annual temperature ranges in the Northern and in the Southern Hemisphere the temperature

[1] The reason why the temperature variation is greater over land than over water will be discussed later (see p. 26).

of the earth as a whole has an annual variation. Table 8 shows the mean temperatures during January and July and during the whole year for each hemisphere separately and for the whole earth. The Northern Hemisphere has a somewhat higher annual mean temperature and a considerably larger

TABLE 8.—MEAN VALUES OF THE TEMPERATURE, DEG F, OF BOTH HEMISPHERES AND OF THE WHOLE EARTH

Part of the globe	Year	January	July	Annual range
Northern Hemisphere.............	59.3	46.5	72.3	25.8
Southern Hemisphere..............	56.0	62.7	49.5	13.2
Whole earth......................	57.7	54.6	60.8	6.2

annual range than the Southern Hemisphere. The higher annual range is due to the much larger continental areas in the Northern Hemisphere. Land surfaces are heated much more strongly in summer and cooled to much lower temperatures in winter than are water surfaces.

The reason for this different behavior of water and land is that the temperature increase due to solar heating extends only to a small depth into the ground while in water the heating effect penetrates to much greater depths, because of the turbulent motion of the water. Moreover, some of the heat received at the water surface is used to evaporate water, while over land the heat loss due to evaporation is mostly smaller.

As an example of the different penetration of the annual heat wave into soil and into water the following examples may be given: At Tiflis, Transcaucasia, the annual temperature range at the earth's surface is 56.5°F. At a depth of 6.5 m, however, the annual temperature range is only 2.7°F. Thus the heat transport from the surface to greater depths, which is effected by molecular conduction, is very slight. The heat received at the ground is largely used to warm the top layers of the soil, which consequently become very hot in summer. In winter, on the other hand, the earth's surface becomes very cold since there is little conduction of heat from greater depths to the surface. As a result of the great variation of the temperature at the earth's surface the air in contact with and near the surface undergoes great temperature changes in connection with the annual as well as with the diurnal period of insolation.

In the Bay of Biscay, on the other hand, the annual temperature range of the surface water is about 14°F; at 25 m depth, about 12°F; at 50 m, 4.5°F. These figures show that the uppermost layers of the sea are not heated so strongly as the land in summer, the reason being the downward transport of heat by turbulent mixing. During the winter the uppermost layers of the sea are not cooled to the same extent as the land because of the transport of warmer water upward from greater depths.

It follows that large annual ranges of temperature with hot summers and cold winters are a characteristic feature of the climate of the interior of the continents. Over the oceans the summers are relatively cool, the winters relatively mild. One speaks, therefore, of continental and oceanic, or maritime, climates.

The annual range of the temperature is thus of great climatic importance. It gives a measure for the extremes of a continental climate, the so-called "continentality." But it should be noted that the yearly range of the temperature increases with increasing latitude over the oceans also, owing to the increase of the yearly variation of the insolation toward the poles. The values of the continentality at different latitudes can therefore not be directly compared by means of their annual temperature ranges. Allowance must also be made for the dependence of the annual temperature range on the latitude.[1]

The fact that the Northern Hemisphere with its larger land masses and with its higher annual temperature range has a higher mean temperature than the Southern Hemisphere would seem to indicate that over the land, the higher the annual temperature range, the higher the temperature on the average for the whole year. This conclusion, however, is correct only for latitudes up to about 35° or 40° north. At higher latitudes the annual mean temperature decreases with increasing annual range, as shown by Table 9. The reason for the opposite variation of the annual mean temperature and the annual temperature

[1] For a discussion see V. Conrad in Köppen-Geiger, "Handbuch der Klimatologie," Vol. 1, B, pp. 138–143, Verlagsbuchhandlung Gebrüder Borntraeger, Berlin, 1936.

range at higher latitudes is that here the summer temperature increases less from the coast toward the interior than the winter temperature decreases.

The higher mean temperature of the Northern Hemisphere is by no means an effect only of its greater land cover. The oceans are also warmer in the Northern than in the Southern Hemisphere. This is partly due to the transport of warm equatorial waters from the Southern Hemisphere into the Northern Hemisphere by the southeast trades, which cross the equator. Another factor favoring the Northern Hemisphere is the partial protection of its oceans from the cold polar waters and from the arctic ice by land barriers, there being no such barrier between the Antarctic Ocean and the southern oceans in more equatorial latitudes.

TABLE 9.—MEAN ANNUAL TEMPERATURE AND TEMPERATURE RANGE

	34°N lat.		47°N lat.	
	Beyrouth	Baghdad	Nantes	Budapest
Long.	35°E	44°E	2°W	19°E
Ann. mean temp.	70.0°F	73.0°F	52.0°F	49.8°F
Ann. temp. range	26.6	45.5	24.9	42.1

7. The Effect of Land and Sea on the Temperature Distribution at the Earth's Surface. The discussion of the latitudinal mean temperatures in the preceding section has already made apparent the great effect of water and land on the temperature distribution. This influence is clearly revealed in Plates I and II,[1] which show the sea-level isotherms during the months of January and July. Over mountainous regions the temperatures have been reduced to sea level, assuming an average lapse rate of 3.3°F per 1000 ft. If such a reduction were not applied, the isotherms over mountain regions would follow the contours closely. This reduction to sea level is naturally quite schematic, and isotherms showing the actual temperature not reduced to sea level are of great practical importance. But such a representation of the actual temperature without elimination of the effect of the altitude would obscure the other influences that are to be discussed here.

[1] The plates will be found in a pocket at the end of the book.

Over large parts of the globe the January isotherms (Plate I) do not run parallel to the circles of latitude. In the Northern Hemisphere the isotherms are deflected toward lower latitudes over the continents, indicating that the continents are colder than the oceans. The coldest region is found in Siberia, where Verkhoyansk has a mean temperature of −58°F during January. This region is often called the "cold pole" of the earth, a designation that is, however, somewhat misleading, for the July temperature is relatively mild, 59°F. A similar July temperature is found, for example, at St. John's, Newfoundland, and the annual mean temperature of Siberia is not extremely low.

Another striking feature of the temperature distribution in the Northern Hemisphere during January is the northward sweep of the isotherms over the eastern part of the North Atlantic, especially along the west coast of Europe, owing to the relatively high temperature of this region. This higher temperature is due to a combination of the effects of the North Atlantic Drift, which brings warmer water from more southerly regions northward, and of the prevailing westerly winds, which transport relatively warm maritime air to northwestern Europe.

The prevailing winds everywhere in temperate latitudes are westerly. Consequently, during the winter the west coasts of the temperate zones are warmer than the east coasts, as can be seen from a closer inspection of the isotherms. The 40°F isotherm, for example, crosses the west coast of the North American continent at about 45° latitude but intersects the east coast only a little above 30° north. The northward and southward displacement of the isotherms in crossing from land to water is much less in the tropics and subtropics, where the annual temperature range is small.

In the Southern Hemisphere, January is a summer month. Hence, here the continents are warmer than the oceans, as is shown by the closed isotherms over the interiors of South America, South Africa, and Australia. On the whole the isotherms in the Southern Hemisphere follow the parallels of latitude more closely than in the Northern Hemisphere since the earth's surface is more uniform, consisting largely of water.

The distribution of the sea-level temperature during July is represented by Plate II. Now the continents of the Northern

Hemisphere are warmer than the adjoining oceans. Over North America the 60°F isotherm extends far north over the Northwest Territories and the Yukon. Over North Africa, the Arabian peninsula, and southwestern Asia, regions surrounded by closed 90°F isotherms are found. In the Southern Hemisphere, which now has its winter, the isotherms again have a much more regular shape than the isotherms over the Northern Hemisphere owing to the smaller continental areas.

Many details of temperature distribution can be recognized even in Plates I and II, in spite of their small scale, such as the low temperature along the Atlantic coast of Morocco, which is due to the cool Canaries Current (see page 331). Such details, however, are more appropriately discussed in Part B on Regional Climatology.

Another feature that is clearly brought out even by a casual inspection of Plates I and II is that the meridional temperature gradient is much greater during the winter than during the summer. This variation has already been pointed out in connection with the discussion of the mean temperatures of the latitude circles (Table 7, page 24).

8. The Annual Variation of Temperature. The maps of the temperature distribution in January and July show that the annual range of the temperature, *i.e.*, the difference between the mean temperatures of the warmest and of the coldest month, is greater at higher latitudes and, at one and the same latitude, greater over the continents than over the oceans. Plate III shows the distribution of the annual range of temperature over the globe. The increase of the annual temperature range with increasing latitude can clearly be seen over the Pacific Ocean. Much more striking, however, is the great increase of the annual temperature range toward the interior of the continents, especially over the large continental masses of North America and Asia. In particular, over eastern Siberia the high latitude and the vast dimensions of the continent combine to give ranges well above 100°F. But even in the interior of the continents in tropical latitudes the yearly temperature range is much larger than over the adjoining oceans, as is shown, for example, by the great temperature variation from summer to winter over the Sahara in Africa. The larger values of the annual range over the continents as compared with those over the oceans are due to the different reaction of land and water surfaces with respect to heating by the sun, as explained previously (page 26).

Besides the yearly mean value and the range, the annual temperature variation is characterized by the position of the maximum and minimum of the monthly mean temperature and by the length of the time for which the annual temperature is above and below the mean, the so-called "asymmetry" of the

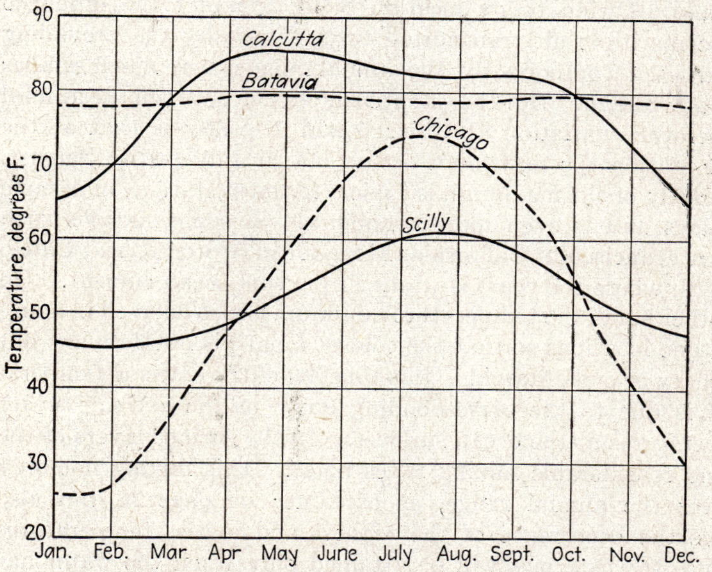

FIG. 5.—Types of the annual temperature variation: 1, continental type—Chicago; 2, maritime type—Scilly; 3, monsoon type—Calcutta; 4, equatorial type—Batavia.

curve. Four typical examples for different forms of the annual march of the temperature are given in Fig. 5, but it should be emphasized that it would be possible to divide the great variety of annual temperature curves into many more types.

a. The Continental Type. At Chicago (Fig. 5) the monthly minimum occurs in January, the maximum in July, both in the month after the solstice. The temperature curve is essentially symmetrical with respect to the maximum and the minimum. A curve of this type is characteristic for the course of the temperature on the continents, especially in temperate latitudes. But even in subtropical and in polar regions an annual variation of this form is frequently encountered in continental locations. Hence, this type with its symmetry and with its extremes occurring in January and July represents the normal continental type. Nevertheless, the annual mean value and the annual temperature range vary considerably within this type.

b. The Temperate-maritime Type. A different behavior is illustrated by the example of Scilly, off the southwest coast of England. Owing to its location Scilly is under the influence of the maritime air transported to the coast by the prevailing westerlies. Consequently, the annual range is here much smaller than at Chicago. But the essential difference between the annual temperature variation at Chicago and at Scilly is that at the latter station the maximum occurs not in July but in August. This delay of the maximum is typical for most stations in oceanic positions and is even more pronounced at some stations such as San Francisco in California, Mogador in Morocco, and Callao in Peru, where the coast is washed by a cool ocean current. On the other hand, the temperature minimum is also delayed in many maritime locations so that the coldest month is not January but February or even March. This time lag of the extreme temperatures is due to the corresponding delay of the extreme water temperatures at the ocean surfaces. In the surface layers of the oceans considerable mixing takes place. This mixing not only reduces the annual range, as explained on page 26, but also delays the occurrence of the highest and lowest temperatures because the extremes will be attained only when the water has been cooled or heated to such a depth that even mixing does not replace cooled or heated surface water by warmer or colder water from below. It must be added here that, like the air temperature, the water temperature in a given region depends not only on the solar radiation but also on the advection of water from other regions so that ocean currents may also influence the temperature.

c. The Monsoon Type. In some respects a counterpart to the temperate-maritime type is that represented by the temperature curve for Calcutta, where the highest temperature is reached in May, with a very slight secondary maximum in September. The premature end of the temperature rise in May is here due to the beginning of the southwest monsoon. The monsoon winds bring greater cloudiness and precipitation with the advection of maritime air, thereby terminating the normal temperature rise, which would otherwise continue until July. While this type is an effect of the monsoon regime, it does not follow that every region under the influence of a monsoon circulation has this type of annual temperature variation.

d. The Equatorial Type. In the tropics the annual variation is very slight, as shown in Fig. 5 by the example of Batavia. Thus the main feature of the annual march of temperature of the equatorial type is its very small annual range. Since the sun passes through the zenith twice during the year in the tropics, many stations have two maxima and two minima. However, this double maximum of the annual temperature curve does not occur at every place where the annual variation of the solar radiation has two peaks. Especially near the boundaries of the tropical zone the time difference between the two maxima of the insolation is so small that no double maximum of the temperature exists. On the other hand, the appearance of two peaks in the annual temperature curve is not necessarily due to the existence of two peaks in the radiation curve but may be caused by a summer maximum of cloudiness and precipitation.

9. The Diurnal Variation of Temperature. The diurnal temperature change is produced by the daily period of the incoming solar radiation. This will be discussed briefly.

During the daytime the temperature rises rapidly as the altitude of the sun increases. The temperature increases for 1 to 3 hr after the sun has reached its highest altitude, for even though the incoming radiation decreases after noon it is still for some time greater than the outgoing radiation. After the temperature has reached its daily maximum, it falls; and this fall continues throughout the night when the sun is below the horizon. But the physical cause of the nocturnal part of the temperature curve is different from that of the part when the sun is in the sky. When solar radiation is absent, the curve is affected only by the radiative cooling from the ground and from the air. Hence, the ideal diurnal temperature curve is not symmetrical. The minimum occurs just about sunrise, and the increase from the minimum to the maximum occupies a shorter time than the fall from the maximum to the minimum. A typical diurnal temperature variation on clear days is represented by the full curve in Fig. 6, which gives the daily mean temperature at Pavlovsk, Russia, during the summer months[1] on clear and cloudy days.

The range of the diurnal temperature variation depends in the first place on the state of the sky. When the sky is overcast, the temperature maximum is smaller because of the reduction of the insolation. The daily minimum temperature is

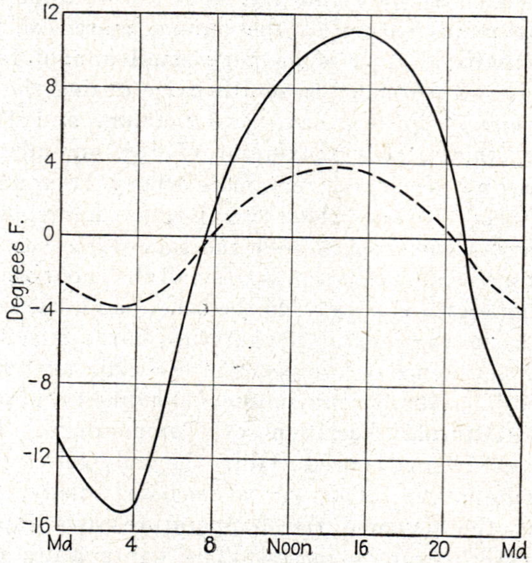

FIG. 6.—Diurnal temperature variation at Pavlovsk in summer. Full curve: clear days; broken curve: overcast days.

higher than on clear days since the outgoing radiation from the ground and from the atmosphere is largely absorbed by the clouds and radiated back toward the ground. As a result, the daily temperature variation on overcast days is smaller than on clear days (see Fig. 6, broken curve).

Furthermore, the daily temperature maximum depends on the stability of the air. If a temperature inversion exists near the ground, the air, which has been heated in contact with the ground, will ascend only to the inversion level, so that a much smaller mass of air is heated than in the absence of the inversion, when the air can ascend to higher levels. Conse-

[1] MEINANDER, R., *Soc. Fenn. Sci. Comm. Phys.-Math.*, **9**, No. 3, 1936.

quently the diurnal temperature range is larger when an inversion exists, and the effect is particularly noticeable if the inversion is low. This phenomenon is in general more of meteorological than of climatological interest. But Sverdrup[1] has shown that in the arctic regions it is also significant from a climatological viewpoint. According to observations on board the "Maud" the monthly mean of the diurnal temperature range increases from February to April because of the increase in the insolation. In May, however, the diurnal range drops again. The explanation is to be found in the height variation of the cold layer next to the ground, which is limited by an inversion. The height of this

TABLE 10.—DIURNAL TEMPERATURE RANGE AT MARITIME AND CONTINENTAL STATIONS
(After Conrad)

	Mean latitude, °N	Annual mean of diurnal range, deg F
5 coastal and island stations	28.6	9.0
4 continental stations	28.4	19.4
6 coastal and island stations	8.6	9.5
3 continental stations	8.4	14.9

inversion increases considerably from April to May so that during the latter month a much larger mass of air is being heated than during April.

Another factor influencing the magnitude of the daily temperature variation is the nature of the earth's surface. Over the oceans the diurnal temperature range—similarly to the annual range, and for the same reason—is much smaller than over land. But the maximum occurs here earlier than over land, only about ½ hr after the sun has attained its greatest altitude. About this time the balance between incoming and outgoing radiation is already reached owing to the smaller amount of heating over water. The effect of oceanic location on the diurnal range of temperature is clearly shown by Table 10, after Conrad.[2]

[1] SVERDRUP, H. U., "The Norwegian North Polar Expedition with the 'Maud' 1918–1925," Meteorology, I, Discussion, Geofysisk Institutt, Bergen, 1933.

[2] CONRAD *op. cit.*, p. 160.

Since the diurnal temperature variation is an effect of the diurnal period of the insolation, the greatest temperature ranges should be expected at low latitudes where the sun reaches a high position in the sky at noon. In many instances an increase of the daily temperature amplitude with decreasing latitude can be found. But the latitude is by no means a dominant factor. The examples in Table 10 show that the contrast between maritime and continental position at the same latitude brings about considerable differences between the diurnal temperature amplitudes at different stations. On the other hand, the difference between the amplitudes at higher and lower latitudes is very small for the selected maritime stations, while for the continental stations the amplitude even decreases toward the equator. Regional differences in the daily periods of cloudiness, water vapor, and dust content often mask the effect of latitude.

At many coastal stations the moderating influence of the water is particularly strong because of the sea breeze, which reduces the temperature during the noon hours. Where the sea breeze blows regularly, it is clearly of great importance since it moderates the temperature during the hottest part of the day.

10. The Temperature Distribution at Higher Altitudes. The temperature distribution at higher altitudes has not as yet been studied in such great detail as at the earth's surface. For Europe and North America considerable material has been collected, but for large parts of the world temperature observations at higher altitudes are still lacking. Thus, any attempt at a discussion must be based to a considerable extent on generalizations that may be proved wrong by more complete observations. Furthermore, because of the scarcity of upper-air data the statistical values are not so reliable as the corresponding surface values.

These facts should be kept in mind in considering Fig. 7. This figure represents a meridional cross section of the vertical temperature distribution in the Northern Hemisphere, during winter on the right-hand side, during summer on the left-hand side. It has been adapted with some changes from a figure first constructed by Ramanathan[1] and later redrawn with some modifications, based on observations at Abisco in northern

[1] Ramanathan, K. R., *Mem. Ind. Met. Dept.*, **35** (5), 1930.

Sweden, by Palmén.[1] The changes in the present figure concern the highest latitudes where observations have become available in Russian territory.[2] For the present figure the data from Fridtjof Nansen Land and from Novaya Zemlya have been combined into mean values and have been used in redrawing the cross section although some modifications have been necessary in the lowest 4 km. Novaya Zemlya extends from about 70° latitude northward, and Fridtjof Nansen Land is just north of 80° latitude,

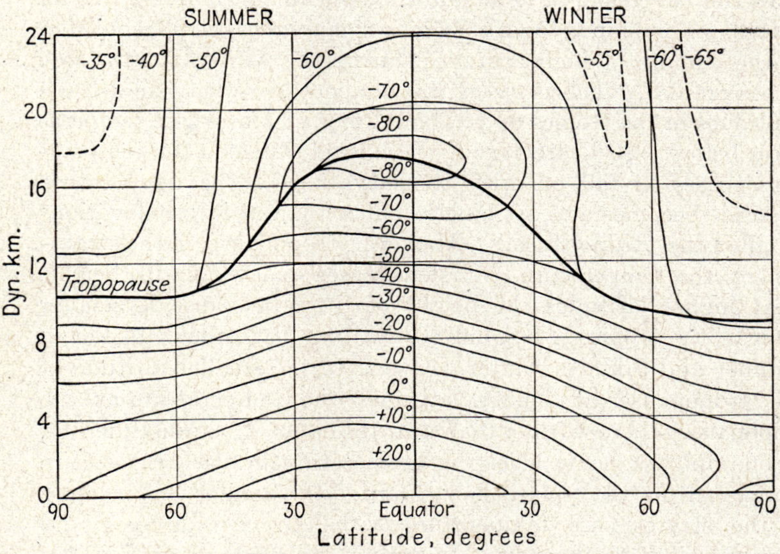

FIG. 7.—Vertical temperature distribution over the Northern Hemisphere.

both island groups being about 60° east of Greenwich. It can hardly be expected that the data from these islands are representative means for the whole latitude belt between 70° and 80° north. Nevertheless, they have been used in constructing Fig. 7 for lack of more complete data. It may be pointed out that at the surface the mean temperatures of Fig. 7 at higher latitudes do not agree with the averages for the mean temperatures at the surface given in Table 7. This discrepancy is not surprising since the present data are obtained from very few stations and represent means for a short time. Despite these

[1] PALMÉN, E., *Met. Z.*, **51**, 17, 1934.

[2] MOLTCHANOV, P. A., *Bull. Am. Met. Soc.*, **18**, 322, 1937.

deficiencies Fig. 7 will give some idea about the temperature variation with altitude and latitude although it will doubtless require considerable correction as more data become available.

Throughout the troposphere the temperature decreases with elevation, approximately at the rate of 6°C/1000 m in the lower troposphere and 7°C/1000 m in the higher troposphere. Only in polar regions during the winter does the temperature in the lowest layers increase so frequently with the altitude that the inversion of temperature appears even in the mean values. This skin layer of cold air is a characteristic phenomenon of the climate of high latitudes during the winter months (see also page 35). The drop of the temperature with the altitude continues in the tropics up to about 16 or 17 km, where temperatures below −80°C are found. The level at which the temperature ceases to fall off with the elevation or the temperature decrease becomes at least much smaller than at lower elevations is called the "tropopause." Above the tropopause, in the stratosphere, the temperature at first increases somewhat with height.

At higher latitudes the height of the tropopause is smaller than in the tropics. Beyond 60° latitude it is about 10 km in summer and 9 km in winter. Owing to the smaller altitude of the tropopause at higher latitudes the temperature at the tropopause is here warmer than near the equator. Consequently, at one and the same level the temperature in the stratosphere increases from the equator northward. In summer this increase of the stratosphere temperature with the latitude seems to continue right to the pole. In winter the stratosphere appears to be warmest between 50° and 60° latitude. Poleward beyond 60° the temperature decreases again.

Similarly to the sea-level temperature the temperature at higher levels depends not only on the geographical latitude but to a certain extent also on the position relative to the continents and oceans. But so far only over Europe and North America do aerologic data give some indication of the temperature effects at higher levels. Over North America the temperature decreases, at least up to 4 km, and along latitude 40° north, from the interior to the east coast, the interior at about 95° west longitude being about 2° to 3°C warmer than the coast. However, the same difference is found, not only in summer, when it could be regarded as an effect of continental heating, but also in winter. It seems, therefore, that the difference is due, not so much to the continental and maritime effect, as to advection. It may be pointed out in this connection that at 40° north latitude along the Atlantic seaboard, according to the mean 3-km isobars for July, west-northwest is the prevailing wind direction, while in the interior of the continent southwest winds are more frequent. It is, of course, not to be expected that the difference between continent and ocean would show up very clearly since the maritime effect at an east coast is slight in a region of prevailing westerly winds.

Over Europe the maritime and continental influences in the free atmosphere have not yet been studied. At Soesterberg, on the Dutch coast, the temperature is always higher than at Lindenberg, about 8° longitude farther east in central Germany, except in the lowest 500 m during July. But the example is not conclusive since Lindenberg is still under the maritime influence, although to a lesser degree than Soesterberg. How strong the effect of continentality is at higher levels under more favorable conditions will have to be studied with the aid of more complete material.

11. Lapse Rates of Temperature. In connection with the vertical temperature distribution the lapse rates of temperature may be considered in somewhat greater detail in view of their importance for atmospheric stability and instability and hence for the formation of clouds and precipitation in the atmosphere. Table 11 shows the temperature lapse rates for stations in North America, Europe, and India and over the oceans. These lapse rates are expressed in degrees centigrade per 1000 m. They are computed from monthly mean values for January and July. In order to eliminate the influence of the earth's surface the lowest 500 m or, where that was not practicable, the lowest 1000 m have been disregarded. Only the temperatures up to 4000 m—or up to 3000 m where sufficient data are not available from higher levels—have been used because the higher layers of the troposphere have in general less effect on the production of hydrometeors.

In comparing these lapse rates for different months it should be remembered that convective phenomena such as cumulus clouds and showers depend, not only on the lapse rate, but also on the moisture content of the air, which is greater in summer

TABLE 11.—LAPSE RATES OF TEMPERATURE IN THE LOWER TROPOSPHERE

	Latitude	Longitude	January, deg C /1000 m	July, deg C /1000 m
North America				
West Coast				
1. Nome	64°N	165°W	4.2	5.6
2. Ketchikan	55°N	132°W	5.7	5.5
3. Seattle	48°N	122°W	5.4	5.4
4. Oakland	38°N	122°W	4.9	3.4
Inland				
5. Fairbanks	65°N	148°W	3.1	7.0
6. Minneapolis	45°N	93°W	2.6	5.6
7. Oklahoma City	35°N	98°W	3.2	5.5
8. San Antonio	29°N	98°W	3.4	5.3
East Coast				
9. Portland	44°N	70°W	3.3	5.3
10. Charleston	33°N	80°W	4.1	5.7
11. Miami	26°N	80°W	4.7	5.5
Europe				
West Coast				
12. Hamburg	54°N	10°E	4.7	5.4
13. De Kooij	53°N	5°E	5.1	5.0
Inland				
14. Pavlovsk	60°N	30°E	5.0	6.0
15. Lindenberg	52°N	14°E	4.6	5.6
16. Vienna	48°N	16°E	4.3	5.4
Oceans				
17. Atlantic Station I	About 37°N	57°W	5.4	5.3
18. Atlantic Station II	About 37°N	47°W	5.6	5.0
19. Coco Solo	9°N	80°W	5.0	5.2
20. Swan Island	17°N	84°W	5.0	5.2
21. Pearl Harbor	21°N	158°W	4.5	4.8
India				
22. Agra	27°N	78°E	5.4	5.4

than in winter and greater at lower than at higher latitudes. Thus, more convective activity can be expected at Miami (11) than at Fairbanks (5) even though the lapse rate is considerably steeper at the latter station. Furthermore, the pseudo-adiabatic rate of temperature change is less with higher temperatures. An average lapse rate of 5.5°C/1000 m at 25°C represents more favorable conditions for convection than a lapse rate of 7°C/1000 m at 5°C. The effect of the temperature on the value of the pseudo-adiabatic rate of temperature change should also be kept in mind when lapse rates at one and the same station during different months are compared. Before condensation has begun, the degree of stability depends on the difference between the actual lapse rate of temperature and the dry-adiabatic rate of temperature change. After condensation has set in, the intensity of the convective phenomena is a function of the difference between the actual lapse rate and the pseudo-adiabatic rate of temperature change. Hence, a steeper temperature lapse rate during the winter does not necessarily imply that the convective phenomena are better developed in winter than in summer.

The inland and east coast stations of North America (5 to 11) and the inland stations of Europe (14 to 16) have steeper lapse rates in summer than in winter so that summer is the season of greatest convective activity, especially if the remarks of the preceding paragraph are taken into account. The smaller stability of the summer stratification is a result of the solar heating, which is strongest next to the ground. During winter, on the other hand, the cooling of the ground and, consequently, of the layers next to the ground is very pronounced so that small lapse rates or even inversions tend to develop. The development of stable lapse rates over the continents during winter is further favored by subsidence that takes place in the continental anticyclones.

Over the oceans (17, 18) and at the majority of the west coast stations (1 to 4, 12, 13) in extratropical latitudes the lapse rates are higher in January than in July. It is largely the more intense cyclonic activity during winter that is responsible for these steeper lapse rates. In summer, on the other hand, cyclones are less frequent and less intense. The decidedly stable stratification at Oakland during summer is due to anticyclonic subsidence (see Sec. 17).

It is of interest to compare the lapse rates at Swan Island (20) and Pearl Harbor (21). The latter station is in the eastern part of the Pacific High, the former in the western part. Swan Island has always steeper temperature lapse rates than Pearl Harbor, showing that the stability of the air is less in the western part of the subtropical anticyclones than in the eastern part (see also Sec. 18).

The station Agra (22) in India has its highest lapse rate in April. The hot season occurs here from March to May, the months from June to September being less warm on account of the monsoon rains (Sec. 8). Hence a decrease of the lapse rate from April to July can be expected. The exceptional decrease of the lapse rate must, however, be attributed to the intense condensation processes during the monsoon season, which tend to reduce the lapse rate to the pseudo-adiabatic value.

TABLE 12.—TEMPERATURE LAPSE RATES IN THE ATLANTIC TRADE-WIND REGIONS AND IN THE DOLDRUMS

Lat.	North, deg C/100 m			South, deg C/100 m		
	6–200	200–500	500–1000 m	6–200	200–500	500–1000 m
0°– 5°	1.16	0.81	0.54	1.08	0.90	0.51
5°–10°	1.18	1.04	0.62	1.40	0.90	0.60
10°–15°	1.17	1.00	0.55	1.37	1.12	0.76
15°–20°	1.35	1.00	0.66	1.45	1.17	0.60

Especially large temperature gradients are found in the trade-wind regions and in the doldrums. Detailed studies of the trades in the Atlantic have been undertaken by Ficker,[1] and Table 12 shows the temperature lapse rates in degrees centigrade per 100 m for different latitudes. In the layer from the sea surface up to 500 m the lapse rate of temperature is often greater than the dry-adiabatic rate. To appreciate the importance of this fact for the atmospheric circulation it may be pointed out that the trade-wind regions and the zone of doldrums comprise together about 40 per cent of the ocean surface of the earth. The high lapse rates are due to the strong heating of the equatorward-flowing air in contact with the warm water. With

[1] FICKER, H. VON, *Veröffentlich. Met. Inst. Univ. Berlin*, **1**, 4, 1936.

increasing altitude the values of the lapse rates decrease rapidly, and the layer is finally terminated by an inversion. The height of this inversion is about 2000 m near the equator and decreases toward the north and south. The lowest altitudes are found near the coast of Africa, where the elevation of the inversion at 15° to 20° north and south latitude is less than 400 m.

CHAPTER III

WIND AND PRESSURE

The temperature differences between various parts of the earth produce pressure gradients that give rise to air motions. But the relations between temperature, pressure, and wind distribution are complicated by the fact that the air motions in turn affect the temperature field. For example, as was pointed out in the discussion on page 29, in the zone of west winds the west coasts of the continents have a more moderate, oceanic climate with cooler summers and warmer winters than the east coasts, owing to the advection of maritime air. Thus, it is not possible to regard the temperature distribution entirely as the cause of the wind distribution. The relation between the two is more complicated than that between cause and effect.

Since the pressure and wind fields are closely related, a summary of the wind distribution over the globe is best supplemented by an account of the pressure distribution, especially at higher altitudes, where direct wind observations are scanty. Therefore, the pressure and wind distribution will be discussed together in this chapter even though the pressure is not strictly a climatological element.

12. Survey of the Wind and Pressure Distribution at the Earth's Surface. As a basis for the later, more detailed discussions it will be useful to give a brief description of the wind and pressure distribution at the earth's surface.

In the subtropical regions easterly winds with components toward the equator prevail, the northeast trade winds of the Northern Hemisphere and the southeast trade winds of the Southern Hemisphere. They are separated near the equator by a zone of prevailing calms, the doldrums. Frequently, however, the doldrums are absent in one part or the other of the equatorial zone. Then the northeast trades and the southeast trades meet directly at a surface of discontinuity on which strong rain squalls may develop. Neither the two trade-wind belts nor the zone of doldrums is strictly symmetrical with respect to the equator. Like the belt of highest temperature the zone of doldrums is largely north of the equator. However, over some parts of the earth it extends into the Southern Hemisphere, as can be seen from Plates IV and V, which show the mean pressure distribution at sea level during the months of January and July. Where the zone of doldrums is displaced particularly far north from its symmetrical position at the equator, the southeast trade winds may reach some distance into the Northern Hemisphere. In that case they become deflected into southwest winds under the influence of the Coriolis force.

One of the most striking features of the trade-wind regions—at least over the oceans, where the trades are best developed—is

TABLE 13.—MONTHLY CHARACTERISTICS OF THE SOUTHEAST TRADES IN THE INDIAN OCEAN BETWEEN LAT. 10°–20°S AND LONG. 80°–90°E
(After Gallé)

Item	Jan.	Feb.	Mar.	Apr.	May	June	July	Aug.	Sept.	Oct.	Nov.	Dec.	Mean
Direction, deg. . .	123	111	128	126	137	147	143	147	140	133	129	134	135
Velocity, m/sec. .	5.66	4.32	4.06	6.11	6.64	6.34	8.24	7.98	8.09	7.75	6.96	6.67	6.55
Persistence, per cent.	83	72	69	84	90	84	89	87	88	90	87	90	85

their great regularity. To illustrate this regularity, the "persistence" of the wind may be introduced. This is defined as the ratio of the mean vectorial wind velocity to the mean velocity without regard to direction. If the wind has the same direction during the whole time interval, the persistence would be 1, or 100 per cent. If the wind directions are distributed entirely irregularly, the mean vectorial wind velocity vanishes, so that the expression for the persistence becomes zero.

Table 13 shows the mean direction, velocity, and persistence of the southeast trades in a 10° square in the South Indian Ocean, from 10° to 20° south and from 80° to 90° east, according to Gallé.[1] The wind direction is counted from north (= 0°) over east (= 90°), etc. These figures should be compared with those of Table 14 for a region in the westerlies.

Beyond the trade-wind regions of each hemisphere are the zones of "prevailing westerlies." They are better developed in

[1] GALLÉ, P. H., *Meddel. Verh. Kon. Neder. Met. Inst.*, **29**, A, 1, 1924.

the Southern than in the Northern Hemisphere, since in the latter the presence of the great continental surfaces causes many modifications. These zones of westerlies are the regions of the migrating cyclones and anticyclones. In connection with these moving pressure centers, winds from other directions than west are observed frequently. Hence, the persistence of the west winds in these regions is much lower than the persistence of the trade winds. Table 14 shows the wind directions and persistences in a certain region of the prevailing westerlies in the Indian Ocean, after Gallé. The persistence in this region is very considerably lower than in the trade-wind belt even though this particular 10° square is located within that part of the west-wind belt which is known as the "roaring forties" owing to the frequency and intensity of the west winds.

TABLE 14.—MONTHLY CHARACTERISTICS OF THE WESTERLIES IN THE INDIAN OCEAN BETWEEN LAT. 35°–45°S AND LONG. 70°–80°E
(After Gallé)

Item	Jan.	Feb.	March	April	May	June	July	Aug.	Sept.	Oct.	Nov.	Dec.	Mean
Direction, deg.	331	331	294	304	285	281	288	285	283	286	294	308	296
Persistence, per cent	33	30	31	38	52	49	56	56	58	57	54	51	47

Poleward from the region of westerlies, beyond 65° to 70° latitude easterly winds with a component toward the equator prevail. But the persistence of these polar easterlies is small.

The motion of the air, as outlined in the preceding paragraphs, is closely connected with the general pressure distribution (see Plates IV and V for the mean pressure distribution). Near the equator but, like the doldrums, not quite symmetrical to it, a zone of low pressure is found. It is followed in both hemispheres by zones of high pressure, the so-called "subtropical" high-pressure belts. Especially in the summer hemisphere these high-pressure belts do not extend continuously around the earth but are interrupted over the continents since the land is hotter than the oceans, as will be discussed in greater detail in the next section. The central regions of the high-pressure belts, which coincide practically with the polar limits of the trade-wind regions, have frequent calms as is to be expected in the center of anticyclones. These central regions of the subtropical highs received the name "horse latitudes" in the era of sailing ships, when regions of frequent calms were a particular nuisance in navigation.

In the regions of the westerlies even the mean pressure distribution is rather irregular, owing to the migrating pressure centers. On the whole, the pressure decreases here toward the pole, as it should, according to the gradient wind relation, in a region with prevailing westerly winds.

In the surface layers of the arctic and of the antarctic zone, at least over the glaciated continental areas of Greenland and Antarctica, the pressure distribution is anticyclonic, because of the low temperature of the air in the surface layers. Since Greenland and Antarctica are not symmetrical with respect to the poles and since, on the other hand, the cooling will be stronger here than over the polar sea, the zones of the easterly winds in the polar regions cannot be symmetrical to the poles.

TABLE 15.—SUMMARY OF THE WIND AND PRESSURE BELTS AT THE EARTH'S SURFACE

(After Hann and Conrad)

Lat.	75°N		60°N		30°N		10°N		Equ.		10°S		30°S		60°S		80°S
Wind		ENE		WSW	Calms	NE		ENE	Calms	ESE		SE	Calms	WNW		ESE	
Press., mb.	1014		1010		1016		1010		1010		1012		1018		990		997

The foregoing brief survey of the wind and pressure distribution over the globe is summarized in Table 15 after Hann and Conrad.[1] It may be emphasized again that the whole foregoing discussion gives only a very generalized survey of the wind distribution over the earth.

It is helpful in visualizing the importance of the zones of doldrums, trade winds, westerlies, etc., for the climates of the earth to consider the size of the areas covered by the various wind systems. Since most of the wind systems are best developed over the oceans, the area covered is given in percentages of the total ocean area (see Table 16, after Lütgens[2]). Table 16 shows

[1] CONRAD, V., in Köppen-Geiger, "Handbuch der Klimatologie," Vol. 1, B, p. 263, Verlagsbuchhandlung Gebrüder Borntraeger, Berlin, 1936.

[2] LÜTGENS, R., *Ann. Hydrographie*, **39**, 265, 1911.

clearly the great importance of the trade-wind, doldrum, and monsoon regions, which cover together about 46 per cent of the water surface of the earth. For the sake of comparison it may be mentioned that North and South America together are equal to about 10.5 per cent of the area covered by the oceans.

TABLE 16.—OCEAN AREAS COVERED BY THE WIND SYSTEMS
(After Lütgens)

Wind System	Percentage of the Whole Ocean Surface
Doldrums	9
Trade winds	31
Monsoons	6
Subtropical regions	20
Westerlies	24
Polar zones	10

13. The Effect of Land and Sea on the Wind and Pressure Distribution. In Chap. II the unequal heating of land and water and its causes were discussed, and in Sec. 12 it was mentioned that this unequal heating of water and land produces considerable modifications of the simple scheme of pressure and wind distribution just outlined. These deviations are most noticeable during the extreme months of January and July, when the one hemisphere has winter and the other summer. The modifications to be expected are indicated schematically in Fig. 8, which is based on Hettner. This figure shows an idealized continent, which extends from about 60° north latitude to about 50° south latitude. It has its greatest width in temperate latitudes of the Northern Hemisphere and tapers off toward the south. In shape and extent it thus resembles the actual continents, especially South America and Africa. The left side of Fig. 8 represents conditions during July, when the Northern Hemisphere has summer and the Southern Hemisphere has winter. The large continental mass of the Northern Hemisphere is then much warmer than the surrounding oceans. Hence, a low pressure is found over the continent. This low interrupts the subtropical high-pressure belt, which, however, is still maintained over the oceans. The continental low is in direct connection with the equatorial low-pressure zone of the doldrums. The Southern Hemisphere has winter during July so that here the land is colder than the water. But the land area in the Southern Hemisphere is small so that the disturbance of the pressure distribution is much less pronounced.

Owing to the interruption of the subtropical high-pressure belt over the continents in the Northern Hemisphere during July the northeast trade winds are here absent. Instead, winds are found that blow around the thermal low with a component toward its center. The inward component of the winds is strongest on the equatorial side of the low because the deflecting

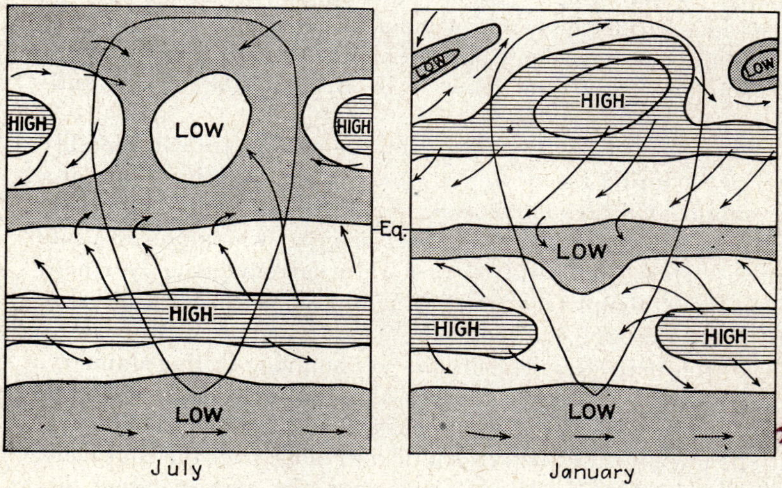

FIG. 8.—Schematic pressure distribution over a continent. (*After Hettner, A, "Die Klimate der Erde," Geograph. Schriften, Vol. 5, B. G. Teubner, Leipzig,* 1930.)

force of the earth's rotation is here smallest. The southwest winds that are observed are a continuation of the southeast trades of the Southern Hemisphere, which are deflected toward the southwest as they cross the equator. The whole circulation around the continental low represents a typical summer monsoon circulation. The deflection of the southeast trades into the southwest monsoon is especially pronounced over India and the Indian Ocean (compare Chap. XII).

The effect of the continents on the pressure and wind distribution during January is shown in the right part of Fig. 8. Since the Northern Hemisphere now has winter and the Southern Hemisphere summer, the conditions are by and large opposite to those of July, except for modifications due to the asymmetrical shape of the continent with respect to the equator. In the Northern Hemisphere, where the continent is now much colder than the adjoining oceans, a strong high has developed. In conjunction with the anticyclonic pressure distribution the wind distribution over the continent is anticyclonic with an outblowing component. On the south side of the continental anticyclone, for example, the winds are northeasterly. A typical example for these winds is the northeast monsoon of Thailand and Indo-China in winter (Plate IV). The northeast monsoon, which prevails over India during the winter, cannot be considered as an outflow from the Asiatic cyclone since the mountain ranges of Tibet cut off completely any direct motion of air from the surface anticyclone to India.

The centers of low pressure over the oceans are considerably intensified during the winter so that cyclonic activity is now at a maximum. On the north side of the continental anticyclone the winds are westerly, while light, irregular winds prevail in its center. Thus, the belts of westerlies surround the Northern Hemisphere completely, but they are displaced considerably northward over the continents.

The Southern Hemisphere has its summer during January. Consequently, the pressure over the continents is now lower than over the adjoining oceans. As a result, the zone of doldrums with its low pressure shows a southward bulge over the continent. But the effect is far less marked than the corresponding displacement during July in the Northern Hemisphere, owing to the smaller size of the southern continent. For the same reason the interruption of the southern subtropical high-pressure belt by the continents during January is only slight. Owing to the southward displacement of the equatorial low pressure across the equator the northeast trades may be drawn somewhat south across the equator and become northwest winds. The southeast trades are deflected toward the warmer land and become east winds—in places, even northeast winds—but the deflection is less than in the Northern Hemisphere.

All these features of the pressure and wind distribution that are indicated schematically in Fig. 8 can be found in Plates IV and V, which represent the actual pressure and wind distribution during January and July. Considering first Plate V for the pressure field during July, it will be seen that the equatorial low-pressure zone is connected with the thermal lows over the continents, which extend far to the north, especially over the vast Asiatic continent, where the lowest pressure is found over the Persian Gulf and Iran. Similarly, the low-pressure trough over the Pacific extends up to 40° north latitude over the western part of the United States. Thus, the subtropical high-pressure belt of the Northern Hemisphere is confined, during the summer, to two cells, the Atlantic and Pacific anticyclones. In the Southern Hemisphere, on the other hand, the high-pressure belt surrounds the earth almost without interruption. In connection with this pressure distribution the southeast trades of the South Indian Ocean blow from the subtropical high pressure of the Southern Hemisphere across the equator and into the Asiatic low. In crossing the equator the winds are deflected into southwest winds and become the southwest monsoon of India. The northeast trades are consequently completely missing over the Indian Ocean.

In January (Plate IV), when the Northern Hemisphere has winter, the pressure is very high over the Asiatic continent and, to a smaller degree, over North Africa and North America, while the pressure gradient in the belt of westerlies is much stronger than during summer. In the Southern Hemisphere the belt of high pressure is now interrupted by the low over each continent. Especially over Africa and over South America the southward displacement of the equatorial trough of low pressure can clearly be seen. A comparison with the map of the pressure distribution during July (Plate V) shows, however, that the displacement of the low pressure southward from the equator is not so pronounced as the displacement northward in July over the Northern Hemisphere. An interesting detail of the pressure distribution during January is the low pressure over the Mediterranean. This center of low is an indication of the fact that a considerable number of the European cyclones pass through this region during the winter owing to the relatively high temperatures, which favor the passage of cyclones.

In consequence of the pressure distribution the northeast trades blow almost in an unbroken belt all around the Northern Hemisphere. Along the east coast of China strong northwest winds are found, which are the winter monsoon winds of this region. Similar winds are observed along the east coast of North America, but they are much less regular so that one can hardly speak of more than a monsoon tendency.

From Plates IV and V it is apparent that the wind velocity and direction have at many places a pronounced yearly period. The best known example of the annual variation of the wind direction is the seasonal shift of the wind where a monsoon regime prevails, as will be discussed in the next section. But also in other regions very pronounced seasonal variations of the wind, especially of the wind velocity, take place. Over the northern North Atlantic and North Pacific, for example, the winds are stronger in winter, when the cyclones are more frequent and intense, than in summer. Owing to the great number of seasonal variations of the wind velocity that are encountered at different stations it is impossible to summarize these yearly periods according to a few types.[1] Instead, they have to be described in connection with the climate of each region.

14. Monsoon Circulation. As was pointed out in the preceding section the disturbing influences of land and water are due to the unequal heating of the continents and the oceans, which makes the land warmer than the water in summer and colder in winter and produces a low over the land in summer and a high in winter. The circulation due to this cause, with the characteristic seasonal variations of the prevailing wind directions and of other meteorological elements, constitutes what are called "monsoons." The system of the monsoon winds around a continent is shown schematically in Fig. 9, but the actual wind distribution over each continent may differ considerably from this simple scheme.

Monsoonal tendencies can frequently be found in the seasonal wind shifts although the shift may not be pronounced enough to be considered as a true monsoon. For example, such a monsoonal tendency is to be found in the winds along the east coast of the United States, as shown in the upper part of Fig. 10. The distances of the end points of the lines radiating from the center of the circle represent the percentage frequency of the different wind directions. In winter, winds from the northwest prevail, in summer from the southwest, so that the monsoon change that

[1] Brose, K., *Deutsches Reich. Reichsamt Wetterdienst. Wissen. Abhandl. (Berlin)*, **I**, 4, 1936.

appears in the frequency percentages consists only in a turning through 90°. For comparison, the lower part of Fig. 10 shows the percentage frequencies of the eight wind directions in northeastern Asia. Here the most frequent wind direction turns through 180° from summer to winter; a true monsoon prevails. The absence of a complete monsoonal change of the wind direction and the greater variability of the winds along the east coast of North America are in part due to the smaller size of the conti-

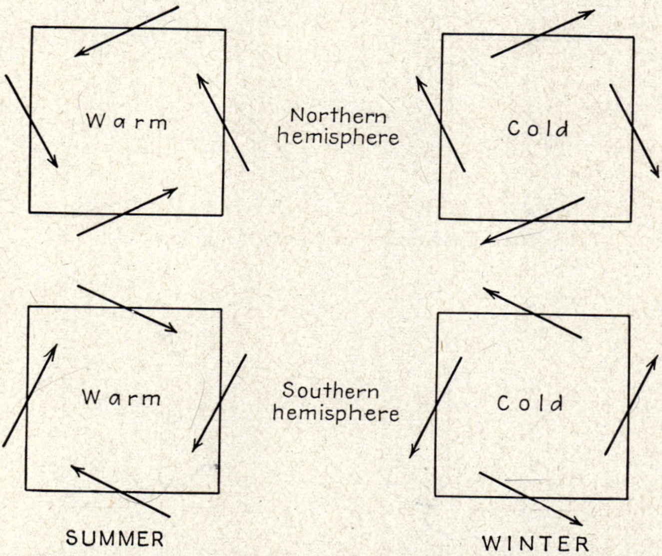

FIG. 9.—Schematic representation of the monsoon circulation. (*After V. Conrad.*)

nent. Another factor that prevents a better development of the monsoon characteristics along the east coast of North America is the lack of an orographic barrier, which would separate the tropical maritime air from the Gulf of Mexico and the polar continental air from the interior of the continent. Therefore, these two air masses can meet without any obstacles, giving rise to frequent intense cyclones so that the wind direction is very variable.

It may finally be pointed out that the monsoon is by no means typical for low latitudes. Its development depends on the unequal heating of water and land during the different seasons.

Since the seasonal temperature changes are very small at low latitudes, monsoonal winds will here be less likely to develop than at higher latitudes, other things being equal.

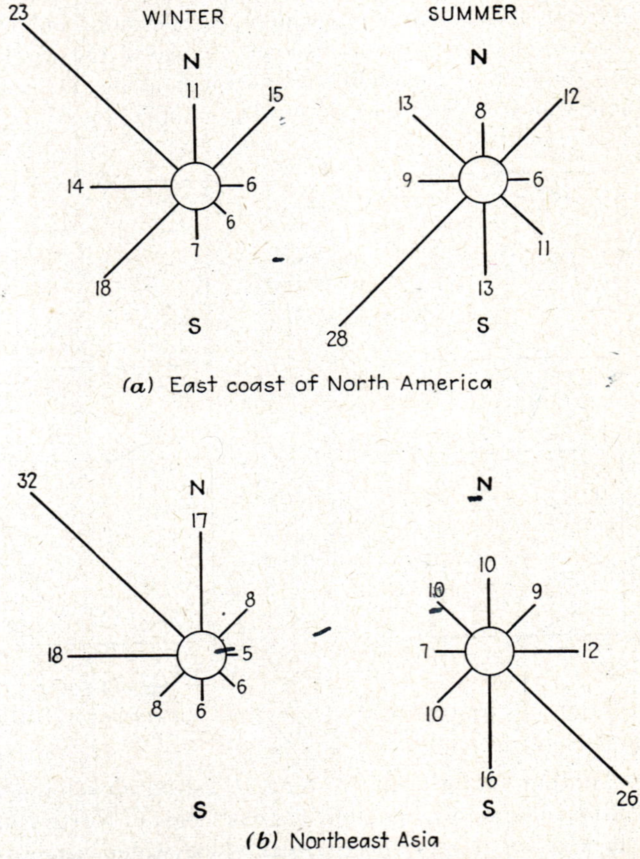

FIG. 10.—Percentage frequencies of wind direction in North America and northeastern Asia showing the development of the monsoon.

15. The Climatic Effects of Ocean Currents. A considerable analogy exists between the currents in the atmosphere and in the oceans. Next to insolation, the ocean currents are the principal factors controlling the water temperature, and the water temperature has a large effect on the climate. Hence it is necessary to make some general remarks about the distribution of ocean currents. A more detailed discussion of ocean currents and their effects on the climates of the continents will be found in Part B. A map of the surface currents in the oceans during February-March is shown in Plate VI. The surface-water temperatures during February and August are given in Plates VII and VIII.

Ocean currents are mainly produced by the prevailing winds. They have not the same direction as the generating air currents but are somewhat deflected, to the right in the Northern Hemisphere, to the left in the Southern Hemisphere, if one faces in the direction of the wind.[1] Consequently, one finds on the west coasts of the continents, from 40° north and south toward the equator, currents that are moving toward the equator (see Plate VI). Since these currents move from colder to warmer regions and since the water temperature, even though it rises while moving toward the equator, lags behind the temperature of its new surroundings, these currents are relatively cool (see Plates VII and VIII). Moreover, owing to the prevailing wind direction there is as a rule a component of motion away from the coast, causing upwelling of colder water from greater depths, which makes the west coasts of the continents at these latitudes still cooler than they would be otherwise. The important climatological effects of the upwelling of cold water from greater depths, especially in the production of the cool and humid coastal deserts, will be discussed later (page 142). As the ocean currents from the north and from the south converge toward the equator, they move westward as the North Equatorial and the South Equatorial currents. They are separated almost throughout the whole extent of each ocean by a countercurrent, which sets toward the east. The region of the countercurrent coincides roughly with the zone of the equatorial calms. Owing to the origin of their water masses the Equatorial currents are relatively cool in the eastern part of each ocean, but they acquire more and more the normal temperature of the environment as they move westward. The Equatorial currents turn north in the Northern and south in the Southern Hemisphere as they approach the west coast of the oceans. Since they are now moving from warmer to colder regions, their water is warm

[1] See, for example, Sverdrup, H. U., "Oceanography for Meteorologists," Chap. VII, Prentice-Hall, Inc., New York, 1942.

relative to the surroundings. At about 40° north and south they begin to move in an easterly direction across the ocean under the influence of the prevailing westerly winds of these latitudes and become again deflected southward in the trade-wind regions.

In the North Atlantic the eastward current is particularly well developed owing to the shape of the coast line which produces a northward deflection of a considerable part of the South Equatorial Current north of Cape St. Roque along the coast of Brazil. The northward current washes the east coast of North America first as the Florida Current and later as the Gulf Stream. Then it continues as the North Atlantic Current across the ocean between Iceland and the British Isles. Finally it flows past the North Cape of Norway and along the north coast of Russia into the Arctic Ocean. The mild winters of northwestern Europe are due to this warm current, whose climatic effectiveness is further enhanced by the westerly winds that carry the air from the warm water well into the interior of Europe. Even along the coast of northern Russia the climatic effect of the North Atlantic Current is of practical significance. Thus, Archangel is blocked by ice for 8 months during an average year. On the other hand, at Aleksandrovsk, 300 miles farther north on the Murman Coast the harbor is rarely icebound longer than 5 months as a result of the warming effect of the North Atlantic Current.

The Equatorial currents of both hemispheres together with the currents in the meridional direction and the easterly drift of temperate latitudes form large anticyclonic eddies. The ocean currents beyond the zones of the west winds in the Northern and in the Southern Hemisphere, on the other hand, form cyclonic eddies. The cold Labrador Current, for instance, is a part of the cyclonic eddy formed by the current system of the northern part of the North Atlantic.

For an exhaustive treatment of ocean currents the reader should refer to the textbooks on oceanography. The purpose of this section is to give only a highly schematized background for later discussions. More detailed reference to individual ocean currents will be made later, where necessary.

16. Wind and Pressure Distribution at Higher Levels. For the upper atmosphere, wind data are few since only pilot-balloon observations can give direct information about wind velocity and direction, although some conclusions can be drawn from cloud observations. The main disadvantage of pilot-balloon data is that, in the present state of meteorological technique, they can be obtained only with clear skies or up to levels which are unobscured by clouds.

An orientation about the mean wind distribution aloft is therefore best obtained by considering the mean pressure distribution at higher levels. Since the wind in the free atmosphere does not deviate much from the geostrophic wind, the wind field can be deduced with sufficient accuracy from the pressure field. The pressure field at higher levels can be computed from the surface pressure field and from the vertical temperature distribution. In this manner maps of the pressure distribution in the higher atmosphere have been constructed.

Before these maps are discussed a summarizing preview of the pressure and wind distribution may be undertaken that stresses mainly the changes in meridional direction.

Figure 11 shows the altitude of various isobaric surfaces from 1000 mb to 100 mb at different latitudes, during summer and winter. For each of the curves representing an isobaric surface a reference level has been given in the center of the figure. Thus, for the two curves showing the height of the 800-mb surface in summer and winter, 1800 m is the reference level. On the left side of the figure, where the summer conditions are shown, the extension of the 1800-m line will be found as the short horizontal line slightly below the 800-mb isobar at the North Pole. Thus the 800-mb isobar is everywhere above 1800 m. On the right side of the figure the extension of the reference line of 1800 m intersects the isobar between 60° and 70° north. The other short horizontal lines intersecting the isobaric curves are all drawn at distances of 100 m. Thus the heights of the various isobaric surfaces at different latitudes can be estimated from the figure. The distances of the isobaric surfaces from each other are arbitrary. It is not possible to draw them to the same scale since such a procedure would make the figure too large. But the inclinations of the different surfaces are all comparable. Consequently, the relative intensity of the geostrophic wind velocity at different levels can be judged from the inclination of the curves at different points because the geostrophic wind velocity is proportional to the inclination of the isobaric surface and does not depend on the density of the air.[1] Figure 11 is

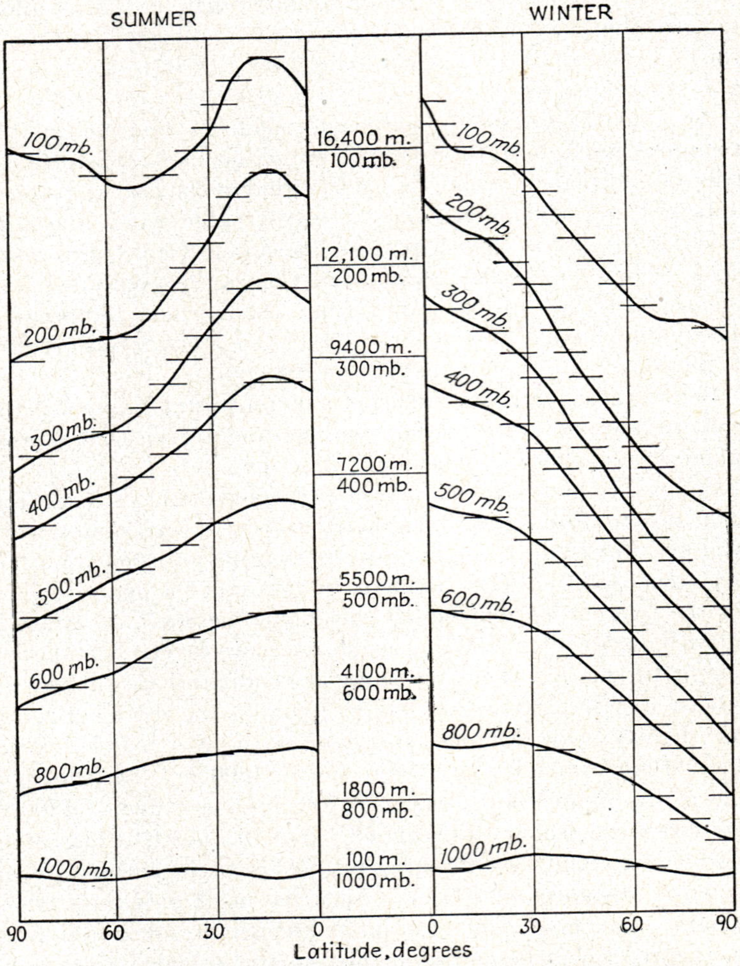

FIG. 11.—Zonal topography profiles in the Northern Hemisphere.

based on observations in the Northern Hemisphere. Hence the following statements refer to the Northern Hemisphere only.

[1] See, for example, HAURWITZ, B., "Dynamic Meteorology," p. 147, McGraw-Hill Book Company, Inc., New York, 1941.

The pressure profiles for the *summer season* with the resulting wind distribution may be considered first. The height profile of the 1000-mb surface corresponds to the meridional pressure distribution at sea level and shows the features discussed in Sec. 12, mainly the pressure increase from the equatorial doldrum belt to the subtropical zone of high pressure, the subsequent decrease northward, through the zone of westerlies, and the slight increase farther toward the pole in the zone of the polar easterlies. In the profiles of the 800-mb and 600-mb surfaces the subtropical highs are hardly indicated, even though upper-air data show that the subtropical highs and the tropical east winds extend to these and higher levels. The apparently poor development of the subtropical highs at the 800-mb and 600-mb levels is a result of the inadequacy of the observational material.

At lower pressures, at 500 mb and less, the pressure maximum is situated at about 10° north. The position of the pressure maximum in the upper troposphere at this latitude during the summer is caused by the meridional temperature distribution. According to Fig. 7, the tropospheric mean temperature has its summer maximum here. Since the pressure decreases less rapidly with elevation at high temperatures than at low temperatures, the highest pressure must be found at about 10° north at upper levels. It follows that at higher altitudes the belt of high pressure is displaced southward from its sea-level position. There is also a westward displacement with elevation on account of the higher temperatures in the western portions of the anticyclones. Consequently, the axes of the subtropical highs have a southwesterly inclination.

The dominant feature of the meridional pressure distribution is the decrease of the pressure from south to north, which can also be explained by the less rapid vertical pressure decrease in the warm air of low latitudes than in the cold air of high latitudes. This meridional pressure gradient and hence the west winds of extratropical latitudes increase in intensity up to the stratosphere. In the stratosphere the direction of the meridional temperature gradient is opposite to the direction in the troposphere. Consequently, the pressure decrease toward the pole becomes smaller in the stratosphere, and at sufficiently high altitudes the pressure increases presumably toward the pole, resulting in easterly winds. The reduction of the pressure gradient in the stratosphere is already indicated by the 100-mb surface, where the total pressure decrease from low to high latitudes is less than in the upper tropopause and the lower stratosphere.

It may finally be pointed out that the highest pressure in polar regions is very shallow according to Fig. 11, so that the polar easterlies, which are poorly developed even at the earth's surface, are to be found only at the lowest levels of the atmosphere. The rapid disappearance of the polar high with the elevation is due to the low temperature in these latitudes, which causes the pressure to decrease more rapidly with the elevation than in the warmer zone of the westerlies. Consequently, the zone of westerlies extends right to the pole, except in the surface layers.

In *winter* the pressure decrease from the equator to the North Pole is much stronger than in summer because the meridional temperature gradient is now much more intense. As a result, the west-wind circulation is considerably more pronounced in winter. Since the highest temperatures are no longer found at about 10° north latitude but at the equator, the pressure maximum at higher levels has shifted toward the equator. In conjunction with the southward displacement of the subtropical anticyclone at the ground the region of equatorial easterlies becomes thus considerably smaller than in summer. Again it must be pointed out that the pressure profiles do not show these details owing to the inadequacy of upper-air data.

The picture of the wind distribution in the upper atmosphere as represented by Fig. 11 is, of course, rather schematic since it cannot show differences in longitudinal direction. Over the Indian Ocean with its monsoon, for example, the atmospheric circulation is completely different from the simple scheme suggested by Fig. 11. Again, if the meridional cross section passed through the Aleutian or the Icelandic Low, the polar easterlies would appear much better developed. However, the predominant feature of Fig. 11, *viz.*, the strong westerlies of higher levels in middle latitudes, can be observed even on the daily maps of the upper air. On these maps the isobars are closed much less frequently than on the sea-level pressure maps. Instead, they generally show sinusoidal forms with troughs of low and wedges of high pressure.

The greater prevalence of the westerlies aloft than at the ground can clearly be shown by the persistence of the winds. Table 17 gives the direction and the persistence of the wind over six stations in the United States at 3000 m elevation. The figures were evaluated by Miss M. Whitcomb and are based on 5 years of observations. A comparison of these figures with those for the "roaring forties" in Table 14 shows how much better the westerlies are developed at higher levels than at the surface. The lowest persistence in Table 17, at Rapid City during July, is equal to the greatest persistence in Table 14 during September. In winter, when the persistence of the upper westerlies at most stations is highest, it is not markedly smaller than the persistence of the southeast trades shown in Table 13.

Table 17.—Persistence of the Westerlies at 3000 m Elevation in the United States

Item	Spokane		Le Havre		Rapid City		Fargo		Sault Ste. Marie		Boston	
	Jan.	July	Jan.	July	Jan.	July	Jan.	July	Jan.	July	Jan.	July
Direct., deg.....	260	243	293	272	306	264	306	287	...	292	291	283
Persist., per cent.	60	72	89	76	82	58	82	76	...	65	87	73

To give a more complete picture of the pressure and wind distribution in the upper atmosphere some maps of the upper pressure distribution may finally be considered. Figures 12 and 13 show the mean pressure distribution at 4 km in winter and summer over the Northern Hemisphere. The winter map (Fig. 12) shows the tendency of the pressure to decrease toward the pole. But instead of a single low-pressure area surrounding the pole there are two centers of low, one over northern Canada, the other over northern Siberia, both in regions where the temperature is very low (see Plate I). In the regions where the Icelandic and the Aleutian Lows are situated at sea level the 4-km isobars do not show any pressure centers. However, the lows over the Canadian archipelago and over northwestern Siberia at 4 km elevation may be regarded as the vertical extensions of the sea-level lows. The vertical axes of these lows are thus inclined toward the west owing to the low temperature of the land. The subtropical highs over the Atlantic and the Pacific, on the other hand, can still be recognized at 4000 m elevation since they are not thermally but dynamically produced. It will be seen by comparison with the sea-level pressure map (Plate V) that the centers of the high are displaced southwestward, as explained previously.

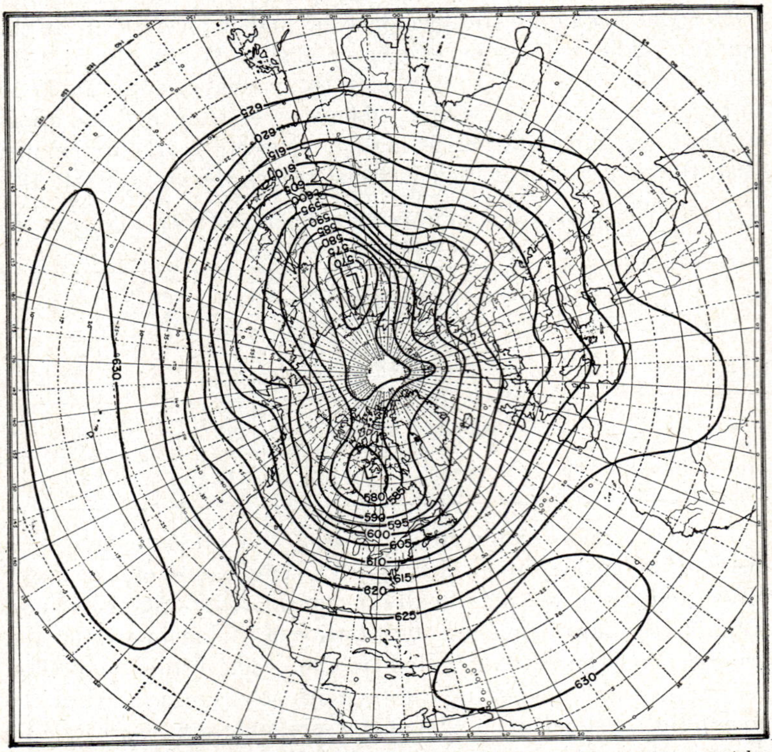

FIG. 12.—Mean pressure distribution over the Northern Hemisphere at 4 km in January. (*After Byers, H. R., and Starr, V. P., Monthly Weather Rev., suppl., No.* 47, 1941.)

In summer (Fig. 13) the meridional pressure gradient, and hence the speed of the westerlies, is considerably smaller than in winter as a result of the smaller temperature contrast between pole and equator. The Pacific anticyclone is not shown on this map because data are lacking to fix its position with any degree of accuracy. Since the air is now warmer over land than over water, high pressure is found over the land at upper levels. Good examples are the high pressure over the Sahara and over

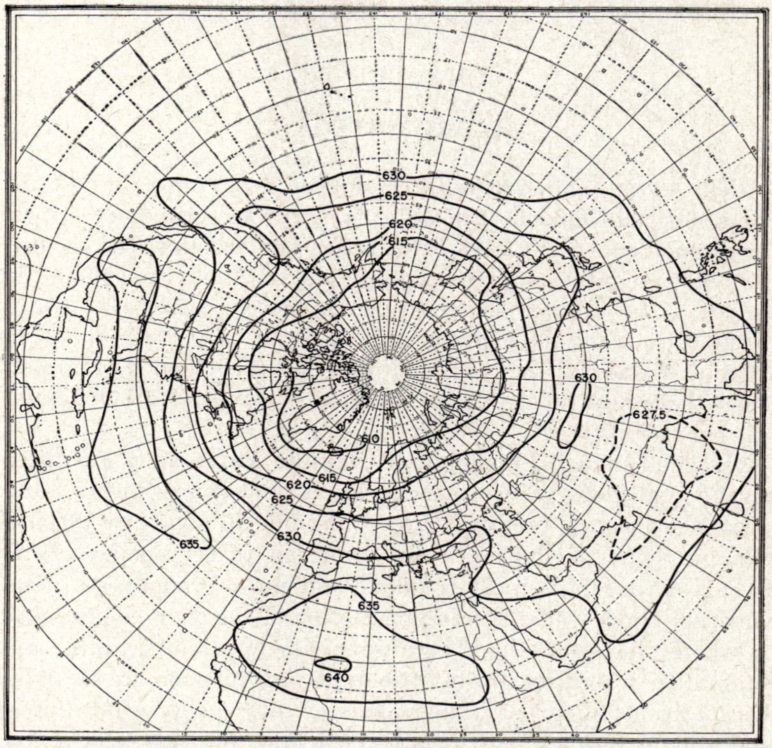

FIG. 13.—Mean pressure distribution over the Northern Hemisphere at 4 km in July. (*From Petterssen, S., "Weather Analysis and Forecasting."*)

the southern United States and northern Mexico. At sea level the pressure is low in these two regions.

CHAPTER IV

HYDROMETEORS

Next to the temperature, precipitation is the most important climatological element since water is one of the prime necessities for animal and plant life. The water and ice that fall to the earth as precipitation have been brought into the atmosphere by evaporation. Hence, evaporation is closely connected with precipitation and is logically discussed in the same chapter. Rain and snowfall are always preceded by cloud formation. Thus the study of clouds and related subjects also belongs in this chapter.

17. The Annual Precipitation over the Earth as a Function of the Latitude. Since condensation and precipitation are almost exclusively produced by cooling due to adiabatic ascent, the rain belts of the earth are found in those regions where ascent of the air is most frequent and most pronounced. On the other hand, where descending motion of the air predominates or where the atmosphere because of its low temperature can retain little water, regions of little or no precipitation are located. The positions and yearly migrations of these belts of precipitation and of dryness will be discussed first, for they will serve as the guiding principle through the intricacies of the distribution of precipitation.

From the discussion of the wind field in Sec. 12 it is apparent that one of the principal zones of precipitation is near the equator, where the northeast and the southeast trades converge into the zone of the doldrums. At the polar limit of the trade-wind regions, within the subtropical high-pressure belts the air descends as in every region with anticyclonic pressure distribution. Therefore, the subtropical high-pressure belts have generally a rather dry climate with little cloudiness, favoring the formation of deserts and semideserts. Poleward beyond the subtropical high pressure, in the zones of the westerlies of the Northern and Southern Hemispheres where the general circulation is broken up into numerous cyclones and anticyclones, precipitation occurs frequently, mainly owing to the ascending motion of air on frontal surfaces between different air masses. This region will be referred to as the "polar-front zone" or the "zone of westerlies." Farther toward the pole the polar easterlies are found. Here the motion of the air, at least in the lowest few kilometers, is mainly descending, and for this reason alone the precipitation in the circumpolar region might be expected to be small. Another reason for the light amounts of precipitation in the polar regions is the low temperature. This makes

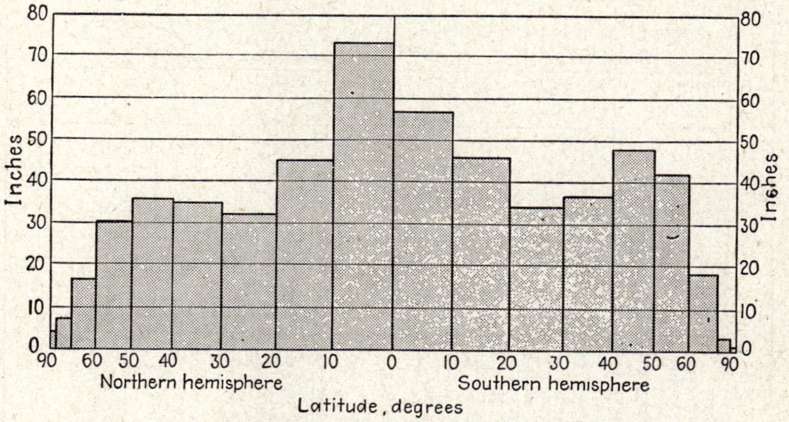

FIG. 14.—Zonal distribution of the annual precipitation.

very small the amounts of water vapor that can be present, even in the case of saturation.

That the theoretical considerations of the foregoing paragraph concerning the zonal distribution of precipitation over the globe are substantially correct is shown by Fig. 14, which illustrates the zonal distribution of precipitation over the earth after Meinardus.[1] This figure gives the distribution of the annual amount of precipitation in inches for latitude zones of 10° width. The abscissa is the latitude on a sine scale to allow for the decrease of the latitude belts toward the poles. In this scale the area of each block is proportional to the total volume of the precipitation falling annually over each latitude belt.

[1] MEINARDUS, W., *Met., Z.*, **51**, 345, 1934.

Figure 14 shows that the height of the annual rainfall is greatest near the equator and falls off as the subtropical high-pressure belts are approached. Beyond the subtropical high-pressure belts the height of the annual rainfall increases again toward the zones of the westerlies, where strong frontal activity takes place. Farther toward the poles the precipitation drops rapidly to low values.

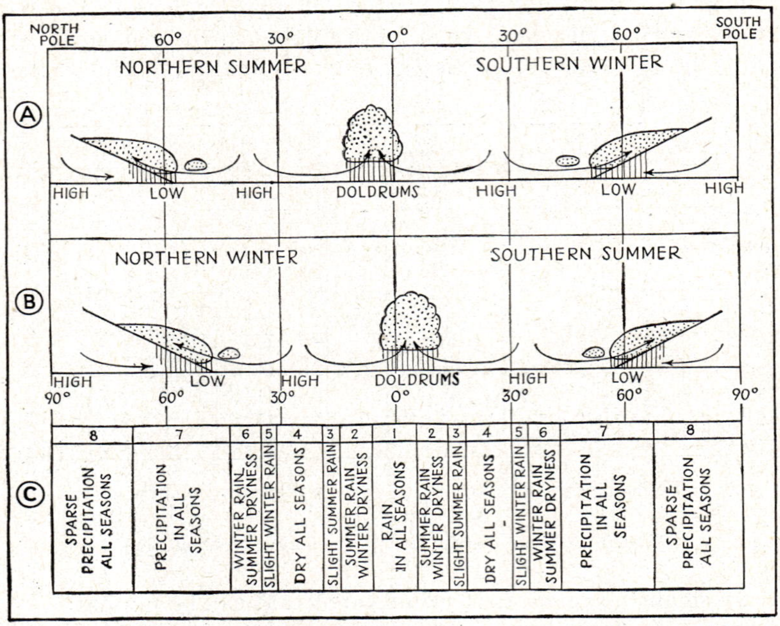

FIG. 15.—Explanation of the precipitation zones. (*After Petterssen, S., "Introduction to Meteorology."*)

This very simple picture of the global distribution of precipitation is disturbed by a number of factors, especially by (1) the seasonal shift of the zones of doldrums, subtropical high pressure, and westerlies; (2) the distribution of water and land; and (3) the presence of mountain ranges.

With respect to the first of these factors it has been pointed out in Sec. 12 and it can be seen from Plates IV and V for the pressure and wind distribution in July and January that these zones which produce rain and dryness follow the sun in its motion to the Northern Hemisphere during the northern summer and to the Southern Hemisphere during the southern summer. The shift of the zones of high and low precipitation produces additional belts of prevailing summer and winter rains.

Thus, one can distinguish eight precipitation zones in each hemisphere from the equator to the pole. These are shown in Fig. 15, after Petterssen. They may be described as follows:

Zone 1. In the equatorial low-pressure belt with its pronounced convergence, high temperature, and strong convective activity, the precipitation is heavy. Where the doldrums occupy the same position throughout the year the rainfall will be distributed uniformly throughout the year. Where an appreciable seasonal shift takes place, two maxima and minima may occur. But, in general, this zone is characterized by *heavy rains throughout most of the year.*

Zone 2 is under the influence of the doldrums in summer. Thus, is has *heavy summer rain and is dry in winter.*

Zone 3 is affected by rain from the doldrums only during the summer and then only slightly. Thus, there is only *slight summer rain and pronounced dryness in winter,* when the zone is under the regime of the subtropical high-pressure belt.

Zone 4. This zone is dominated by the conditions of the subtropical high-pressure belt throughout the year. Hence *dryness prevails throughout the year.*

Zone 5. The polar-front zone extends its influence to the poleward boundary of this zone in winter so that this is a region with *slight winter rain and pronounced dryness during the remainder of the year.*

Zone 6. Still farther away from the equator, the duration of the winter rain will be longer than in zone 5. But in summer the zone is under the regime of the subtropical high-pressure belt. The zone has *winter precipitation and summer dryness.*

Zone 7 is dominated practically throughout the whole year by the prevailing westerlies with their frontal ascent of air. *Precipitation occurs during all seasons.*

Zone 8. In this region polar anticyclones predominate. However, cyclones and fronts occur occasionally. In view of the low temperatures and of the anticyclonic circulation, *precipitation is slight.*

The average annual amount of the precipitation is almost the same in the Northern and in the Southern Hemisphere, according to Meinardus 39.2 in. and 39.6 in., respectively, in spite of the difference in the zonal distribution in both hemispheres, shown in Fig. 14. The most outstanding of these differences is that the precipitation is considerably higher from 0° to 10° north latitude than from 0° to 10° south latitude. This can be explained by the fact that the doldrums are largely north of the equator all the year round. At higher latitudes the secondary maximum of precipitation in the zone of the westerlies from about 40° to 60° is more pronounced in the Southern than in the Northern Hemisphere. The higher precipitation in the zone of the westerlies in the Southern Hemisphere is essentially an effect of the much larger percentage of ocean surface in this zone in the Southern Hemisphere. In the Northern Hemisphere only 45 per cent of the area between 40° and 60° is taken up by the oceans; in the Southern Hemisphere, 98 per cent. Since the oceans represent the greatest reservoirs of moisture, it is natural that the precipitation is greater where the earth's surface is largely covered by water. That the total amount of precipitation is almost equal in both hemispheres, in spite of the much larger extent of the continents in the Northern Hemisphere, is due to the fact that almost 50 per cent of the total annual precipitation falls on a belt extending from 20° north to 20° south of the equator. Here the percentage of land area is almost equal in both hemispheres. In this belt, in fact, the precipitation is larger in the Northern than in the Southern Hemisphere. On the other hand, in the zone of the westerlies the precipitation is more abundant in the Southern Hemisphere than in the Northern Hemisphere. Thus, the mean precipitation is the same in both hemispheres.

The precipitation over the oceans and over the continents—whose distribution is the second factor influencing the amount of precipitation—can be compared with the aid of Table 18. This table gives, according to Meinardus, the zonal distribution of the annual precipitation for the continents and the oceans separately and the means, which are weighted according to the percentage of land and sea area within each latitude. Figure 14 is based on the weighted means.

As just stated, the precipitation in the same latitude zone is as a rule greater over the oceans than over land. Only in the belt from 0° to 10° south is the annual amount of the precipitation greater over land than over water. In every other latitude belt more rain falls over the oceans than over the continents. The secondary maximum in the polar-front zone is much greater over the oceans than over the continents.

For the whole earth the average height of the annual precipitation is 26.4 in. over land and 44.9 in. over water. According to its mass, 19 per cent of the precipitation falls over land and 81 per cent over water, while the percentages of land and water are 29 per cent and 71 per cent, respectively.

TABLE 18.—ZONAL DISTRIBUTION OF ANNUAL PRECIPITATION, IN., OVER THE EARTH

Latitude	90°–80°	80°–70°	70°–60°	60°–50°	50°–40°	40°–30°	30°–20°	20°–10°	10°–0°
Northern Hemisphere									
Oceans	4.4	8.4	26.9	44.2	53.2	46.3	35.3	49.3	78.4
Continents	4.4	5.7	12.0	19.2	20.2	23.2	26.6	32.1	55.3
Weighted mean	4.4	7.6	16.4	30.2	35.9	34.8	32.0	44.8	73.1
Southern Hemisphere									
Oceans	1.8	4.0	19.2	42.0	48.0	38.6	36.2	46.6	55.7
Continents	0.7	3.1	6.7	38.1	31.4	22.2	26.0	42.8	60.6
Weighted mean	0.7	3.3	18.1	42.0	47.6	36.8	33.9	45.7	56.7

18. The Effect of Continents and Oceans on the Distribution of Precipitation. It now becomes necessary to consider the actual distribution of the annual precipitation in order to see how the simple picture of the latitudinal distribution of precipitation is modified by the distribution of land and water. Plate IX shows such a map, largely based on work by Schott and Meinardus. In studying this map it is, of course, important to keep in mind that the amounts recorded for precipitation over the oceans and over the unsettled parts of the continents are not wholly reliable.

It has been suggested by Wüst[1] that the figure of Meinardus for the precipitation over the ocean is too high since it is 1.7 times larger than the mean precipitation over the land. While it can be expected that more precipitation occurs over water

[1] WÜST, G., *Z. Ges. Erdkunde (Berlin)*, **47**, 35, 1922.

than over land, the factor 1.7 appears doubtful. The suspected overestimate may be due, at least in part, to the fact that the maps from which Meinardus obtained his rainfall data for the ocean areas are mainly based on information gathered at island stations. Over islands the precipitation will be greater than over the free ocean owing to orographic and convective effects. The question of the overestimate of the precipitation over oceans will be considered again on page 85 in connection with evaporation.

The great equatorial rain belt can easily be recognized in Plate IX. The precipitation, of course, is not the same everywhere within this belt. But with a few exceptions, for instance along the dry east coast of equatorial Africa, it is higher in this zone than farther north and south, where it decreases with the latitude as one approaches the dry belts of the two hemispheres. North of the equator the dry belt extends from Iran and Afghanistan through the Arabian Desert and the Sahara far out into the North Atlantic. In the Western Hemisphere the dry regions of northern Mexico and the southwestern United States form a part of this dry belt. In the Southern Hemisphere the driest parts of the dry belt are found over the western parts of Australia, South Africa, and South America. As in the case of the equatorial rain belt the degree of dryness is not the same everywhere in this zone. The two subtropical dry zones do not form continuous dry belts surrounding the globe. They are interrupted by zones of higher precipitation along the east coasts of the continents and over the western parts of the oceans because in the subtropical highs the air has an ascending component during its westward motion and becomes more unstable. Moreover, the trades with their easterly wind components prevail so that the east coasts are exposed to moist winds from the ocean.

Farther poleward, beyond the dry belts, the precipitation increases again as the zones of frontal disturbances are approached. In these latitudes the west coasts of the continents have higher precipitation than the east coasts since the prevailing wind direction is here from the west. The precipitation map shows clearly the decrease of the annual amount toward the interior of the land. Over the North and South American continents this decrease from the sea inland is greatly influenced by the effects of the mountains, which will be discussed in the next section. Over Eurasia no such orographic obstacles exist, the Ural mountains being too low to have a pronounced influence. Consequently, the decrease of the precipitation inland is much more gradual. At high latitudes, beyond the zones of the westerlies, the annual amount of precipitation decreases again owing to the low temperature of the atmosphere.

To assess the importance of a certain amount of rainfall for the climate and for the vegetation it is necessary to compare it with other climatic elements. The same amount of rainfall may produce a desert or steppe climate at low latitudes where the temperature and hence the evaporation are high, while in a cool climate it may be adequate to permit the existence of forests or of glaciers. This question will be discussed in greater detail in Chap. VI, when the different climatic types are considered.

Another important factor is the variability of the rainfall. In regions where the precipitation is abundant the variability is in general small and, moreover, of little practical importance, since there will be as a rule sufficient water for vegetation even if the amount of precipitation is below normal. But in regions where the annual precipitation is just barely sufficient for agriculture in normal years its variability is very significant. Years with precipitation below normal bring crop failure with serious economic dislocations, as, for example, in the western United States, or even famine, as in China or India.

19. Precipitation and Altitude. The world map (Plate IX) shows clearly the effect of the third factor (page 66), the altitude of the land, on the distribution of precipitation. On the windward side the amount of precipitation is considerably increased by the forced ascent of the air. Notable examples for such orographically produced areas of high precipitation are found on the west side of the Rocky Mountains in North America, on the west side of the Andes in the temperate regions of South America, and along the west coast of Norway. The rains of the southwest monsoon during the Indian summer are greatly intensified on the Ghats along the west coast of the peninsula and again on the southern slope of the Himalaya. Where the prevailing winds are from an easterly direction, the maximum precipitation, of course, is found on the east side of the mountain range. Such is the case, for example, in the subtropics along the east coast of Madagascar. The leeward slopes of all these mountain chains, on the other hand, are in the rain shadow and have much less precipitation. Under such conditions even deserts may develop. A good example of such a desert on the lee side of a mountain range is the Patagonia desert along the east coast of southern South America.

In the greater and more extended mountain systems each range has its windward and leeward slopes, with higher and lower precipitation, respectively. The total precipitation also decreases on each range as one proceeds through the mountain system to the lee side. For example, heavier precipitation occurs over the western than over the eastern Rocky Mountains of North America.

The amount of precipitation increases with the altitude. But at least in the tropics this increase continues up to a certain elevation only. This is not surprising since the amount of condensed water during pseudo-adiabatic ascent of the air is greater at higher than at lower temperatures. Moreover, at greater elevations there are more gaps in the mountain ridges than at lower levels so that part of the air can pass around the barrier instead of ascending over it. Conrad[1] quotes an example for the zone of maximum of orographic rain in the tropics that is reproduced in Table 19.

TABLE 19.—RAINFALL AND ALTITUDE IN CENTRAL JAVA

Altitude, ft.	6,925	8,235	10,105
Rainfall, in.	105.1	118.1	83.9

In the mountains of temperate latitudes such a level of maximum precipitation does not seem to exist according to the most recent measurements. However, Conrad points out that these data are not quite conclusive since the precipitation in the higher mountains of temperate latitudes falls largely as snow, which covers the mountain slopes very loosely and might easily be carried to higher and higher elevations by the strong winds.

The question of the dependence of precipitation on the altitude is of special interest in considerations involving the potential

[1] CONRAD, V., "Fundamentals of Physical Climatology," Harvard University Press, Cambridge, Mass., 1942.

water power of a region.[1] Water or subsequently melting snow that falls on high ground will flow to lower levels. It represents therefore a store of potential energy. The potential water power of a region depends thus on two factors, the amount of precipitation and the height at which the precipitation reaches the surface of the earth. Even if the whole potential water power could not be utilized for the generation of energy owing to losses by evaporation and penetration into the ground, it would give an important indication of the amount of available water power in a given region.

20. The Annual Variation of Precipitation. In addition to the annual total of the precipitation its seasonal variation is of great importance for a characterization of the climate. The usefulness of precipitation for agriculture is much greater when it occurs during the growing season of the plants than when it occurs during the period when the vegetation is dormant, as during the cold season of temperate latitudes.

There are a great number of different types of the yearly variation of precipitation. In the following discussion only six will be distinguished (Fig. 16).

1. *The equatorial type* has two maxima shortly after the equinoxes, *i.e.*, shortly after the time when the sun passes the zenith (Fig. 16, Yaoundé in the Cameroons). The origin of the two rainfall maxima is easily understood. In the first place, the equatorial rains are caused by convergence of the air flow in the zone of the doldrums, and the zone of the doldrums follows the sun in its yearly shift north and south of the equator. Consequently, the zone of maximum rains follows the sun also. Moreover, vertical convection, which contributes considerably to the strength of the equatorial rains, is strongest when the sun is highest or shortly thereafter, *i.e.*, at the equator, at the time of the equinoxes.

The equatorial type of annual variation of precipitation is found between about 10° north and south of the equator. With increasing distance from the equator the two rainfall maxima come closer and closer together and finally combine into one single maximum. In general, the two maxima are of different intensity. Moreover, not every station near the equator actually has two maxima. On the other hand, even stations outside the

[1] Penck, A., *Sitz.-Ber Akad. Wiss. (Berlin), Phys.-Math. Kl.*, **9**, 135, 1930.

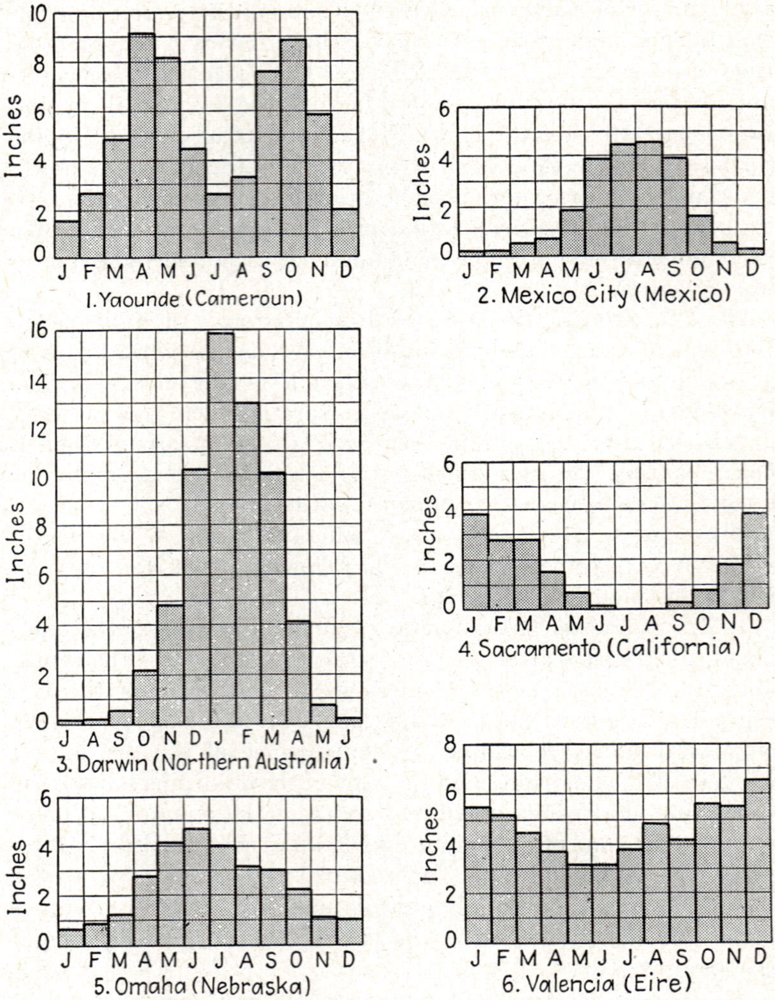

FIG. 16.—Types of annual variation of precipitation. 1, equatorial type—Yaoundé; 2, tropical type—Mexico City; 3, monsoon type—Darwin; 4, subtropical type—Sacramento; 5, continental type—Omaha; 6, maritime type—Valencia.

equatorial zone may have two rainfall maxima, but in such cases the double maximum is due to other causes, frequently to regular seasonal wind changes.

2. *The tropical type* has one rainfall maximum (see Fig. 16, Mexico City); it occurs in summer at or shortly after the time when the sun has reached the zenith, because the zone of the doldrums is then closest to the station and the convective activity is greatest owing to convergence. During the time when the sun is in the other hemisphere, corresponding to the winter of higher latitudes, the precipitation is small. There may even be a dry period since the zone of the doldrums is then far away and since the station may at this time be under the influence of the subtropical high-pressure belt with its dry weather and cloudless skies.

While in the tropics the rains are in general most intense during the season when the sun is highest, there are rare exceptions to this rule, caused by orographic conditions. Tropical regions with highest rainfall in winter may be found where the trade winds from the ocean are forced to ascend over a mountain range. Since the trades are stronger in winter than in summer, the maximum occurs here during this season, contrary to what is experienced elsewhere in the tropics. But as a rule the rainfall during the remainder of the year is quite abundant, too, and a season of dryness or near dryness does not exist.

3. *The monsoon type* is represented by Darwin in northern Australia (Fig. 16). Since in the Southern Hemisphere the seasons are opposite to those in the Northern Hemisphere, the graph for the monthly precipitation at Darwin starts with July in order to facilitate the comparison with the other types of yearly rainfall distribution shown in Fig. 16. Darwin lies at about 12.2° south latitude. The sun passes here through the zenith at the end of October and about the middle of February. The maximum rainfall, however, occurs in January, when the northwest monsoon is strongest and masses of moist oceanic air are often found over the northern part of the Australian continent. This type resembles the preceding one in that the rainfall maximum occurs during the summer. But the latitude of the station is sufficiently low so that two rainfall maxima could be expected if it were not for the effect of the monsoon winds.

In summer the prevailing winds are from the ocean to the continent so that the precipitation is more abundant than if the monsoon were absent. In winter, on the other hand, the prevailing wind direction is from the dry and relatively cool continent to the ocean. Hence precipitation is small or completely absent. The monsoon rains of the Northern Hemisphere fall from June to August or September in India and along the coast of eastern Asia. The monsoon type is thus closely related to the continental type with prevailing summer rain, to be discussed.

4. *The subtropical type* has winter rains. As discussed in the preceding section, the subtropical high-pressure belts with their dry regions and the zones of westerlies with their frontal precipitation follow the seasonal motion of the sun. Consequently, along the polar margins of the subtropical belts in both hemispheres summer is the dry season and winter the wet season. As an example the monthly rainfall at Sacramento, Calif., is shown in Fig. 16. On the east coasts of the continents this type does not occur, largely owing to the monsoon effect. The best known examples of the subtropical type of rainfall occur along the west coast of California and in the Mediterranean region; this type is therefore also called the "Californian" or "Mediterranean" type.

5. *The continental type* has summer rains. In the interior of the continents the summer months have a heavier precipitation than the winter months (see Fig. 16, where the monthly precipitation at Omaha, Neb., is given as an example). The smaller amount of precipitation in winter than in summer is partly due to the lower temperature of the air, which reduces the possible water-vapor content. Moreover, the lower temperatures in winter favor the development of anticyclones, which tend to reduce the amount of precipitation. On the other hand, in summer the convective activity will be considerably greater than in winter.

A comparison of the tropical, monsoon, and continental types (curves 2, 3, and 5, Fig. 16) shows that they are quite similar in that the maximum precipitation occurs in summer. The main difference between the continental type and the other two types is that the drier period during winter is less pronounced over the continents in temperate latitudes.

The form of the annual-precipitation curve is the same throughout the longitudinal extent of the European and Asiatic continent from central Europe to the east coast of Asia. The maximum of precipitation occurs during the summer; the minimum during the winter. As a result, there is a continuous transition from the continental type of precipitation to the monsoonal type.

6. *The Maritime Type.* The cyclonic activity over the oceans is most intense during winter. Furthermore, the maritime air masses are warmer and their moisture content, consequently, larger during late autumn rather than midwinter. Therefore, the precipitation over the oceans and the adjoining coastal regions is greatest during the early winter, while in the interior of the continents the frequent presence of anticyclones and the lower temperature reduces the precipitation considerably. As an example of this type, Valencia, the Irish Free State, is shown in Fig. 16.

Along the more northerly part of the west coast of North America this type represents a continuation of the subtropical type that is found in California. It differs from the latter in that a dry period during the summer is absent. The maritime type is best developed along the west coasts in temperate latitudes; the east coasts, especially the east coast of Asia, have a monsoon regime with summer rains, as explained previously.

On the east coast of North America a monsoon is absent, and the region is visited by cyclones in summer as well as in winter. Therefore, the monthly amount of precipitation is here rather uniform. It would thus be possible to introduce a different type of annual variation of precipitation. The small seasonal variation along the Atlantic seaboard is shown in Table 20.

TABLE 20.—PRECIPITATION, IN., ALONG THE ATLANTIC SEABOARD

Place	Jan.	Feb.	Mar.	Apr.	May	June	July	Aug.	Sept.	Oct.	Nov.	Dec.	Year
Portland, Maine	3.9	3.6	3.7	2.9	3.0	3.0	3.0	3.0	2.8	3.5	3.3	3.7	39.3
Boston	3.5	3.7	3.5	3.4	3.1	2.9	3.4	3.5	3.2	3.1	3.4	3.4	40.1
New York	3.3	3.3	3.4	3.3	3.5	3.5	4.1	4.3	3.4	3.4	3.3	3.3	42.1
Washington, D.C.	3.2	3.0	3.5	3.3	3.6	3.9	4.4	4.0	3.1	3.1	2.5	3.1	40.5

The discussion of the various types of the seasonal variation of rainfall cannot be concluded without emphasizing that these six types represent a schematization and simplification of the great variety of types which actually do occur on the earth.

Obviously, many modifications and transitions between these types occur. The foregoing division is to be regarded as a first aid and a guide for the student of climatology and not as a rigid scheme into which all annual rainfall curves can be fitted.

21. The Diurnal Variation of Precipitation. Two main types of the diurnal variation of the precipitation can be distinguished. The *maritime type* has its maximum during the night or in the early morning; the *continental type* in the early or late afternoon (for examples, see Table 21). It should be added, as in the

TABLE 21.—DIURNAL VARIATION OF THE PRECIPITATION AT BATAVIA
(After Braak[1])
Per cent of the daily total

Season	Hours					
	0–4	4–8	8–12	12–16	16–20	20–24
Rainy (Jan.–Feb.)	22	18	13	14	16	17
Dry (July–Sept.)	12	7	6	23	33	19

[1] Quoted after V. Conrad, in Köppen-Geiger, "Handbuch der Klimatologie," Vol. 1, B, p. 521, Verlagsbuchhandlung Gebrüder Borntraeger, Berlin, 1936.

case of the annual variation, that all sorts of deviations from these two types occur. Thus, many continental stations have actually two maxima, one in the early morning, one in the afternoon. Some continental stations, for example, Bern (Switzerland) and Irkutsk (Siberia) have their main maxima at night and thus belong to the maritime type. At other continental stations the continental type prevails in summer, the maritime type in winter.

The continental type with its precipitation maximum in the afternoon is evidently due to the effect of convection, which over land is strongest in the early afternoon. Thus, a tendency exists to produce a precipitation maximum about this time of the day. Moreover, convection is stronger during the summer than during the winter, explaining why at some stations the diurnal-precipitation curve changes from the maritime type in winter to the continental type in summer. The most pronounced afternoon maxima of precipitation occur in the tropics since here the intensity of the solar radiation and consequently the convective activity is greatest. However, even in the tropics the maritime type with its maximum during the night or the early morning is found in those localities which have a maritime exposure. In those regions with a seasonal shift of the prevailing winds the type may change from one season to the other, as shown in the preceding example for Batavia. In order to explain the origin of the maritime type of the diurnal variation of the precipitation it must be remembered that, in contrast to the surface of the land, the surface of the ocean has hardly any daily temperature variation (see page 35). Consequently, the temperature of the surface layers of the atmosphere also changes very little. Somewhat higher up, on the other hand, the air temperature is largely controlled by radiation and becomes warmer during the day and cools off during the night. Thus, the vertical temperature gradient is greater at night and smaller during the day. Consequently, over the ocean the convective activity is greater at night than during the day.

It follows that the difference of the diurnal period of convection over land and sea is in the right direction to explain the daytime maximum of precipitation over the land and the nocturnal maximum over the sea. Specifically, frontal precipitation, because of the smaller stability over the sea, will be greater at night than in daytime. A further factor that tends to make the precipitation greater at night than in daytime is the lower temperature. Since the air cools during the night, it is more predisposed to condensation. Where the maritime type of the diurnal variation of precipitation occurs inland, this factor must be the most influential.

For a more detailed analysis of the difference between the two types of the diurnal-precipitation period it is important to consider the three quantities that characterize precipitation, duration, intensity, and amount. The amount is the product of the intensity and the duration. Köppen[1] has compared the variation of these three quantities for three typical stations (Table 22). Valencia is a station whose daily variation of the amount of precipitation is of the maritime type, while at Pavlovsk the continental type is found. San José, Costa Rica, is a tropical station in Central America. While the diurnal range of the amount of rain differs strikingly at the first two stations,

[1] Köppen, W., "Die Klimate der Erde," p. 67, Walter de Gruyter & Company, Berlin and Leipzig, 1923.

the variations of the duration and of the intensity of the precipitation are parallel. The duration of rain has its maximum at the two extratropical stations during the late night or the early morning, while the most intense rains occur even at such a typically maritime station as Valencia in the afternoon, when convective activity is greatest. In the tropics, where most rains are of convective origin, amount, duration, and intensity of the precipitation all have their maximum in the afternoon.

TABLE 22.—DIURNAL VARIATION OF THE PRECIPITATION AT THREE TYPICAL STATIONS

(After Köppen)

(A = amount in thousandths of the total, D = duration in thousandths of the total duration, I = intensity in thousandths of inches per hour of rain)

Station		Hours					
		0–4	4–8	8–12	12–16	16–20	20–24
Valencia, Irish Free State, 52°N	*A*	181	*183*	160	149	162	165
	D	186	*213*	154	128	145	164
	I	38	33	41	*46*	44	40
Pavlovsk, U.S.S.R., 60°N	*A*	148	156	144	*201*	185	157
	D	174	*220*	152	146	149	156
	I	16	14	19	*27*	24	20
San José, Costa Rica, 10°N	*A*	13	6	44	342	*485*	84
	D	126	56	54	224	*343*	191
	I	35	31	23	108	*131*	40

22. Snow. Until now no distinction has been made in this chapter between the different hydrometeors of which rain and snow are the most important, hail, dew, and frost being rather negligible from a climatological point of view.

Snow, however, is of considerable practical importance. Where it falls during a considerable part of the year or where large masses may fall during a short time without melting soon after reaching the ground, it has a considerable effect on traffic and transportation. Severe snowstorms may result in the blocking of roads and railroad lines.[1] If snowfalls are frequent, they

[1] PALMER, A. H., *Monthly Weather Rev.*, **47,** 698, 1919. Also BROOKS, C. F., *Monthly Weather Rev.*, **43,** 2, 1915; **45,** 271, 1917.

necessitate the erection and maintenance of snow fences and snow tunnels and the use of equipment to clear highways and railroad tracks. Sudden warm spells after large snowfalls result in disastrous floods.

For the vegetation a permanent snow cover during the winter is important for two reasons. In the first place it provides a protection for the underlying ground and seeds and plants that it contains. Furthermore, it permits the storage of precipitation. Rain falling during the winter has disappeared by the time the growing season begins. But as the snow melts during the spring, the water filters into the ground, moistening it to great depths. This storage of water in the form of snow is especially important in regions where precipitation is not large because it ensures a sufficient supply of water at the beginning of the growing season.

The melting of the snow in spring delays the increase of temperature, for a considerable amount of the solar radiation is used, not to heat the ground, but to change snow of 32°F into water at the same temperature.

Since snow reflects most of the incident solar radiation, the snow surface and the adjacent air are cooled to very low temperatures while the temperature at greater elevations is considerably higher. In this manner the typical temperature distribution of polar continental air masses is produced, with its very low surface temperature and with an inversion aloft.[1] Hence the continental regions with permanent snow covers during the winter, such as Canada and northern Siberia, are source regions of polar continental air.

Data on the frequency and amount of snowfall and the duration of the snow cover are still very scanty for large parts of the globe. An indication as to the occurrence of snow and the persistence of a snow cover can be obtained from the mean monthly temperature maps. According to Köppen the snow will remain on the ground for an appreciable time where the mean temperature of the coldest month is 27°F or below. But the estimates of the extent of the snow cover based on such a criterion are, of course, not very reliable. Therefore, only a

[1] See for example, Petterssen, S., "Weather Analysis and Forecasting," p. 169, McGraw-Hill Book Company, Inc., New York, 1940.

few general remarks will be made here about the world-wide occurrence of snow.

At greater elevations snow falls at lower latitudes than at sea level—at sufficiently great heights, even at the equator. The snow, once it has fallen, remains longer on the ground at higher altitudes, and at sufficiently great heights it persists the year round. Some examples of the altitudes above which the snow does not melt from year to year are given in Table 23, after Paschinger,[1] together with the annual mean temperatures and the summer temperatures. These examples show that permanent snow occurs even in regions where the temperature of the warmest

TABLE 23.—HEIGHT OF THE SNOW LINE (*H*), ANNUAL MEAN TEMPERATURE (*T*) AT THE SNOW LINE, AND SUMMER TEMPERATURE
(After Paschinger)

Region	Latitude	*H*, ft	*T*, deg F	Summer temperature, deg F
Norway	68°N	3,500	21	
Northern Tirol	47°N	9,200	25	June–Sept. 38
Himalaya, northern slopes	27–34°N	18,700	27	Summer 43
Himalaya, southern slopes	27–34°N	16,200	33	Summer 45
Kilimanjaro	3°S	18,400	21	

months is appreciably above freezing. To understand this fact it must be remembered that the existence of a permanent snow cover depends on two factors. In the first place the temperature must be reasonably low, and in addition the precipitation must be heavy. If the snowfall is large, the amount of heat received during the warmer season is not sufficient to remove the snow cover completely. As an example, the snow conditions on the northern and southern slopes of the Himalaya may be considered. On the southern slopes of the Himalaya even the annual mean temperature is slightly above freezing. But the amount of snow that falls here is so great that a permanent snow cover can exist slightly above 16,000 ft throughout the year. On the northern slopes of the Himalaya the snow line is considerably higher, 18,700 ft, because these slopes are in the lee of the monsoon winds and therefore receive considerably less precipitation than the southern slopes. Consequently, the

[1] PASCHINGER, V., *Petermanns Mitt.*, Ergänzungs-Heft 173, 1912.

snow cover is less deep and can be melted to a considerably greater altitude than on the south side.

The example of the Himalaya shows clearly the great importance of the amount of precipitation for glaciation and the development of ice ages. It is not sufficient and, on the other hand, not even necessary for the mean temperature to drop below the freezing point. More important is the amount of the snowfall.

23. Evaporation. The water that is suspended in the atmosphere in the form of water vapor, liquid water, or ice and that is returned to the earth as precipitation has been brought into the air by evaporation. The evaporation from a water surface depends in the first place on the difference between the saturation vapor pressure over the water surface and the actual vapor pressure of the air. The greater this difference, the more intense the process of evaporation. If the vapor pressure of the atmosphere is higher than the saturation vapor pressure with respect to the water surface, the opposite process takes place: water vapor condenses on the water surface. This may happen when warmer air passes over a colder water surface. Such situations frequently occur, for example, over the Grand Bank off Newfoundland, especially during spring and summer. They induce the formation of fog (advection fog).

The amount of evaporation increases with the wind velocity. If the air remains stagnant over the water, its vapor pressure will increase, and thus the difference between the saturation vapor pressure with respect to the water surface and the vapor pressure of the air will decrease. Consequently, the evaporation becomes gradually smaller. On the other hand, when a wind is blowing, the air in contact with the water is always replaced by less moist air, at least over small bodies of water and near the shore if the wind is offshore, so that the atmospheric vapor pressure does not reach such high values as in calm air. Furthermore, with higher wind velocities the turbulence is greater, and the water vapor is more rapidly transported upward from the surface. This turbulence effect, of course, is also active in the central parts of large bodies of water where a direct transport of fresh air from land is out of the question. Finally, the increase of evaporation with higher wind velocities may partly be due to ocean spray, which brings water droplets into the air and thus increases the effective surface.

The evaporation increases also with higher temperature, other things being equal.

It has already been pointed out that condensation rather than evaporation may take place if the air temperature is higher than the water temperature. This case will arise, however, only if the vapor pressure of the atmosphere is higher, which is not necessarily always true when the air is warmer than the water. For example, if the air temperature is 77°F and the relative humidity only 50 per cent, the vapor pressure of the atmosphere is 15.8 mb. If the water temperature at the same time is 68°F, the saturation pressure with respect to water is 23.4 mb and evaporation can still take place. In this example, only if the relative humidity rises to 68 per cent or higher will evaporation cease.

On the other hand, if the water temperature is higher than the air temperature, evaporation can always take place, for the saturation vapor pressure with respect to the water will then always be greater than the vapor pressure of the atmosphere. Moreover, in this case the air is heated from below by contact with the warmer water. Hence, the stratification is unstable, turbulence large, and evaporation high.

Evaporation can be determined in different ways. The most direct method is observation of the evaporation from a vessel filled with water. However, the results are not entirely reliable, especially when such observations have to be made on board ships, the ship itself disturbing the normal course of evaporation. Climatological mean values of the evaporation over the oceans can be computed from energy considerations.[1] Over land evaporation is often determined by equating the difference between precipitation and evaporation of an area to the amount of water that is removed by rivers and streams. In this case it is assumed that no water is lost or gained by net transport from or to the area because of the air currents. This assumption is sufficiently well satisfied for large areas. Finally, relations between evaporation and the meteorological elements can be derived from the turbulence theory.

The data on evaporation are still very scanty, and it cannot be claimed that the figures which are given below have more value

[1] SVERDRUP, H. U., "Oceanography for Meteorologists," p. 65, Prentice-Hall, Inc., New York, 1942.

than that of a first orientation. Wüst[1] has estimated that the average evaporation for the whole earth may be about 29.2 in. per year.

The average annual amount of precipitation, on the other hand, is 39.4 in. for the whole earth (see page 68), which is considerably larger than Wüst's figure. The average values for evaporation and for precipitation should, of course, agree. Wüst's figures for evaporation are probably too low, but the discrepancy cannot be accounted for entirely by assuming that they are too small. It is very likely that the figure for precipitation is somewhat too high since the precipitation on the oceans is to a large extent estimated from observations on islands, where convection gives larger precipitation, as explained on page 70.

Table 24 gives the amount of evaporation over the continents and oceans and mean values for various latitudes after Wüst. In considering these figures it should be remembered that they will

TABLE 24.—ZONAL DISTRIBUTION OF EVAPORATION, IN.
(After Wüst)

Latitude	60°–50°	50°–40°	40°–30°	30°–20°	20°–10°	10°–0°
Northern Hemisphere						
Continent	14.2	13.0	15.0	19.7	31.1	45.3
Ocean	15.7	27.6	37.8	45.3	47.2	39.4
Mean	15.0	20.1	28.0	35.8	42.9	40.6
Southern Hemisphere						
Continent	(7.9)*	(19.7)*	20.1	16.1	35.4	48.0
Ocean	9.1	22.8	35.0	44.1	47.2	44.9
Mean	8.7	22.8	33.5	39.0	44.5	45.7

* Estimated from ocean measurements.

be on the whole too small. However, they give at least an idea of the order of magnitude and of the relative size of the evaporation at different latitudes. It should be added here that the figures for evaporation over land are based on zonal values of precipitation and of run-off so that they represent actual evaporation better than figures which could be obtained from pan observations. The latter would rather represent the so-called

[1] WÜST, *op. cit.* See also MEINARDUS, W., *Petermanns Mitt.*, **80,** 1, 1934.

"potential evaporation," *i.e.*, the amount of water that would be evaporated if a continuous water supply were available.

The higher evaporation over the oceans is mainly due to the fact that here an unlimited supply of water is available for evaporation, while over land in many regions, especially in deserts and on steppes, the water supply is very small or nil, so that the actual evaporation is very small even though the potential evaporation is large. Between 10° north and 10° south evaporation is actually larger over land than over water. In this equatorial zone the continents are humid, vegetation is dense, and the water supply never becomes exhausted. The evaporation maxima over the oceans are not at the equator but at 10° to 20° northerly and southerly latitude. The occurrence of the maxima in these zones rather than at the equator may be due to the strong trade winds that reach from higher latitudes down to these regions. Nearer the equator, on the other hand, the wind velocities are much lower and the relative humidities higher. These remarks do not apply to the continents since the trade winds are not well developed here, at least during a great part of the year. Consequently, over the continents the evaporation maximum is nearer the equator.

The data concerning the regional and annual variation of the evaporation are very scanty. Over the continents in temperate latitudes the maximum of the evaporation occurs during the warm season, but over the oceans the evaporation is very small during the same period. The reason is that in summer the air temperature becomes higher than the water temperature. Hence, as explained previously, evaporation is reduced considerably and may even be changed into the inverse process, condensation. Over small lakes the annual variation of evaporation resembles that over the continents with a maximum in summer, because the temperature range of such small bodies of water is much greater than that of the ocean and because these lakes are often crossed by dry air masses. Maps of the mean daily evaporation over the North Atlantic and the North Pacific have been published by Jacobs.[1]

24. Atmospheric Humidity. The atmospheric humidity—*i.e.*, the water-vapor content of the air—can be expressed in a great number of different ways, as is explained in textbooks on meteor-

[1] Jacobs, W. C., *Ann. New York Acad. Sci.*, **44**, 19, 1943.

ology. In the following discussion, reference will be made to the water-vapor pressure and the relative humidity, for which more data are available than for the mixing ratio or the specific humidity. It is easy to infer from the behavior of these two variables the distribution of the other quantities expressing atmospheric humidity even though the actual computation requires as a rule a knowledge of the air pressure and the temperature. Thus, the specific humidity or the mixing ratio is with sufficient accuracy proportional to the water-vapor pressure and inversely proportional to the pressure of the air. Since the air pressure varies hardly more than 10 per cent in horizontal direction, it is permissible to identify changes of water-vapor pressure with changes of specific humidity.

a. Vapor Pressure. Maps of the distribution of the vapor pressure over the earth in January and July have been published by Száva-Kováts.[1] From these maps, values of the vapor pressure for each tenth degree of latitude have been determined. The zonal distribution that is thus obtained is plotted in Fig. 17. For the latitude a sine scale has been used as before in order to show the relative size of the different latitude belts. To give an approximate idea of the magnitude of the mixing ratio or the specific humidity a scale for the mixing ratio is added on the right-hand side of the figure, assuming an average pressure of 1000 mb. This mean value is somewhat low for sea level, but it seems adequate for the purpose.

The full curve, called "mean," is the average of the January and July values and may be regarded as representing the yearly means, with obvious reservations. It is highest at the equator and decreases toward the poles, as is to be expected, since its distribution must follow that of the temperature quite closely. For the same reason the vapor pressure at each latitude is higher in summer than in winter, and the highest value is found somewhat north of the equator in July, somewhat south in January. The northward displacement in July is larger than the corresponding southward displacement in January owing to the corresponding asymmetrical distribution of the temperature (see page 23). The slight irregularities of the curves may in part be caused by the inadequacy of the data over large parts of the earth. But to a large extent they are doubtless real, showing

[1] Száva-Kováts, J., *Ann. Hydrographie,* **66,** 373, 1938.

the deviations of the actual distribution from the ideal zonal symmetry due to the actual arrangement of water and land.

The great dry belts of the earth do not show in the curves of Fig. 17. On the maps published by Száva-Kováts the dry regions stand out as having considerably lower vapor pressure than the oceans at the same latitude. But the difference is not great enough and the area covered is not large enough to affect the zonal mean values noticeably. It may be pointed out in this connection that the moisture content of desert air is by no means

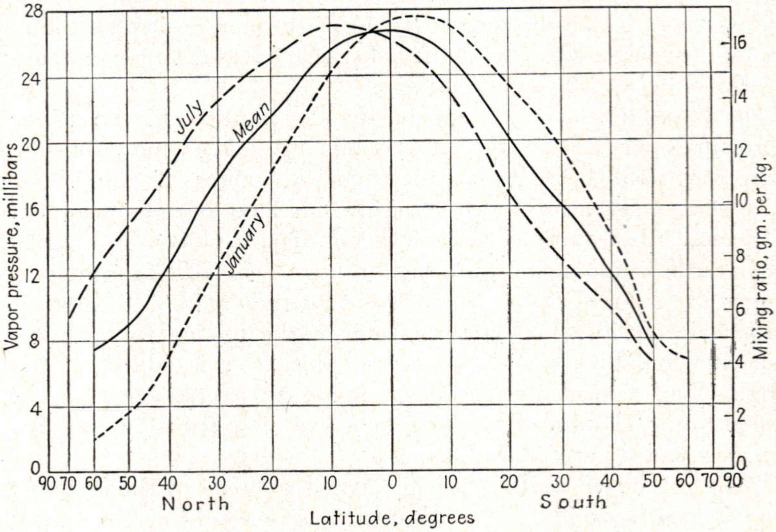

FIG. 17.—Zonal distribution of the vapor pressure.

small when compared with that of the air in more humid climates at higher latitudes. In In-salah in the Sahara, for example, the temperature during the hottest month is about 99°F on the average, the relative humidity 25 per cent. Hence, the specific humidity is about 11 g/kg. This amount is greater than the specific humidity of polar maritime surface air over western Europe in summer. The regions of lowest water-vapor content are found at high latitudes where the temperature is low.

Since the source of the atmospheric water-vapor content is at the earth's surface, the humidity decreases upward. However, no specific statements can be made concerning the rate of decrease of the vapor pressure with altitude.

The annual variation of the vapor pressure is in most regions parallel to that of the temperature. The vapor-pressure maximum occurs in the season of highest temperature, the minimum in the season of lowest temperature. Only in those tropical regions where the annual temperature range is small and where a pronounced rainy season exists does the vapor-pressure maximum coincide, not with the highest temperature, but with the heaviest precipitation.

The diurnal variation of the vapor pressure is different over land and sea. Over the ocean the vapor pressure follows the temperature, with a minimum in the early morning hours, a maximum in the early afternoon. Over land, two maxima and two minima occur. The main minimum is reached about the time of the temperature minimum. Then the vapor pressure rises to a maximum in the forenoon, but it decreases again to a secondary minimum in the afternoon about the time of the temperature maximum. The main maximum occurs in the evening hours.

To understand the origin of the diurnal variation of the vapor pressure it is necessary to realize that two factors determine the water-vapor content of the surface air. The rising temperature causes greater evaporation and hence tends to increase the water-vapor content. The turbulent mass exchange, on the other hand, transports water vapor from the surface of the earth to higher altitudes, in the direction of decreasing water-vapor content. The turbulent mass exchange is greatest in daytime because the wind velocity and the convection are then strongest. Thus the turbulent mass exchange tends to produce a minimum of the water-vapor content near the surface in daytime, a maximum at night. Over the oceans the temperature effect is strongest because the ocean surface is not heated greatly in daytime and the stratification of the air is rather more stable during the day than at night. Over land the influence of the daily period of the turbulent mass exchange predominates, at least during the warmer season when convection and turbulent mixing are very pronounced. Consequently, around noon the supply of water vapor by evaporation from the ground becomes less than the amount that is transported upward by turbulence and convection, and a minimum occurs. Later in the afternoon when the turbulent transport becomes again smaller, the influence of evaporation is strongest. The addition of water to the surface air by evaporation from the ground is now greater than the transport to higher levels. Hence, the water-vapor content increases again until it reaches the evening maximum.

In very dry regions, such as steppes and deserts, the effects of turbulent mixing are particularly strong, and the amount of water available for evaporation is particularly small. Hence, the afternoon minimum may here predominate completely while the forenoon minimum disappears so that the diurnal variation becomes opposite to that over the oceans, with a maximum in the morning and a minimum in the afternoon.

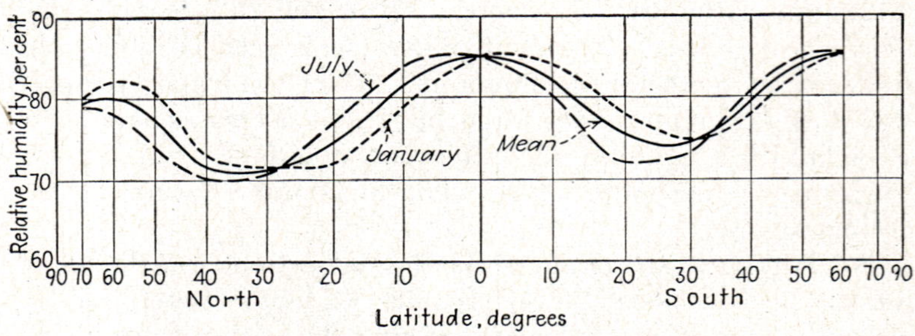

FIG. 18.—Zonal distribution of the relative humidity.

At one and the same station the oceanic type may prevail at one season and the continental type at another. Such a change of the type of the diurnal variation of the vapor pressure is due to the annual period of convection intensity and to the state of the soil. If the ground is moist or snow-covered in winter but dry in summer, the oceanic type may be observed during the winter, the continental type during the summer.

b. Relative Humidity. The zonal distribution of the relative humidity is shown in Fig. 18. The data represented in this diagram are based on two maps published by Száva-Kováts.[1] The "mean" curve gives again the averages for January and July. Considering this curve first, it will be seen that the largest maximum occurs at the equator. From there, the relative humidity decreases toward the regions of the subtropical high-pressure belt where the minima of the relative humidity occur. Toward higher latitudes the relative humidity increases, even though the vapor pressure decreases continuously from the

[1] *Ibid.*

equator poleward. This increase of the relative humidity toward the zones of the westerlies is obviously a result of the decrease of the temperature. Beyond 60° latitude, at least in the Northern Hemisphere, the relative humidity seems to decrease again slightly. The minimum of the relative humidity is at a higher latitude in the Northern than in the Southern Hemisphere because the dry regions of the large continental masses of North America and Asia extend to higher latitudes than the dry regions in the Southern Hemisphere. Moreover, the minimum values are lower in the Northern Hemisphere than in the Southern Hemisphere because the relative humidity is lower over land than over water and the continental influence is naturally stronger in the Northern than in the Southern Hemisphere. For the same reason, the secondary maximum is smaller in the Northern than in the Southern Hemisphere.

In January, the equatorial maximum of the relative humidity is somewhat south of the equator. Similarly, the subtropical minima are displaced southward so that the driest zone in the Northern Hemisphere is closer to the equator, the driest zone in the Southern Hemisphere farther away from it. This shift of the belts of highest and lowest relative humidity and the analogous shift during July agree with the corresponding motions of the thermal belts and of the zones of precipitation and dryness.

Between the equator and about 30° north and south the average relative humidity for every latitude is higher in summer than in winter. At higher latitudes the relative humidity is higher during winter. This behavior may be interpreted in the following manner: The relative humidity is the ratio of the actual to the saturation vapor pressure, and the greater, the latter, the higher the temperature. Thus, even though the actual vapor pressure is mostly higher in summer than in winter, the relative humidity may have a minimum during the warmer season. This type of annual variation is especially found over the continents.

To discuss the different types of annual variation of the relative humidity in some detail four typical examples are given in Table 25. The division into four types is very schematic, for a great number of deviations occur. Over the *continents* the effect of the temperature on the relative humidity is greater than that of the actual vapor content. Consequently, the minimum occurs during the warm season, the maximum during the cold season.

The most characteristic feature of what has been called the *oceanic* type is the very slight range of the variation, although there is a definite tendency for a maximum during the warmer season. The *monsoon* type has a maximum during the summer, when maritime air masses are most frequent because of the winds from the water, and a minimum during winter, when continental air masses predominate because of winds from the continents. In many of the regions where the annual variation of the relative humidity follows the monsoon type, it would have a different seasonal rhythm if the monsoon effect were absent. The *tropical* type is quite similar to the monsoon type since here the maximum of the relative humidity occurs also during the height of the rainy season. Both the tropical and the monsoon types contribute toward the higher values of the relative humidity in summer than in winter at lower latitudes that are observed in Fig. 18.

TABLE 25.—ANNUAL VARIATION OF THE RELATIVE HUMIDITY, PER CENT

Type	Station	Winter*	Spring	Summer	Autumn
Continental .	Salt Lake City, Utah	67	51	37	47
Oceanic.	Thorshavn, Faeroes	83	81	85	85
Monsoon. . . .	Vladivostok, U.S.S.R.	72	75	87	71
Tropical. . . .	Puebla, Mexico	58	53	71	69

* Winter comprises the months of December to February. The other seasons are defined accordingly.

On mountains in temperate latitudes the relative humidity frequently has its maximum in summer, its minimum in winter. The reason for the summer maximum of the relative humidity at higher elevations is the stronger upward transport of water vapor in summer when the turbulent mass exchange is greatest.

The daily period of the relative humidity is in general opposite to that of the temperature. The maximum occurs in the early morning hours, the minimum in the early afternoon. But where a sea breeze sets in during the daytime the relative humidity increases during the day.

25. Cloudiness. Cloudiness is the amount of the sky covered by clouds. It is usually given in tenths of the total area of the sky. If the sky is completely cloudless, the cloudiness is 0. If the sky is completely covered by clouds, it is 10 (tenths). The observed cloudiness does not take into account the thickness of the cloud cover. It is entirely possible, and does indeed happen, that the cloudiness is 10 while the sun is shining.

In interpreting mean values it is necessary to keep in mind that cloudiness can vary only between 0 and 10. Thus, if a station is largely either completely cloudless or completely overcast and the intermediate values occur very rarely, the mean value will be near 5 even though this value may be very infrequent. It would therefore be highly desirable to have frequency distributions of the various grades of cloudiness. However, the following discussion will necessarily be restricted to the mean values.

The latitudinal distribution of cloudiness is fairly parallel to that of the precipitation, with a maximum in the zone of the doldrums and a minimum in the subtropics, as shown in Table 26 after C. E. P. Brooks.[1] The main maximum of cloudiness

TABLE 26.—ANNUAL MEANS, PER CENT, OF THE CLOUDINESS
(After C. E. P. Brooks)

Latitude	90°–80°	80°–70°	70°–60°	60°–50°	50°–40°	40°–30°	30°–20°	20°–10°	10°–0°
				Northern Hemisphere					
Ocean	63	70	72	67	66	52	49	53	53
Land	..	63	62	60	50	40	34	40	52
Mean	..	66	63	62	56	45	41	47	53
				Southern Hemisphere					
Ocean	..	64	76	72	67	57	53	49	50
Land	..	..	..	70	58	48	38	46	56
Mean	..	..	..	72	66	54	48	48	52

occurs at higher latitudes because here frontal cloud systems—which cover the whole sky—are more frequent, while in tropical regions convective clouds predominate, which often cover only part of the sky.

Maps of the distribution of cloudiness over the earth during January and July are shown in Plates X and XI. They show, apart from the zonal distribution demonstrated by Table 26, that cloudiness on the whole is greater over the oceans than over land and how it varies from summer to winter.

[1] BROOKS, C. E. P., *Mem. Roy. Met. Soc.* (*London*), **1,** No. 10, 1927.

In tropical regions with pronounced rainy seasons the cloudiness has a maximum at the same time as the precipitation, a minimum during the dry season. The same is true in a great many other climates. Thus, where a monsoon regime prevails, the cloudiness is greatest during the summer when the monsoon rains fall and lowest during the dry winter. Along the Mediterranean and the California coast, on the other hand, where winter is the rainy season, summer the dry season, the cloudiness is greater in winter than in summer. In the interior of the continents, however, the annual variations of cloudiness and of precipitation are often opposite. The precipitation has here its maximum in summer (page 76), while the maximum of the cloudiness occurs during the winter. The reason for this perhaps somewhat surprising fact is that in summer convective clouds of the cumulus type and convective showers predominate, while in winter frontal clouds of the stratiform type and frontal precipitation are more frequent. As just mentioned, the latter clouds cover a greater part of the sky and persist for a longer period than convective clouds. On the other hand, frontal rain is less intense than convective showers so that it gives less precipitation even though it lasts longer. Thus, the duration of precipitation and the amount of sky covered by clouds actually have a maximum in winter, but the amount of rainfall is greatest in summer. It must, however, be emphasized that not everywhere in the interior of the continents are the annual variations of cloudiness and precipitation opposite. In eastern Siberia, for example, where anticyclonic pressure distribution predominates during the winter, the maxima of cloudiness and of precipitation occur during the summer. There are many types of the seasonal variations of cloudiness.

The daily variation of the cloudiness is also governed by the differences between the clouds of the cumulus and the stratiform type. Since the former develop during the daytime owing to convection, they have a maximum in the early afternoon hours. The formation of stratiform clouds, on the other hand, is favored by a stable stratification so that their maximum occurs in the early morning hours, their minimum later during the day. As a result, the maximum of cloudiness occurs during the early morning hours where the stratiform clouds are most frequent, as at many stations in temperate latitudes during winter. In summer when convection is strong the same station may have a maximum of cloudiness during the early afternoon when the formation of cumulus clouds is greatest, or there may be two maxima, one in the early morning hours and the other during the early afternoon.

It may be mentioned here that the number of hours of sunshine is closely related to the cloudiness even though an accurate mathematical relationship between the two elements obviously cannot exist since the sun may shine even if the sky is cloudy or since the sun may be obscured even if the sky is not completely overcast.

26. Fog. Fog is a cloud that is directly in contact with the ground. One speaks of fog when the visibility in the horizontal direction is less than 1000 m. If the visibility is between 1000 and 2000 m, the atmospheric condition is referred to as "light fog." The standardization of the term "fog" has only fairly recently come into general use. Earlier investigations about the frequency and occurrence of fog in different regions can often hardly be compared with each other owing to the different definitions attached to the term.

The various processes by which fog may be formed are similar to those which form clouds, with some important differences arising from the fact that the fog originates near the ground. A complete discussion of the various factors that may give rise to the formation of fog belongs in textbooks on meteorology. Here only a few facts need be recapitulated in order to understand the distribution and the annual period of the frequency of fog, to be discussed.

The formation of fog may be due to one of two causes, (1) evaporation of water into air and (2) cooling of air. The formation of fog by evaporation is likely to occur when colder air comes into contact with warmer water. Then the saturation vapor pressure with respect to the water is higher than the saturation vapor pressure of the air. Hence, the water vapor that evaporates into the air will condense and form fog. In the atmosphere this may occur when warm rain falls through colder air as happens at frontal surfaces (frontal fog) or when cold air flows over warm water (steam fog, arctic sea smoke).

When fog is formed by cooling of the air, one may distinguish, according to the cause of the cooling, between radiation fog, advection fog, and upslope fog. Radiation fog is produced by radiational cooling, which is most effective on calm, clear nights. Advection fog occurs when the air moves over a colder surface so that it is cooled from below by contact. Over land advective fog occurs fairly often in winter when air from a warmer water surface crosses over the continents. In summer, on the other hand, advection fog occurs rather when warmer air from the warmer land reaches the ocean, especially if it reaches the region of a cold ocean current. Radiation fog forms largely over land and seldom over the ocean because the nocturnal cooling of the water surface is very slight.

Upslope fog is formed when moist air is cooled sufficiently by adiabatic expansion while being forced to move upward along sloping ground. Upslope fog is therefore mainly important in hilly and mountainous regions.

The distribution of fog in winter and summer is shown in Plates XII and XIII. Owing to the somewhat different definitions of "fog" the maps are not so reliable as they should be, but they represent quite well the main features of fog distribution in those areas from which observations are available. A well-known area of great fog frequency is in the western Atlantic in the neighborhood of Newfoundland. These fogs are typical advection fogs, which originate when warmer air passes over the region of the cold Labrador Current. Other regions with frequent fogs are the Kuril Islands in the eastern North Pacific and all the regions along the west coasts of the continents where colder water is found near the surface, due either to cold ocean currents or to the upwelling of cold water from greater depths.

There are two different types of seasonal variation of the fog frequency, one characteristic for continental stations, the other for maritime stations. The continental type has the maximum of fog frequency in winter, the minimum in summer. This distribution of the maximum and the minimum throughout the seasons is to be expected, for radiation fog will largely form in winter when the nights are clear and calm owing to the presence of a large continental anticyclone. In summer, the frequency of anticyclones over the continents is much less and the nights are shorter so that the radiational cooling does not extend over such a long period and is therefore less effective. As far as fogs of the advection type are concerned it has already been pointed out that over land they will mainly occur in winter when relatively warm and moist maritime air penetrates into the continent.

The maritime type of the seasonal variation of fog frequency has a maximum in summer and a minimum in winter. This seasonal variation is easily understood since radiational fog is infrequent over the oceans and advection fog has a maximum over the ocean in summer when the water is cooler than the land.

As always in the discussion of typical annual and diurnal variations it is necessary to add that not at all land stations does the maximum of fog frequency occur in winter and not at all maritime stations does it occur in summer. The formation of fog, especially over land, depends very much on local peculiarities, which may bring about great changes in the fog frequency over small distances.

Since all types of fogs are dissolved with sufficiently strong heating, the fog frequency has a daily period with a maximum in the early morning and a minimum in the afternoon.

27. Thunderstorms. While in general it is quite simple to decide whether a particular storm should be classed as a thunderstorm, difficulties arise when only thunder or lightning is observed. The following discussion of thunderstorms is based on a paper by C. E. P. Brooks.[1] Considerable additional material has been used in plotting Fig. 19 and Plates XIV and XV. A "day of thunder" is defined as a day on which either thunder and lightning or distant thunder has been observed. A report of lightning alone is not considered as indicating a thunderstorm. The observational data on which a climatological discussion of thunderstorms can be based are still far from satisfactory. In the first place, data from a great number of regions, especially from oceans, are either lacking or incomplete. Furthermore, even where observations are available many occurrences of distant thunder are probably not noted, especially during the rainy season in the tropics, where distant thunder may be so common during certain times of the day that it may pass unnoticed. The following should be considered with these reservations.

Figure 19 shows the latitudinal variation of the thunderstorm frequency for land and sea separately and combined into a mean value for the whole latitude. The uppermost diagram represents

[1] Brooks, C. E. P., *Geophys. Mem.*, **3,** No. 24, 4, 1925.

the averages for the half year from April to September, the middle diagram for the half year from October to March, and the lowest diagram for the whole year. For each half year a frequency of 10 per cent indicates that 18 days with thunder occur during

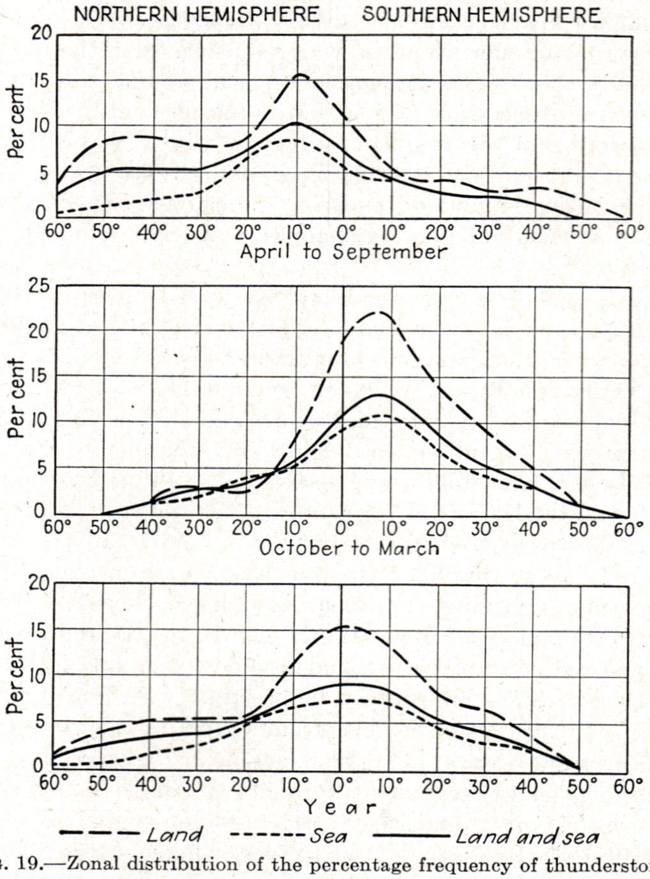

FIG. 19.—Zonal distribution of the percentage frequency of thunderstorms.

the half year, 20 per cent corresponds to 36 days, etc. For the whole year, 10 per cent means 36 days, 20 per cent 73 days, etc.

In general, the thunderstorm frequency is greater over land than over sea. Only at 20° north latitude during the half year from October to March are appreciably more thunderstorms observed over water than over land, for the parallel of 20° north latitude crosses the Sahara, where thunderstorms are very rare. During the other half year, from April to September, the small frequency of thunderstorms over the Sahara is compensated by the greater frequency over other continents, especially over Indo-China, so that the frequency at this latitude is now greater over land than over water.

During the summer half year of the Northern Hemisphere, from April to September, the maximum of the thunderstorm frequency is at about 10° north latitude since the doldrum belt is north of the equator. During the other half year, from October to March, the maximum has shifted into the Southern Hemisphere. The thunderstorm frequency decreases from the tropics toward higher latitudes. The effect of the subtropical high-pressure belts manifests itself in the slight depressions or in the interruptions of the regular decrease of the thunderstorm frequency toward higher latitudes. Only during October to March in the Southern Hemisphere is this effect of the subtropical high-pressure belt not noticeable.

A more detailed picture of the geographical distribution of thunderstorms is given in Plates XIV and XV.

During the summer half year of the Northern Hemisphere, from April to September (Plate XIV), regions of maximum frequency are found over Central America, the West Indies, the southeastern Gulf states and New Mexico, equatorial Africa, northeastern Indo-China, and the East Indies. Over the Sahara and the Arabian Desert, on the other hand, thunderstorms are very rare. Other regions of small thunderstorm activity at low latitudes are found over the South Atlantic, over the Indian Ocean, and over Australia. Beyond 50° south latitude and in the arctic regions of the Northern Hemisphere, thunderstorms are very infrequent phenomena. But even at lower latitudes thunderstorms occur very seldom over large parts of the oceans.

A comparison of Plate XV with Plate XIV shows how the maximum of the thunderstorm frequency is displaced into the Southern Hemisphere during the half year from October to March, when the Southern Hemisphere has summer. In the Northern Hemisphere very few thunderstorms occur during the winter half year beyond 50° north latitude. The very large area with a frequency of more than 10 per cent over North America during the summer has shrunk to a much smaller region of 5 per cent frequency over the Gulf states. Over central and eastern Europe and the Balkans, where the frequency was higher than 10 per cent during the summer, it is only 1 per cent to 3 per cent during the winter. The maximum over equatorial Africa, which was centered in the latitude zone from 0° to 10° north, has now moved about 20° toward south. The maximum over Indo-China has disappeared, and the maximum over the East Indies extends farther south, reaching to northern Australia. In South America the thunderstorm frequency is high over Ecuador, Peru, and a large part of the Amazon basin.

More detailed discussions of the thunderstorm distribution will be given in Part B on Regional Climatology, but a few remarks concerning the annual and diurnal variations of the thunderstorm frequency may be made here. Since thunderstorms are mainly due to atmospheric convection, the monthly maximum of their frequency occurs mostly in summer, while the minimum takes place in winter (see pages 39 to 43, where the variations of the vertical lapse rate of temperature are discussed). Nearer the equator, however, where the difference between the hot and the cold season becomes insignificant, this is no longer true. Southern Asia, North Africa, and the East Indies have two maxima, at the beginning and at the end of the rainy season. In Central America and the West Indies the maximum occurs, not in July, as in the more temperate latitudes of the Northern Hemisphere, but in August, during the hurricane season. Over the continental areas of the Southern Hemisphere the maximum occurs similarly in summer, the minimum in winter. Over the oceans the times of the maxima and minima of thunderstorm frequency show greater variability. In the North Atlantic the maximum is reached in August; in the Indian Ocean north of the equator in May. South of the equator the maximum occurs in all oceans in January to March.

The diurnal maximum of the thunderstorm frequency over land occurs during the afternoon when convection is strongest, the minimum during the night or the early morning; but there are important exceptions to this general rule. Over the sea, where the tendency toward instability is greatest at night, the thunderstorm maximum falls in the early morning hours.

CHAPTER V

AIR MASSES, FRONTS, CYCLONES, AND ANTICYCLONES

Changes of cloudiness, precipitation, temperature, and humidity can generally be subdivided according to their causes into two main classes, (1) weather phenomena that are associated with frontal passages and (2) weather phenomena that are the result of air-mass transformations. Consequently, the factors that influence the climate of any region may be discussed on the basis of known frontal and air-mass activity.

Whereas fronts form essential elements of extratropical cyclones, anticyclones are singularly devoid of fronts because of the frontolytical wind fields. Therefore, any discussion of frontal zones is intimately connected with that of regions of cyclonic and anticyclonic activity. Furthermore, air-mass weather depends partly upon the type of circulation. Hence, fronts and air masses should be discussed in connection with the regions of predominant cyclonic or anticyclonic circulation.

28. Air-mass Classification. Most classification schemes are based upon the geographical location of the various air-mass source regions. The air masses are defined as tropical, polar, continental, maritime, etc. This geographical classification is adequate provided that one is interested only in source-region properties and provided that the characteristics of the source regions are clearly understood. However, over the greater part of the globe air masses are continually subjected to processes that cause a modification of the air-mass properties. A satisfactory classification scheme should encompass these transformations; consequently, it is necessary to expand the simple geographical classification.

The transfer of heat between the air and the underlying surface represents one of the most important processes that cause a modification of air-mass properties. When the atmosphere is heated from below, it becomes increasingly unstable, while it becomes more stable when cooled from below. To take care of these important modifications Bergeron[1] introduced a thermodynamical classification, *viz.:*

1. *K* air masses are colder than the underlying surface.
2. *W* air masses are warmer than the underlying surface.

Stability changes, however, do not depend solely upon the transfer of heat between the air and the underlying surface. Above the earth's surface, horizontal convergence and divergence produce ascending or descending currents, which, in turn, affect the air-mass stratification. Subsiding air becomes more stable, while ascending air is accompanied by a steepening of the lapse rate, *i.e.*, a trend toward instability. These changes affect the air-mass stratification, mainly above the friction layer, and may be independent of surface heating or cooling. For example, air that is being heated from below may not show weather phenomena connected with instability solely because upper-level subsidence restricts the instability to a shallow surface layer. Besides vertical motion, advection of warm or cold air aloft may affect the air-mass stratification. In view of these upper-level factors, Petterssen[2] has suggested an over-all influence classification that indicates directly the stability of the air, *viz.:*

1. *s* denotes a stable stratification.
2. *u* denotes an unstable stratification.

Since surface- and upper-air processes determine the air-mass modifications and since they may act in the opposite sense, it has been decided to adopt a classification scheme that takes into account low-level and high-level changes separately. That is, air masses, wherever desirable, shall be classified not only as *W* or *K* masses but also as *u* or *s* masses. The *u* indicates upper-level instability and occurs in regions under strong cyclonic control or where there is advection of cold air aloft. Conversely, *s* designates upper-level stability and arises either from subsidence within anticyclonic circulations or from advection of warm air aloft. In connection with advection it is commonly observed that instability, arising from the advection of cold air aloft, occurs in regions of cyclonic flow. On the other hand, stability, arising from the advection of warm air aloft, generally takes place in regions of anticyclonic flow. Consequently, on most

[1] BERGERON, T., *Geofys. Pub.* **5,** No. 6, 1928.

[2] PETTERSSEN, S., "Weather Analysis and Forecasting," p. 169, McGraw-Hill Book Company, Inc., New York, 1940.

occasions, one can associate the *s* designator with anticyclonic circulation and the *u* designator with cyclonic circulation.

On the basis of the above considerations the following classification scheme and Plates XVI and XVII have been constructed in cooperation with H. C. Willett.[1] *P* indicates polar, *T* tropical, *m* maritime, *c* continental, *K* heated from below, *W* cooled from below, *s* stable aloft, and *u* unstable aloft. It is noticeable that

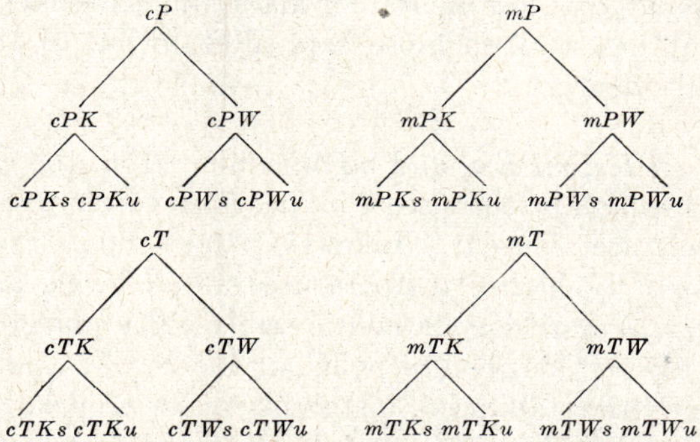

arctic and equatorial air masses are no longer recognized as separate geographical types. Arctic air may be considered as a modified form of *cP* air, while equatorial air is now designated by *mTu, viz.*, the unstable tropical maritime air within the equatorial low-pressure circulation.

The seasonal distribution of air-mass types is indicated by Plates XVI and XVII. Besides air-mass symbols, these charts contain the mean positions of the most important frontal zones and also an outline of the general pressure distribution. The prevailing air flow can be determined by the pressure field. Wherever the *K-W* or the *s-u* modifications are slight, the designator has been omitted; for example, no *K* or *W* designators appear within the equatorial low-pressure areas since the underlying surface is reasonably homogeneous.

The properties of the different air masses may be summarized very briefly as follows:

cPWs a cold, dry continental air mass that is stable in the lower layers through cooling from below and stable aloft through subsidence.

[1] WILLETT, H. C., "Descriptive Meteorology," Academic Press, New York, 1944.

cPKs a cold, dry continental air mass that is unstable in the lower layers through heating from below and stable aloft through subsidence.

cPKu a cold, dry, unstable continental air mass. The instability arises partly from the surface heating and partly from the ascending motion that accompanies strong cyclonic flow.

cPWu a cold, dry continental air mass that is cooled from below but has a steep lapse rate aloft owing to the lack of subsidence.

mPWs, mPKs, mPKu, and *mPWu* differ from the continental polar masses mainly because of higher surface temperatures and higher moisture content. Likewise, the tropical masses are characterized by higher temperatures at all levels and, in the case of tropical maritime air, considerably higher moisture values. As in the case of continental polar air the *K, W, s,* and *u* designators signify the stability or instability of the tropical air masses.

For a detailed description of air-mass properties the reader is referred to standard meteorological texts. The foregoing classification scheme is described in more detail by Willett.[1] Since it is convenient to review the air-mass characteristics in Part B on Regional Climatology no detailed description will be presented in this chapter. The average distribution of air masses over the globe in January and July is shown in Plates XVI and XVII.

29. Fronts, Cyclones, and Anticyclones. The principal frontal zones of the world are indicated in Plates XVI and XVII. In general, the polar fronts represent the mean positions of the boundaries between polar and tropical air masses wherever the wind field is frontogenetical. That is, the mean locations coincide with pressure troughs or regions of cyclonic wind shear. The arctic and antarctic fronts separate either polar maritime and polar continental air masses or decidedly different types of polar maritime air masses. Again these mean fronts occur only in regions where the wind field is predominantly frontogenetical. The equatorial or intertropical front is primarily a zone of converging *mTu* air masses.

[1] *Ibid.*

Extratropical cyclones develop on the main fronts. Consequently, the frequencies, intensities, and paths of the various cyclones depend upon the frontal changes, both in intensity and location. The major anticyclone movements, in turn, are closely connected with the cyclone trajectories. Since these migratory pressure systems behave somewhat differently in various sections of the world, it has been decided to postpone further discussion until Part B on Regional Climatology is reached. There, in each chapter, a reasonably detailed discussion will be undertaken of the air masses, primary and secondary fronts, cyclones and anticyclones, and weather types that accompany the various pressure systems. A discussion of the semipermanent pressure systems, such as the subtropical anticyclones, has been presented in Chap. III.

CHAPTER VI

CLIMATIC ZONES AND TYPES

The climate of a given locality or a given region is determined, not by just one climatic element, but by the combination of the various meteorological variables that make up the climate. Two places may have very similar temperatures, but very different amounts of precipitation; thus their climatic difference becomes apparent only if other climatic elements are considered in addition to temperature. This fact was pointed out in Table 1 (page 2), where the temperature and precipitation were considered at Cairo, Egypt and at Galveston, Tex. At both places the annual mean temperatures and the mean monthly extremes are almost identical. But the climate of the two cities is very dissimilar, Cairo having little precipitation while the rainfall at Galveston is quite plentiful.

Since the climate of a particular region is made up of a great number of different elements of which temperature and precipitation are only the two most important, it is obvious that no two places have exactly the same climate. In view of this great variety it is necessary to have some classification that permits of a first orientation in the matter. Two such classifications will be discussed in this chapter.

30. The Eight Main Climatic Zones. The first, rather crude attempt at a division of the earth into climatic zones was undertaken on an astronomical basis. The amount of incoming solar radiation depends on the latitude. The temperature depends to a large extent on the incoming solar radiation. Hence, there is a certain justification for a climatic division on an astronomical basis even though it is obvious that such a procedure must give rather unsatisfactory results. This astronomical division of the earth into five climatic zones may be considered here briefly since it can serve as a framework for the more elaborate classifications to be discussed later.

The greatest distances of the sun from the equator during its annual course are 23.5° south latitude where it reaches the Tropic of Capricorn about Dec. 21 and 23.5° north latitude where it reaches the Tropic of Cancer about June 21. The zone between the two tropics is the *Torrid*, or *Equatorial*, *Zone*. In this zone the sun is in the zenith at noon twice during the year, except directly at the two tropics, and its height at noon is never smaller than 43°. Moreover, it is never less than 10.5 hr above the horizon. Consequently, the seasonal variations of the temperature are small, much smaller than the diurnal variations, although there is a tendency toward two maxima and two minima of temperature throughout the year, especially near the equator. Of much greater importance than the annual temperature period is the change between dry and wet seasons.

Poleward from the Equatorial Zone the *Temperate Zones* of the Northern and Southern Hemispheres follow. These zones extend to the Arctic and Antarctic Circles, which are at 66.5° northerly and southerly latitude, respectively. Astronomically these polar circles are distinguished by the fact that the center of the sun just reaches the horizon on the shortest day (if the effect of the atmospheric refraction is disregarded). Throughout these zones the annual mean temperature decreases toward the poles while the annual temperature range increases in the same direction. Beyond the polar circles are the *Polar Zones*, where the summer is generally short and cool so that the heat received is not sufficient for most plants, especially for trees.

Some characteristics of the distribution of the climates are represented fairly well by this astronomical classification, for the most important single factor determining the climate is the latitude. But, just as the astronomical division of the year into four seasons requires modifications for meteorological purposes, it is necessary to change the limits of these astronomical regions considerably in order to bring them into better agreement with the climatological facts.

It should be pointed out that for the maintenance of plant life it is not the value of the annual mean temperature that is important but the existence of a sufficiently warm season. Yakutsk, Siberia, has an annual mean temperature of 13.3°F. Angmagssalik on the east coast of Greenland has an annual mean temperature of 28.0°F. But while the coast of east Greenland is barren, forests grow around Yakutsk and even some crops are raised, although the ground thaws only superficially during the summer. The reason for the climatic difference between the two localities is to be found in the mean temperature of the warmest month, 43.2°F at Angmagssalik, 66.2°F at Yakutsk.

It follows that the limits of the Temperate Zone should be defined by the temperature of the coldest and of the warmest month of the year. Köppen has chosen a mean temperature of 64.4°F (18°C) for the coldest month as the equatorial boundary of the Temperate Zone because this represents approximately the upper limit of comfortable mean temperatures. As the polar boundary of the Temperate Zone he selected the isotherm 50°F (10°C) for the warmest month because this isotherm coincides approximately with the polar boundary of the regions in which trees grow. The North Temperate Zone would thus be bounded on its equatorial side by the January isotherm of 64°F and on its polar side by the July isotherm of 50°F, since in general January and July are, respectively, the coldest and warmest months. The boundaries of the South Temperate Zone are analogous. Plates I and II show where these meteorologically defined limits of the Temperate Zones agree with the astronomically defined limits and where they disagree. The 64°F isotherm of January in the Northern Hemisphere runs on the whole somewhat north of the Tropic of Cancer, while the 50°F isotherm of July is mostly south of the Arctic Circle, except over Alaska, the Yukon, and Siberia. In the Southern Hemisphere the position of the January isotherm of 50°F should be especially noted in this connection. This isotherm is well north of the Antarctic Circle, at latitudes of about 50° south and even closer to the equator.

This system of five main climatic belts, which has its roots in the old astronomical division of the earth, has to be supplemented by three more climatic zones, which are typical even though they do not surround the earth completely: a northerly and a southerly dry belt at the boundaries of the Torrid Zone, in the subtropical high-pressure belt; and the boreal tree belt, with cold winters, which is found on the continents at higher latitudes in the Northern Hemisphere between the Temperate Zone and the Polar Zone. In the Southern Hemisphere this climate is not found, since no continent extends sufficiently far south. The name "snow-forest climate" has been given to this climate by Köppen since it is characterized by the existence of a snow cover lasting for a considerable part of the winter and by the growth of forests due to sufficiently high summer temperatures. The dry belts, on the other hand, are found in both hemispheres. However, they are interrupted on the east side of the continents and over the adjoining oceans since the trade winds with their easterly components bring moist air from the water over the land and since the subsiding motion is here, on the west side of the subtropical highs, much less pronounced or absent (page 70).

The foregoing attempt at a division of the earth into different climatic *zones* should not be confused with a classification of the various climatic *types*. When the climate of a large region, preferably one that surrounds the earth, is uniform, the region may be called a *climatic zone*. The same climatic type, on the other hand, may be found at different, unconnected regions. The remainder of this chapter will be taken up with a discussion of climatic types and their classification.

31. General Remarks Concerning the Classification of Climates. The main purpose of a classification of climates is to provide a reasonably easy and logical guide that permits a comprehensive survey of the great variety of climatic types. It would therefore be most desirable from a meteorological standpoint to have a genetical classification based on the physical causes of the climates. Such a classification has not yet been devised. Moreover, it appears doubtful that a purely genetical classification would be entirely satisfactory since it might include too many different climates in one group. It would be possible, for example, to speak of "monsoon climates," but such a grouping would combine the climates of large parts of India and those of the east coast of Asia, which differ appreciably from each other.

The classification that will be chiefly discussed in this chapter and that will be used later when a climate has to be briefly characterized is the one developed by Köppen.[1] It defines the climatic types according to the values of the climatic elements, especially temperature and precipitation, but without regard to the geographical location of the regions where these climatic types are found.

[1] Köppen-Geiger, "Handbuch der Klimatologie," Vol. 1, C, Verlagsbuchhandlung, Gebrüder Borntraeger, Berlin, 1936.

In order to have precise and easily applicable criteria for the determination of the limits between different climatic types Köppen chooses certain numerical values of the temperature and precipitation. A consequence of this precise definition is that in the marginal regions, where one climatic type changes gradually into another, the notation of the climate sometimes seems unsatisfactory. Such difficulties are unavoidable, however, wherever a boundary has to be drawn through a region in which actually a continuous change takes place. The climatic boundaries are in reality zones of gradual transition, and their diffuseness should not distract the attention from the fact that in the greater part of a region the climatic type characterizes the climate quite well. It can rather be considered as a distinct advantage of Köppen's classification that each climate can be assigned to a different type without subjective judgment.

The numerical values of the different boundaries are selected mainly according to their effect on plant growth because the vegetation cover of the earth is of the greatest economic importance to civilization. Temperature and precipitation are chosen as the two principal elements of the classification. This choice is dictated, first, by the circumstance that for these two variables the observations are most complete. Second, heat and water are the two most important factors dominating the organic and the inorganic world. It is, of course, impossible to characterize a climate completely with a few figures referring to temperature and precipitation; and it is not implied that other climatic elements, like evaporation, humidity, radiation, or wind, are not also important. But at least it is possible in this manner to find expressions that permit the combination of regions with the same climatic effects and that facilitate the differentiation between regions where the effects on plants, animals, and human life are different.

Moreover, the distribution of the other climatic elements is reflected in the distribution of temperature and precipitation, at least, if both these elements are suitably combined. Separately, they cannot give any satisfactory idea of the climate, as has already been shown by the example given in Table 1 (page 2). The same amount of precipitation may have a very different effect, depending on whether it falls in a hot climate with high evaporation or a cold climate with low evaporation. Precipitation amounts of 8 to 12 in. per year that fall in some parts of the Sahara are sufficient for the growth of trees in eastern Siberia and for the maintenance of glaciers at Spitsbergen.

From the discussion in the previous chapters it is apparent that for a characterization of the climates it is not sufficient to give only annual mean values; these are even less satisfactory in the case of the temperature than in that of precipitation. The seasonal temperature range has to be included in some form among the factors distinguishing the different climates. In view of the scarcity of other statistical data, the system of climates has to be based on annual and monthly means. Köppen chooses, therefore, the temperatures of the warmest and coldest month as criteria for the boundaries between different climates, as will be seen in the next section. In making a classification of the climates, it is important to keep in mind that for plant growth in temperate latitudes the heat during the summer vegetation period is much more important than the temperature during the winter, when the plant hibernates. At lower latitudes a season of dryness has a similar effect on the vegetation as a season of low temperature at higher latitudes. It is furthermore important to know whether the temperature remains below freezing during the whole winter or not. In the first instance, even a light snowfall is sufficient to maintain plant growth during spring, for the water coming down in the form of successive snowfalls is stored at the ground until spring. If thaws occur during winter, on the other hand, the moisture will disappear in the soil and will not be available at the beginning of the growing season.

Since the plains cover a much larger area of the continents than the mountainous regions, the former are mainly considered in Köppen's classification.

In order to obtain a sufficiently detailed description of the different climates that is simple enough to be grasped easily, Köppen introduces only a few main divisions. These, however, can be subdivided, the subdivisions being denoted by qualifying letters; thus a "climate formula" is obtained that can be extended if necessary. The climate formula gives a brief description of the climate, at least as far as the air temperature and the precipitation are concerned. It includes also the seasonal trend of these two elements.

32. The Principal Climatic Groups and Their Subdivisions. Köppen's classification has been changed slightly during its evolution. The form presented here is the final shape in which it has been published by Köppen in the "Handbuch der Klimatologie," Vol. 1, C, edited by Köppen and Geiger.

Köppen distinguishes 11 principal climatic types, which can be combined in 5 climatic groups denoted by capital letters. A small letter, *f*, *s*, or *w* indicates the absence of a dry period (*f*) or the presence of a dry period in summer (*s*) or winter (*w*). One obtains thus the following scheme arranged according to geographical latitude (Table 27):

Table 27.—Principal Climatic Types
(After Köppen)

Climatic group		Dry period	Degrees of dryness and cold
Tropical rainy climates	*A*	*f* (*s*) *w*	
Dry climates	*B*	....	*S* *W*
Warm temperate rainy climates	*C*	*f* *s* *w*	
Cold snow-forest climates	*D*	*f* (*s*) *w*	
Polar (snow) climates	*E*	....	*T* *F*

The parentheses indicate that *As* and *Ds* hardly occur. An *As* type of climate would imply a tropical rainy climate with prevailing winter rain. But the zone of the doldrums where the maximum of precipitation occurs shifts with the sun, as explained in Sec. 17, so that the season of high sun (summer) in the tropics is the season of greater rain, if a seasonal variation of precipitation does exist. *As* types of climates, therefore, hardly occur. At higher latitudes over the continents greater amounts of precipitation occur during the summer, when the temperature of the air and hence its moisture content are greater than in winter. Consequently, as also discussed in Sec. 17, summer must here be the rainier season, winter the drier season, and *Ds* climates are found only over small areas.

In the dry climates *B* two different degrees of dryness may be discerned, *BS* (steppe climate) and *BW* (desert climate). In the polar climate it is similarly practical to separate the climates *ET*, which still permit the growth of a tundra vegetation, from the climates *EF*, which are characterized by eternal frost.

The climatic groups *A*, *C*, and *D* may be called the "tree climates" because trees can grow only in regions with these climatic types.

33. The Numerical Values of the Boundaries between the Five Climatic Types. The five climatic types are separated from each other by the values of the extremes of the monthly mean temperatures or by a combination of precipitation and temperature. As the boundary between the tree climates *A*, *C*, and *D* on the one hand and the polar climates *E* on the other hand the mean temperature of the warmest month is chosen. Where the mean temperature of the warmest month is below 50°F, the polar climate *E* is found; where the mean temperature of the warmest month is above 50°F, one of the tree climates exists.

The boundaries between the climates *A*, *C*, and *D* are determined by the mean temperature of the coldest month. The tropical rainy climates *A* are separated from the warm temperate rainy climates *C* by the temperature 64.4°F[1] of the coldest month. The lowest mean temperature of the coldest month in the warm temperate rainy climates is 26.6°F or above. Where the mean temperature of the coldest month is less, the cold snow-forest climate *D* prevails. The temperature of 64.4°F has been chosen as the boundary between the *A* and the *C* climates because this temperature is considered as the optimum for human activities. Thus, in the tropical *A* climates the mean temperature is always above the optimum. The value of 26.6°F, according to Köppen, is the highest temperature at which a snow cover can persist for an appreciable period during the winter season. The boundaries of these climatic types and of various subdivisions are shown schematically in Fig. 23 (page 122).

The dry climatic types *B* are distinguished from the tree climates by a combination of the temperature and the precipitation, for the effects of precipitation, especially on the vegetation, depend not only on the amount of precipitation itself but also on the evaporation. The observations of evaporation are still far too scanty to be of direct use in a study of the dis-

[1] Köppen fixed the temperature limits in degrees centigrade. (64.4°F is 18°C, 26.6°F is −3°C, and so on.)

tribution of climatic types. But the effect of the evaporation can at least roughly be taken into account by forming an expression that depends, not on the precipitation alone, but on precipitation and temperature. In the last publication on his classification system Köppen chooses as the condition for a dry climate that the annual precipitation in inches

$$r \leqq 0.44t - 8.5$$

where t stands for the annual mean temperature in degrees Fahrenheit. It is only in this connection that the annual mean temperature is used to define one of the major climatic types. Table 28 shows for some temperatures the corresponding precipitation limits. If the annual mean temperature is 80°F, for

TABLE 28.—LIMITS BETWEEN DRY AND RAINY CLIMATES
(After Köppen)

Temperature, deg F..........	50	60	70	80	90	100
Precipitation, in............	13.5	17.9	22.3	26.7	31.1	35.5

example, a climate would have to be classified as dry if the annual amount of precipitation were less than 26.7 in., while with a lower annual mean temperature the climate would be designated as one of the tree climates. The above inequality for the demarcation of the boundary between dry and moist climates has been derived by Köppen from a study of the relation between temperature, precipitation, and vegetation. He found that the vegetation changes to the steppe type where the rainfall is less than the amount shown in Table 28 for a given temperature.

As indicated in Table 27, the five main climatic types whose limits are discussed in the preceding part of this section are further subdivided. In the driest parts of regions that have a dry climate even the steppe vegetation with its grasses cannot exist and deserts are found. According to Köppen the boundary between the steppe *BS* and desert climate *BW* lies where the precipitation drops to half the value that it has at the limit between dry and moist climates, or where the annual precipitation

$$r = \frac{0.44t - 8.5}{2}$$

These limits for both the steppe and the desert climates require a certain readjustment when the precipitation is not uniformly distributed throughout the year. If the rain falls mainly during the hot season at a time when the evaporation is higher than the yearly average, a larger amount of precipitation is necessary to prevent the formation of a steppe or a desert. On the other hand, when the rain falls mainly during the cool season, the loss of water by evaporation will be much smaller. The following compilation shows the criteria used by Köppen for the boundaries in these two cases:

Precipitation	Boundary between *BS* and rainy climates	Boundary between *BW* and *BS*
Maximum in summer..........	$r = 0.44t - 3$	$r = \frac{0.44t - 3}{2}$
Maximum in winter..........	$r = 0.44t - 14$	$r = \frac{0.44t - 14}{2}$

The letters t and r stand here again for the *annual* mean temperature and for the *annual* mean precipitation, in degrees Fahrenheit and inches, respectively. A comparison of these criteria with those for a climate without considerable seasonal variation of precipitation shows that the rainfall must be at least 5.5 in. larger when its maximum occurs in summer if the climate should not be classed as dry. On the other hand, when the rainfall maximum occurs in winter, the annual total may be 5.5 in. smaller than with an even distribution of rainfall, and yet the climate is not denoted as dry.

If a great number of stations are to be classified, Fig. 20 can be used to decide whether the climate is dry. The figure also permits one to distinguish between the different types of dry climates. If the precipitation shows no pronounced annual variation, the full curves should be used; if the precipitation has a winter maximum, the dotted lines are the proper boundaries; and if the precipitation maximum occurs in summer, the broken curves are to be considered. Suppose that a station has an annual mean temperature of 70°F and an annual rainfall of 10 in. If the precipitation has no annual period or if the maximum is in summer, the station has a desert climate *BW*, but if the rain falls mainly during the winter a steppe climate *BS* prevails. The notations h and k, which refer to the temperature, are explained in Sec. 34 (page 123), where more detailed subdivisions of the climatic types are discussed.

While in the dry *B* climates vegetation is either absent or sparse owing to the lack of water, in the cold *E* climates the low temperature restricts the plant growth. As has already been pointed out, the limit between the tree climates *A*, *C*, and *D* and

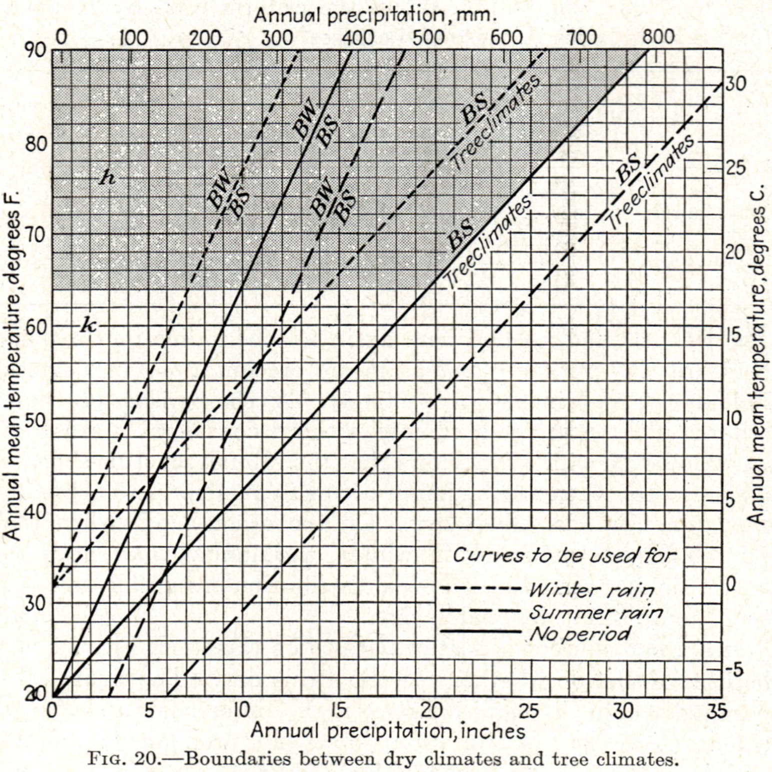

FIG. 20.—Boundaries between dry climates and tree climates.

the polar climates is given by the mean temperature of the warmest month. Where this temperature is below 50°F, an *E* climate prevails. Within the *E* climates, one distinguishes between tundra climate *ET*, where the temperature at least of the warmest month is above 32°F, and the frost, or icecap, climate *EF*, where the mean temperature of every month is below freezing. The frost climate has a permanent snow or ice cover, and vegetation is consequently absent. The tundra climate, as its name implies, has the characteristic tundra vegetation, which consists chiefly of mosses and lichens.

As indicated in Sec. 32 the tree climates *A*, *C*, and *D* are further subdivided according to the seasonal distribution of the precipitation. The criteria for distinguishing between climatic types without a dry season (*f*), with a dry season in winter (*w*), and with a dry season in summer (*s*) are different for the tropical *A* climates, where the temperature difference between the

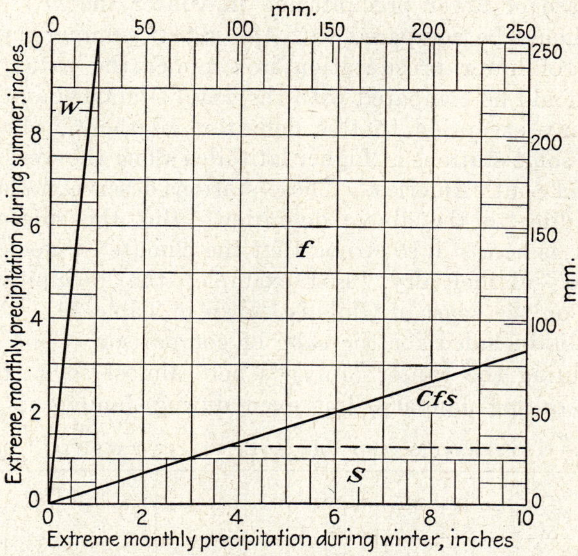

FIG. 21.—Precipitation divisions of the *C* and *D* climates.

seasons is insignificant, and for the *C* and *D* climates, where summer and winter temperatures differ markedly. In the latter two climates Köppen considers the ratio of the precipitation during the wettest month of the warmer or cooler season and of the driest month of the opposite season. If the driest month occurs during the cold season and has only one-tenth or less of the precipitation amount during the wettest month of the warm season, the climate has a dry winter (*Cw* or *Dw*). In view of the strong evaporation in summer, Köppen speaks of a climate with a dry summer when the wettest month of the cold season has three times as much precipitation as the driest month of the warmer season (*Cs*). *Ds* exists only over small areas, for example, over the interior of Oregon. If the difference between the critical months is smaller, the climate is of the type *Cf* or *Df*.

The boundaries between the subdivisions of the *C* and *D* climates according to the precipitation are illustrated in Fig. 21. Here the winter precipitation has been chosen as abscissa, the summer precipitation as ordinate. It is necessary to indicate the meaning of the coordinate axes on the figure in general terms, referring to the "extreme" monthly precipitation. Obviously, in the case of lower precipitation in winter the *driest* winter month should be compared with the *wettest* summer month; in the case of lower precipitation in summer the *wettest* winter month should be compared with the *driest* summer month.

The only exception to this definition of the *Cs* climates is found at some stations at higher latitudes along the west coast of North and South America. These stations deserve the notation *Cs* according to the above definition. But the summer precipitation is here still so strong that the climate cannot properly be called summer dry. For example, the precipitation at Valdivia on the coast of Chile is shown in Table 29. The temperature is included for the sake of comparison. The precipitation during the winter (July) is here almost seven times as great as during January, but even during January it is considerable so that the summer can hardly be characterized as dry.

TABLE 29.—TEMPERATURE AND PRECIPITATION AT VALDIVIA, CHILE (*Cfs*)

Item	Jan.	Feb.	Mar.	Apr.	May	June	July	Aug.	Sept.	Oct.	Nov.	Dec.	Year
Temp., deg F.....	61.9	60.4	57.9	53.1	49.6	45.5	45.7	46.4	48.0	52.0	55.0	59.0	52.9
Precip., in.......	2.6	2.9	5.5	9.3	15.7	17.2	16.3	13.7	8.4	5.5	5.0	4.4	106.3

As a further condition for the *Cs* type, it is therefore required that the amount of precipitation during the driest month is less than 1.2 in. Otherwise, the climate is denoted by *Cf* or by *Cfs* in order to indicate that considerably less precipitation occurs in summer than in winter, even though the climate cannot properly be called summer dry. In Fig. 21 the subdivision *Cfs* has been indicated by a broken line.

In the tropical *A* climates a definition of the limit between winter dry (*w*) and moist (*f*) climates similar to that in temperate climates would not be satisfactory, for the criterion used for the

C and *D* climates would not divide the winter dry regions with their savanna type of vegetation from the tropical rain forests. ("Winter" in connection with tropical climates means, not necessarily the season of lower temperature, but the time when the sun is on the other side of the equator.) Instead, a tropical climate is denoted by *Af* if the rainfall even during the driest month is more than 2.4 in. (more accurately, 60 mm). If the precipitation is less than this amount during the driest month, the climate is of the type *Aw*. The type *As* occurs only over

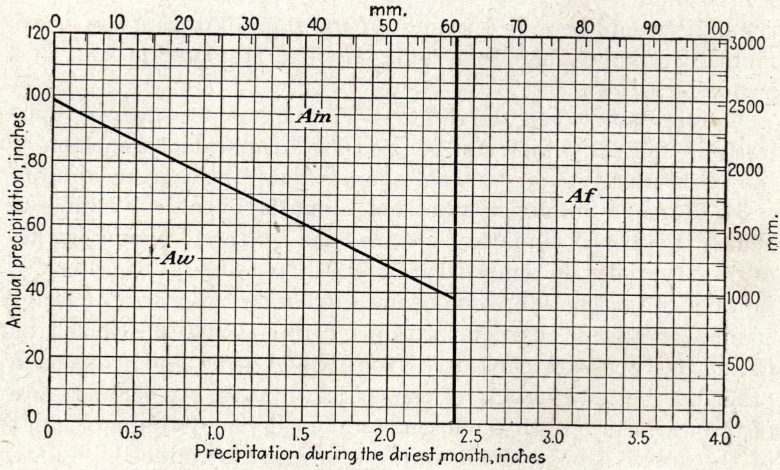

FIG. 22.—Precipitation divisions of the *A* climates.

small areas since, owing to the shift of the rain belts with the sun, summer is always the wettest season in the tropics.

Tropical rain forests with their large water requirements may even grow in regions where a regular seasonal period of little or no precipitation is found. It is necessary, however, that in such regions precipitation during the rainy months be so large that enough moisture remains in the ground to maintain the tropical forest even during the drier months. This type of climate is intermediate between the types *Af* and *Aw*. It resembles *Af* in its amount of precipitation, *Aw* in the seasonal distribution. Since it is thus a mixture of these two types, it is denoted by *Am*. It is also referred to as the "tropical monsoon climate" since a high annual amount of precipitation, in spite of a season of dryness, is characteristic of monsoon regions. The limit between the types *Am* and *Aw* is given by the expression

$$a = 3.94 - \frac{r}{25}$$

where r is the annual rainfall, a the rainfall during the driest month, both in inches. If, with a given annual rainfall, the rainfall during the driest month is greater than the value a obtained from the preceding formula, the climatic type is *Am*; if it is smaller, *Aw*. For example, if the annual rainfall is 50 in., the limit would be $a = 1.94$ in. If the rainfall during the driest month is greater than this value, the climatic type is *Am*; if it is smaller, *Aw*.

The limits between the different *A* climates are shown graphically in Fig. 22, which can be used for a rapid determination of the particular type of tropical climate found at a given station.

34. Further Subdivisions of the Climatic Types. Within the *C* and *D* types of climates that are fixed by the foregoing limits, a great variety of temperatures is still possible. Therefore, a number of further subdivisions are introduced in order to permit a more detailed characterization of the climates. Since the *C* and *D* types are separated from the *A* type and from each other by the temperature of the coldest month, the temperature of the warmest month is chosen as the principal dividing element of the subdivisions. Four subdivisions are introduced. If the summer is hot with a mean temperature of the warmest month above 71.6°F (22°C), the climatic type is denoted by the letter *a*. If the warmest month is below 71.6°F but above 50°F—since otherwise the climate would be one of the *E* types—it is characterized by the letter *b*, provided that more than 4 months have a mean temperature over 50°F, and by *c* if fewer than 4 months have a temperature over 50°F. Finally, if the mean temperature of the coldest winter month is below −36.4°F (−38°C), the letter *d* is added to the climatic type. The limits of these four subdivisions are summarized in Table 30.

If all the combinations of the three types of seasonal variation of precipitation and the four subdivisions according to the temperature were realized in both the *C* and *D* climates, 24 variations would be possible. Actually, only 14 of these combinations are important, as indicated in Table 31, where the types that occur are indicated by an x. The other 10 types either occur not at all or only over very small areas.

The boundaries between the polar and the tree climates and their subdivisions are represented schematically in Fig. 23, except for those subdivisions which are determined by precipitation. The abscissa of this figure is the mean temperature

TABLE 30.—LIMITS OF THE SUBDIVISIONS OF CLIMATES *C* AND *D*

Type	*a*	*b*	*c*	*d*
Mean temperature of warmest month	> 71.6°F	71.6°F to 50°F		
Duration of temperature of at least 50°F	More than 4 months		1–4 months	
Mean temperature of coldest month	≧ −36.4°F			< −36.4°F

of the warmest month, the ordinate the mean temperature of the coldest month because these two quantities are used as criteria for the different climatic types. The upper left-hand corner of the figure is cut off since the temperature of the coldest month cannot be greater than that of the warmest month. Both coordinate scales are interrupted, the ordinate scale (coldest month) between −20°F and +20°F, the abscissa (warmest

TABLE 31.—OCCURRENCE OF THE SUBDIVISIONS OF *C* AND *D*

Type	*C*			*D*		
	w	*s*	*f*	*w*	*s*	*f*
a	*x*	*x*	*x*	*x*	–	*x*
b	*x*	*x*	*x*	*x*	–	*x*
c	–	–	*x*	*x*	–	*x*
d	–	–	–	*x*	–	–

month) between 50°F and 60°F. An unbroken ordinate would have made the figure too large. The break in the abscissa has been inserted in order to have a place for the boundary between the subdivisions b and c of the C and D types that depends, not on the temperature, but on the duration of the temperatures above 50°F during the year. The break in the abscissa is made so that no temperature corresponds to the vertical line that represents the boundary between *b* and *c*. The figure may be

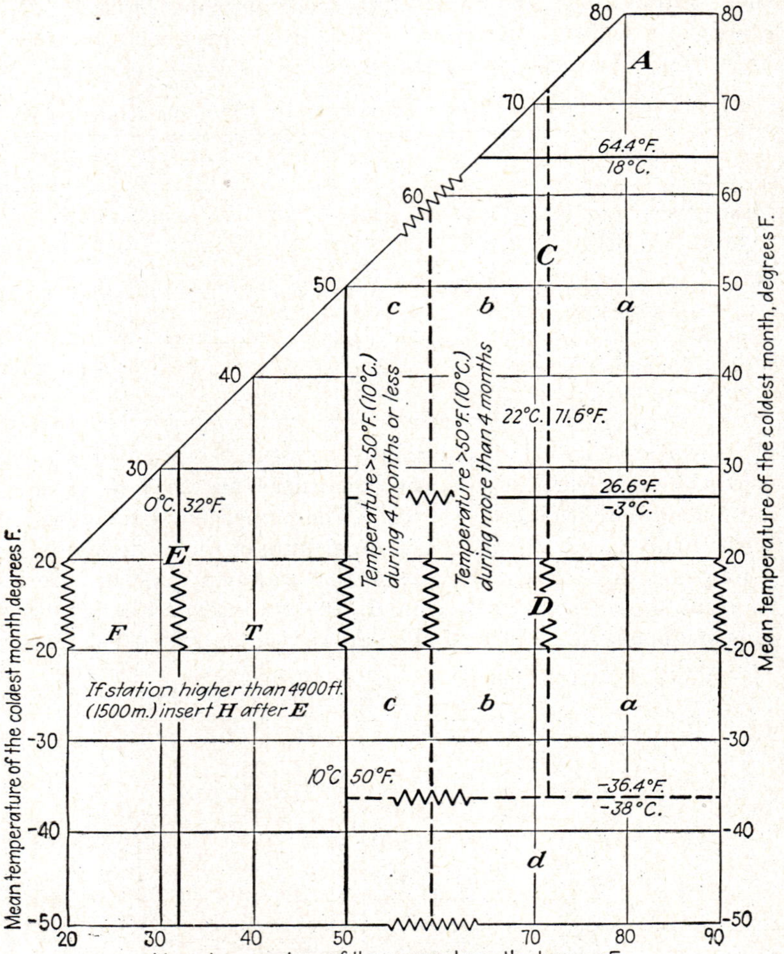

FIG. 23.—Boundaries between polar and tree climates and their subdivisions.

used to determine the climate of a station. The numerical value of each boundary, except that between *b* and *c*, is indicated in degrees centigrade and in degrees Fahrenheit. Figures 20 to 23 permit a typing of the climate of a given station.

In addition to the subdivisions of the *C* and *D* climates Köppen has suggested a great number of other notations that permit of a still more detailed description of the climate of a given region. It is thus possible to characterize a climate very briefly by a symbol consisting of certain letters. Only a few of these notations will be used here.

In the tropical climates near the equator the annual variation of the temperature is, as a rule, very small. If the annual temperature range is less than 9°F, this fact may be indicated by the letter *i* for "isothermal." (In his latest review of his classification Köppen[1] suggests the use of the letter *i* only if the yearly range is less than 1°C. But this limit seems unnecessarily narrow.) It will be seen later that some regions with *C* climates also deserve the notation *i*. In these regions the altitude reduces the temperature so much that the climate belongs to the *C* group though owing to the proximity of the equator the annual temperature range is small. Examples for such climatic types are to be found in the African highlands near the equator and in the Andes of equatorial South America.

For the *B* climates the letter *h* is used to indicate that the annual mean temperature is above 64.4°F. If the annual mean temperature is below 64.4°F, the letter *k* is added. In Patagonia a very cool desert climate prevails. Even the temperature of the warmest month is here below 64.4°F. This Patagonia climate can therefore be considered as a separate subdivision of the *B* climates for which the notation *k'* is used.

If fog occurs frequently in the dry climates, this fact is indicated by the letter *n*. This type is usually found along littorals paralleled by cool ocean currents as, for example, along the coasts of Peru and southwestern Africa. If fog is rare but humidity high, the climate is designated by *n'*. Strictly speaking, this designation is applied only if the summer temperature is below 75.2°F; different notations are used for higher summer temperatures. But these distinctions will be omitted here.

The *E* climates with their low summer temperatures are mainly found in polar latitudes. But in mountainous regions of lower latitudes this type also occurs provided that the elevation is sufficiently high. To indicate that a region has an *E* climate owing not to high latitude but to high elevation the letter *H* is

[1] Köppen-Geiger, *op. cit.*, Vol. 1, C, p. 43.

used if the region is higher than 4900 ft (1500 m). The same notation could be used for the *C* climates that are found at high elevations under the equator, but Köppen does not make use of this possibility.

The 11 principal climatic types and their subdivisions and additional notations as far as they will be used here may finally be listed together.

1. *Af* tropical rainy climate
2. *Aw* tropical savanna climate
3. *BS* semiarid (steppe) climate
4. *BW* arid (desert) climate
5. *Cf* warm temperate rainy climate without dry season
6. *Cw* warm temperate rainy climate with dry winter
7. *Cs* warm temperate rainy climate with dry summer
8. *Df* cold snow-forest climate without dry season
9. *Dw* cold snow-forest climate with dry winter
10. *ET* tundra climate
11. *EF* frost or icecap climate

a	*C* or *D* climate	with hot summer
b	*C* or *D* climate	with warm summer
c	*C* or *D* climate	with cool, short summer
d	*D* climate	with very cold winter
H	polar climate due to high altitude	
h	hot dry climate	
i	small annual temperature range	
k	cool dry climate	
k'	cold dry climate	
m	mixed or monsoon type of tropical climate	
n	dry climate	with frequent fog
n'	dry climate	with high humidity

35. Geographical Distribution of the Climatic Types. The geographical distribution of the climatic types shows certain characteristic features found, with some modifications, on every continent. To bring out these similarities, which help greatly toward an orientation with respect to the multitude of climates, Köppen has constructed a schematic figure showing the distribution of his types over an ideal continent. This continent extends from the North Pole to about 60° south latitude. Its width decreases from 30° north latitude southward to conform with the general pattern of the continents in the Southern Hemisphere.

The distribution of the climatic types on such an ideal continent after Köppen is reproduced, with some modifications, in Fig. 24. The climatic types over the adjoining oceans are also shown, but with less detail. The northernmost part of the

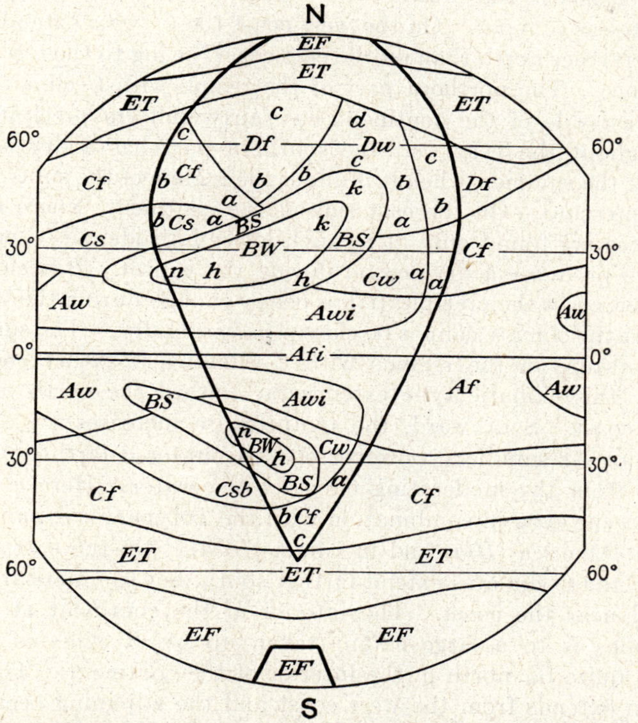

FIG. 24.—Distribution of climates on an ideal continent. (*After Köppen.*)

continent is taken up by an *EF* climate, followed by an *ET* climate. Farther south, *D* climates are found. In the western part of the continent, which is under the influence of westerly winds from the oceans in these latitudes, the annual variation of the precipitation is not very pronounced, but in the eastern part precipitation is markedly less in winter except in a narrow strip along the east coast, which is more under the maritime influence. The temperature within this large region of *D* climates becomes milder toward the south, as indicated by the geographical arrangement of the letters *a*, *b*, *c*, and *d*. The *d* type with very severe winters occurs only in the interior of the continent, far away from the moderating influence of the ocean. It is displaced toward the east from the center of the continent since the westerlies carry the influence of the ocean far inland. Farther south the winters are warm enough to classify the climates as *C* types. On the west coast the *C* types extend very much farther north than on the east coast, owing to the maritime influence. The northern part of the region with *C* climates on the west side of the continent has fairly uniform precipitation throughout the year, but the southern part has a dry season during the summer, when it is under the regime of the subtropical high pressure. On the east side of the continent, where the *C* climates are found only at a much lower latitude, the monsoon regime produces a dry season during the winter. But close to the east coast the precipitation is heavy enough during the winter so that the climate cannot be classed as winter dry. The summer temperature in the regions with *C* climates is rather cool (*c*) where this climatic type extends far toward the north on the west coast. Southward, the temperature increases, of course. The hottest summers, however, are encountered farther inland, away from the moderating effect of the water. Here *a* types may occur even in combination with the *D* type, for example, in North America (*Dfa*) and in China (*Dwa*). On the east coast, where the *C* climates extend farther south, hot summers are also found near the coast. The interior of the continent at lower latitudes is to a large extent taken up by *B* climates, which reach quite far north in the interior of the continent. The dry region extends from the west coast and the adjoining ocean far inland and toward the north, where it is sufficiently far away from maritime influences. The northern part of this arid and semiarid region is cool (*k*), but the southern part is hot (*h*). Where the desert reaches the west coast, fogs are frequent (*n*) (see page 123). As the equator is approached, the desert type of climate changes into the steppe type again, and the steppe type into the savanna type (*Aw*). On both sides of the equator the *Af* type is predominant. On the east side of the continent the dry climates do not reach to the coast, for the easterly winds bring moist air from the oceans. The *C* types of climate change here directly into *A* types. Owing to the prevailing easterlies there are as a rule no winter dry *A* climates along the east coast at low latitudes, or if a pronounced seasonal variation exists because of a monsoon regime the precipitation during the rainy season is so intense that the notation *m* is appropriate.

South of the equator the climatic types are found in about the same order except for the modifications that are produced by the smaller width of the continent. Thus, the region of *B* climates is considerably smaller than in the Northern Hemisphere and does not reach to such high latitudes because the narrowness of the land prevents the development of high-latitude deserts and steppes, except in South America in the rain shadow of the Andes. The *D* climates are entirely absent in the Southern Hemisphere, for in higher temperate latitudes where this climatic type is found the southern continents are too narrow.

The main features of this schematic distribution of the climatic types will be recognized on the world map of the climatic types, which is shown as a Frontispiece. A complete agreement with the foregoing scheme of Fig. 24 is, of course, not to be expected since each continent has its own peculiarities. The geographical distribution of the climatic types will be discussed in connection with the description of the climates of the different parts of the globe, in Part B on Regional Climatology. It is suggested, however, that the reader shall compare the actual distribution of the climates in the Frontispiece with the schematic distribution indicated in Fig. 24 and note the differences the explanation of which in many instances will be quite obvious from meteorological and climatological principles.

Table 32 shows the area of the climatic types[1] for water and land and for the earth as a whole. For the whole earth the tropical *A* climates cover the largest area, more than one-third the whole earth. The *C* climates have the next largest distribution. *A* and *C* climates together cover more than one-half the earth's surface. They are followed in size by the *E* climates, while *B* and *D* climates have the smallest extent. For the oceans, which cover more than 70 per cent of the earth's surface, the relative importance of the various climatic types is substantially the same except that the *B* and *D* climates, which are characteristically continental types, are even less important.

[1] WAGNER, H., *Petermanns Mitt.*, **67**, 216, 1921.

For the continents, on the other hand, the distribution of the areas is quite different. The *B* climates account for more than one-fourth the total land area, the *D* climates for more than one-fifth. The *A* climates and the *C* climates, which take up almost three-fourths the ocean area, extend over only about one-third the land area.

TABLE 32.—AREAS OF THE CLIMATIC REGIONS, PER CENT

Type	Continent		Ocean		Whole earth	
Af	9.4	19.9	28.6	42.7	23.0	36.1
Aw	10.5		14.1		13.1	
BS	14.3	26.3	3.6	4.2	6.7	10.6
BW	12.0		0.6		3.9	
Cw	7.6	15.5	0.4	31.9	2.5	27.2
Cs	1.7		2.9		2.6	
Cf	6.2		28.6		22.1	
Df	16.5	21.3	1.5	1.7	5.8	7.3
Dw	4.8		0.2		1.5	
ET	6.9	17.0	16.0	19.5	13.4	18.8
EF	10.1		3.5		5.4	

The *Dw* climate is almost completely continental. Similarly, the *Cw* climate is found only over a very small ocean area. The maritime extent of these two climatic types is even smaller than that of the desert climate *BW*. The tree climates *A*, *C*, and *D* together cover 57 per cent of the continents, 76 per cent of the oceans, and 71 per cent of the whole earth.

36. Modifications of Köppen's Classification of Climates. The classification of climatic types after Köppen, which has been presented in the preceding sections, has been developed gradually. In particular, the boundaries between the dry *B* climates and the moist tree climates have been changed a number of times until Köppen finally adopted the expressions given on page 114. However, it has been suggested by Russell[1] that this last expression for the boundary between the dry climates and the tree climates should be abandoned in favor of an earlier expression used by Köppen. In the United States this modification would lead to a better agreement between the climatic boundaries and the actual border between desert and steppe vegetation.

[1] RUSSELL, R. J., *Univ. Calif. Pub. Geography*, **5**, No. 1, 1931; No. 5, 1932.

Furthermore, Russell prefers as the boundary between the *C* and *D* type the temperature of 32°F during the coldest month instead of 26.6°F. Russell's limit for the *C* and *D* climate was adopted by Ackerman[1] on his map of the climatic types of North America. While the temperature of 32°F as the freezing temperature of water is doubtless of meteorological and climatological importance, the significance of Köppen's boundary of 26.6°F for the mean temperature of the coldest month consists in its representing quite well the upper limit at which a regular snow cover can be expected in winter. The importance of a snow cover was discussed in Sec. 22.

Similarly, Russell advocates using the same mean temperature of 32°F during the coldest month as the limit between the hot (*h*) and the cool (*k*) desert climates, instead of the annual mean temperature of 64.4°F used by Köppen. The mean temperature of the coldest month is beyond doubt of greater climatic significance than the annual mean temperature, whose value can be very misleading, as pointed out already on page 1. On the other hand, at least for the vegetation, the determining factor in the *B* climates is not the temperature but the lack of sufficient precipitation, so that the choice of a temperature limit is not of great importance.

Köppen has already suggested that it might be desirable to make a further subdivision of the tundra climates *ET*. Figure 23 shows that a station can have a polar tundra climate, even though the monthly mean temperature never drops below freezing. It is necessary only that the temperature of the warmest month remains below 50°F—otherwise the climate would be *C* or *D*—and that the annual temperature range is sufficiently small. The station Evangelist's Island off the west coast of Chile, for example, has a mean temperature of 47.2°F during the warmest month and of 37.4°F during the coldest month, so that the climate is of the type *ET*. Another example is Cape Pembroke on the Falkland Islands (Table 37). It would be possible to differentiate between such oceanic climates with cool summers and mild winters on the one hand and the true *ET* climates with cold winters on the other hand by using as a boundary the temperature of the coldest month. According

[1] Ackerman, E. A., *Geog. Rev.*, **31**, 105, 1941.

to Köppen a monthly mean temperature of 26.6°F during the coldest month would be a satisfactory boundary.

Another possibility of extending Köppen's classification would be a distinction between the regular *C* climates of low altitudes and the *C* climates near the equator which are due to high altitudes. This extension was indicated on page 127.

However, regardless of the merits of the proposed modifications, it is preferable to retain just one classification that has been adopted widely instead of introducing modifications and changes that only make it more complicated instead of easier to gain a clear picture of the various climates of the earth. To facilitate orientation in respect to the great number of terrestrial climates is the main purpose of a classification. Once this goal is attained and the student of climatology has become familiar with the distribution of the climatic elements over the globe, there is little further need for the classification.

37. Thornthwaite's Classification of Climates. Among other attempts at a quantitative classification of climates, Thornthwaite's[1] will be described here, although more briefly. It was pointed out in the discussion of Köppen's classification that the importance of precipitation for vegetation depends not only on its amount but also on the intensity of the evaporation. For if the evaporation is large, less of the same amount of precipitation will be available for plant life than if the evaporation is small. In Köppen's classification this effect is taken care of in a not very satisfactory way by combining precipitation with temperature. Thornthwaite computes the "precipitation effectiveness ratio" (*P-E* ratio) in order to obtain a better expression for that part of the precipitation which is effective in furthering plant growth. The *P-E* ratio is the total monthly precipitation *P* divided by the total monthly evaporation *E*. The sum of the 12 monthly *P-E* ratios is called the "*P-E* index." The difficulty in applying the *P-E* ratios and indexes to climatological investigations arises out of the lack of observations of evaporation over most of the the earth. In order to overcome this difficulty Thornthwaite studied the relation between precipitation *P*, evaporation *E*, and temperature *T* at 21 stations in the western United States

[1] THORNTHWAITE, C. W., *Geog. Rev.*, **21**, 633, 1931; **23**, 433, 1933. See also "Climate and Man," Yearbook of Agriculture, p. 98, United States Department of Agriculture, Washington, D.C., 1941.

where evaporation measurements were made from April to September for periods ranging from 4 to 12 years. He found that the following relation holds:

$$P\text{-}E \text{ ratio} = \frac{P}{E} = 11.5\left(\frac{P}{T - 10}\right)^{10/9}$$

where P and E are in inches and T is in degrees Fahrenheit, so that the P-E ratio can be computed from observations of the precipitation and the temperature that are much more numerous than those of evaporation. The preceding formula is based on observations made in the western United States during the warmer season. Hence, its application to other regions and other times of the year represents an extrapolation that is hardly fully justified, as already pointed out by Thornthwaite in his first paper. Below freezing the precipitation is of no direct benefit to plants. More specifically, empirical studies have shown that the *smallest mean monthly temperature* at which precipitation ceases to be of use to plants is 28.4°F rather than 32°F. At a temperature lower than 28.4°F this value should be substituted for T.

For the sake of convenience P/E is multiplied by a factor 10 so that the

$$P\text{-}E \text{ ratio} = 115\left(\frac{P}{T - 10}\right)^{10/9}$$

$$P\text{-}E \text{ index} = \sum_{n=1}^{12} 115\left(\frac{P}{T - 10}\right)^{10/9}$$

The actual computations of these quantities are greatly facilitated by charts given in Thornthwaite's papers.

By evaluating P-E indexes for a great number of stations and comparing them with the characteristic vegetation, Thornthwaite found that the following humidity provinces can be distinguished:

Humidity province	Characteristic vegetation	*P-E* index
A (wet)	Rain forest	128 and above
B (humid)	Forest	64–127
C (subhumid)	Grassland	32–63
D (semiarid)	Steppe	16–31
E (arid)	Desert	Less than 16

These five humidity provinces are subdivided into the following four subtypes:

r = moisture abundant at all seasons
s = moisture deficient in summer
w = moisture deficient in winter
d = moisture deficient at all seasons

They are determined by the values of the *P-E* index and *P-E* ratio for the season of highest precipitation, but the numerical limits will not be discussed here. Obviously, not all the combinations of humidity provinces and subtypes can occur.

If other factors affecting the plant growth, especially the temperature, are equal in different localities, the type of vegetation is determined by the respective humidity of the localities. In considering climatic regions of the earth, however, the temperature has to be taken into consideration, also. To do this, Thornthwaite has devised a monthly coefficient of thermal efficiency, the *T-E* ratio. He calls the sum of the 12 monthly *T-E* ratios the "*T-E* index."

The thermal efficiency is very low in polar regions, and the *T-E* index is so defined that it becomes zero at the polar limits of the tundra. The most favorable thermal conditions for plant growth are in the tropical regions. Therefore, Thornthwaite defines the *T-E* index in such a manner that its numerical value for the most favorable thermal conditions becomes equal to that of the numerical value for the most favorable humidity conditions, *i.e.*, for the humidity province *A*. An equation was developed empirically that ascribes to the poleward limit of the tundra a *T-E* index of zero and to the poleward limit of the tropical rain forest and savanna a *T-E* index of 128. The latter is the same as the *P-E* index of the drier margin of the rain forest. The equation which was derived states that

$$T\text{-}E \text{ ratio} = \frac{T - 32}{4}$$

and

$$T\text{-}E \text{ index} = \sum_{n=1}^{12} \left(\frac{T - 32}{4} \right)$$

Thornthwaite distinguishes the following six temperature provinces:

Temperature Province	*T-E* Index
A′ (tropical)	128 and above
B′ (mesothermal)	64–127
C′ (microthermal)	32–63
D′ (taiga)[1]	16–31
E′ (tundra)	1–15
F′ (frost)	0

[1] The word "taiga" is the Russian name for the subarctic coniferous forests.

An additional factor to be considered is the seasonal temperature variation. Thornthwaite differentiates between five thermal

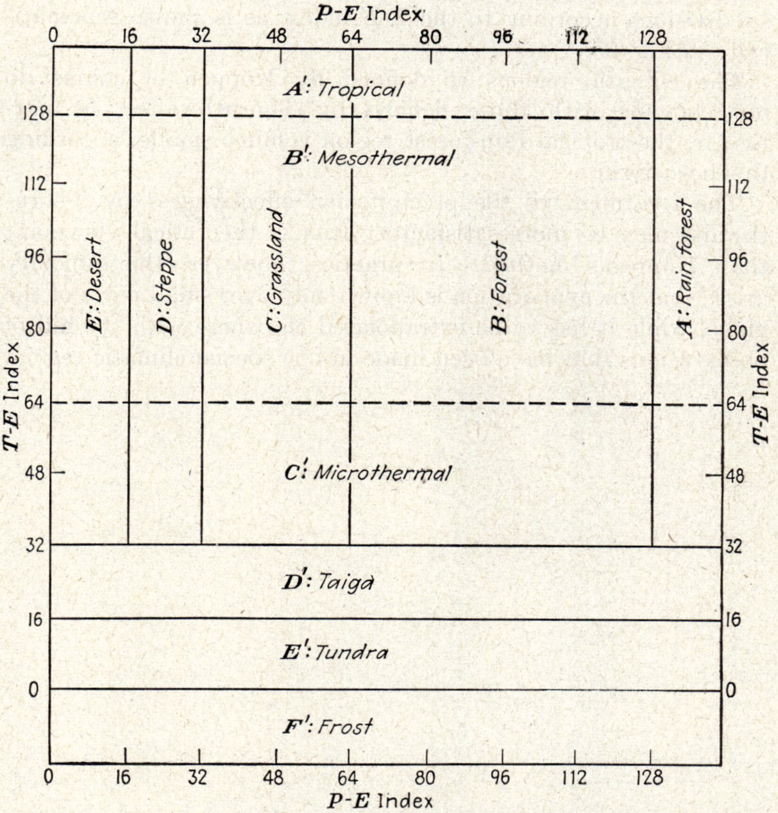

FIG. 25.—Temperature and humidity provinces. (*After Thornthwaite, Geog. Rev.*, **21**, 649, 1931.)

subprovinces according to the degree of summer concentration of temperatures favorable for vegetation. But this refinement will not be discussed here.

The precipitation effectiveness and the temperature efficiency are the two most important climatic factors, as are temperature and precipitation in Köppen's classification. With a temperature sufficiently high to maintain vegetation the *P-E* indexes establish the climatic boundaries. Elsewhere the *T-E* index is the dominant factor. Thus, there are eight major provinces, *viz.*, *A*, *B*, *C*, *D*, and *E*, where the temperature efficiency is adequate and D', E', F' where it is inadequate. In each of the climates where the temperature efficiency is adequate every humidity province is subdivided into three possible thermal subdivisions according to the *T-E* index, as is shown schematically in Fig. 25.

The climatic regions as defined by Köppen of course do not coincide with those defined by Thornthwaite. In particular, the tropical rain-forest region is much smaller according to Thornthwaite.

The treatment of the precipitation effectiveness by Thornthwaite may be more satisfactory from a theoretical viewpoint than Köppen's method. In practice, however, the difficulty arises that the evaporation is known only over small parts of the globe, while it has to be extrapolated elsewhere with the aid of observations that have been made in one special climatic region.

CHAPTER VII

DESCRIPTION OF THE CLIMATIC TYPES

In Chaps. I to IV the different climatic elements have been discussed separately. But in order to describe the climate of a particular region it is necessary to consider the various climatic elements together. As was shown in the preceding chapter, the combination of these climatic elements gives a great variety of climatic types, which have been brought into a logical system by Köppen's classification. The climatic features of these types will be discussed in the present chapter. In this description little emphasis will be placed on their geographical positions. Instead, the common features of one type in different parts of the world will be stressed. A separate discussion of the climates of different regions of the earth will be found in Part B on Regional Climatology.

38. The Tropical Rainy Climates (*A*). According to the definition of the boundary between the *A* and the *C* types of climates the temperature even of the coldest month in the *A* types is at least 64°F[1]; thus the colder season cannot be referred to as winter in the same sense as for higher latitudes. The yearly mean temperature in the *A* climates is as a rule between 70° and 80°F. The annual range of temperature is small, considerably less than the daily variation. At Bolobo in the Belgian Congo, 2° south of the equator, for example, the average daily temperature range is 16°F, the annual range 2°F. In a great part of the regions covered by *A* types of climates the annual temperature range is not more than 9°F so that the notation *i* is appropriate. This small variation of the temperature with the seasons is, of course, due to the relatively small variation of the solar altitude and of the length of the day throughout the year in tropical latitudes.

[1] For the sake of convenience the limits will now be given in whole degrees in the text.

There are aperiodic changes of the temperature too, but they are less strong than in extratropical regions because the horizontal temperature contrasts are much less pronounced than in higher latitudes.

Owing to the absence of a real winter there is no interruption of the vegetation period due to low temperatures as in temperate latitudes. But, in the savanna type (*Aw*), vegetative life has a rest period during the dry season. In those regions with *A* climates where a pronounced seasonal variation of the precipitation occurs the dry season coincides with the winter, and the greatest heat occurs toward the end of the drier season or at the beginning of the rainy season before the rains become too heavy (see the data for Cuiabá as representatives of the *Aw* climate in Table 33).

The amount of the rainfall is quite large, rarely less than 35 in. per year and in most regions with *A* climates considerably more. The rain is mostly of the convective type, but in some regions and during some seasons tropical cyclones are important in causing precipitation. Convection is particularly strong in the tropics because the stratification of the air is largely convectively unstable. The ascending motion of the air is further favored by the convergence of the trade winds in the zone of the doldrums. Since convection depends on the intensity of the solar radiation the precipitation shows a dependence on the height of the sun. At the polar limits of the regions with *A* climates this dependence is indicated by the *Aw* climates with dry winters, but even in the *Af* climates the rains show a seasonal variation as was illustrated by the example of Yaoundé, Cameroons, in Fig. 16. This station is about 4° north of the equator and has two rainfall maxima, in April and October, just after the equinoxes, and two minima just after the solstices, *i.e.*, at the times when the sun has its highest and lowest position in the sky, respectively. The convective origin of the rainfall explains also why the rains in the tropics occur rarely in the forenoon, but rather in the afternoon or, in maritime locations, later in the evening (see Sec. 21). The clouds are also largely of the convective type, and the variation of the cloudiness follows that of the precipitation quite closely.

The amount of precipitation in the tropical climates is much more variable than the temperature. The rainfall varies with the time of the year, with the locality, and from year to year, while the temperature is quite uniform. The annual aperiodic changes of the rainfall, especially in winter dry regions (*Aw*) may be so large that crop failures and famines result.

The Frontispiece shows that the *A* climates are largely found in a belt near the equator. On the east coasts of the continents that are under the influence of winds from the sea the *A* belt extends farthest to the north and south and is followed by *C* climates, approximately at the tropics of Cancer and Capricorn. On the west coasts the *A* climates reach only to latitudes between 5° and 15° and are then replaced by the dry *B* climates.

Climatic regions of the tropical rain-forest type (*Af*) or its modification, the *Am* type, are mainly located near the equator within the equatorial rain belt. Among the regions whose climate is of this type are the Amazon Valley in South America, the Congo basin, the Guinea coast in Africa, and the East Indies. The *Am* variation occurs mainly and in its most pronounced form in coastal regions where a monsoon wind blows against a mountain range during part of the year (see Colombo, Ceylon, Table 33).

The savanna type of climate (*Aw*) differs from the tropical rain-forest type (*Af*) in that it has a dry season during the winter. The yearly amount of the precipitation is therefore in general smaller than in the *Af* climates. *Aw* climates are developed over regions that are reached by the migrating equatorial rain belt only during the season when the sun is high. Hence, these regions are situated on the poleward side of regions with *Af* climates and bounded on the side farther away from the equator by regions with dry climates (*B*). The *Aw* type is found in the llanos of the Orinoco Valley in Colombia and Venezuela, the campos of Brazil, the Sudan of North Africa, in South Africa, and northern Australia. In many locations, especially in India, Burma, and Indo-China, the *Aw* climates are due to the monsoon.

The yearly temperature range in the *Aw* climates is mostly greater than in the *Af* climates owing to the greater distance from the equator at which the regions with *Aw* climates are located. Since the rainy season with its greater cloudiness occurs during the summer, the highest temperatures are usually attained before the sun reaches its highest position in the sky, as for instance at Colombo and at Cuiabá.

TABLE 33.—EXAMPLES OF *A* CLIMATES

Month	Jan.	Feb.	Mar.	Apr.	May	June	July	Aug.	Sept.	Oct.	Nov.	Dec.	Year	Range
Singapore, Straits Settlement (lat. 1°N; long. 104°E; el. 16 ft) *Afi*														
Temp., °F. .	77.9	78.4	79.3	79.9	80.6	79.9	80.2	79.7	79.5	79.7	79.0	78.3	79.3	2.7
Precip., in. .	9.7	7.1	7.3	7.8	6.5	7.0	6.7	7.8	6.9	7.9	10.1	10.4	95.1	
Jupiter, Fla. (lat. 27°N; long. 80°W; el. 30 ft) *Afw*														
Temp., °F. .	64.6	66.9	70.2	72.3	76.5	80.1	81.7	81.7	80.6	76.6	77.5	66.4	74.6	17.1
Precip., in. .	3.3	2.9	3.0	2.5	4.9	6.9	5.2	5.7	9.2	10.1	3.1	2.6	59.4	
Colombo, Ceylon (lat. 7°N; long. 80°E; el. 24 ft) *Ami*														
Temp., °F. .	79.0	79.7	80.6	81.5	82.1	80.4	80.0	80.5	80.4	79.3	79.5	79.2	80.2	3.1
Precip., in. .	3.5	1.9	4.5	7.8	13.0	7.7	6.5	3.0	5.9	13.1	11.0	5.2	83.1	
Cuiabá, Brazil (lat. 16°S; long. 56°W; el. 541 ft) *Awi*														
Temp., °F. .	80.8	80.6	80.8	80.4	78.4	75.0	74.8	78.1	81.3	82.6	81.7	81.0	79.7	7.8
Precip., in. .	9.6	8.9	8.1	4.1	2.0	0.2	0.2	1.1	2.0	4.4	6.0	7.9	54.5	

39. The Dry Climates (*B*). The common characteristic of the *B* types of climates is their aridity. The potential evaporation—*i.e.*, the amount of evaporation which would occur if a permanent water supply were available—is greater than the precipitation. According to the degree of dryness and the resulting type of vegetation the semiarid steppe type and the arid desert type are distinguished, the steppe type representing the transition from the desert climates to the more humid climates. Since the aridity depends not only on the smallness of the precipitation but also on the magnitude of the potential evaporation, the amount of precipitation alone cannot be used as a boundary between the dry and the humid climates. The same amount of precipitation falls in some areas in the Sahara and in the northern parts of Siberia and North America. But in the one location an arid desert or a semiarid steppe is found, and the other two locations are taken up by the tundra, which may even be waterlogged in summer since the ground at greater depth remains frozen and since little water evaporates. In order to take the evaporation into account in determining the boundary between dry and moist climates Köppen uses a combination of temperature and precipitation, as explained in Sec. 33.

Dry climates are found at widely different latitudes. Hence, few general statements about their temperature can be made, and, as discussed previously, a subdivision into hot and cool dry climates was introduced by Köppen (page 123). The annual temperature range in the dry climates is large in comparison with the average temperature range in the same latitude because the *B* climates occur predominantly in the interior of the continents, often on the lee side of mountain ranges, which prevent any moderating oceanic influences. Similarly, the diurnal temperature ranges are high owing to the clear skies and to the relatively small water-vapor content of the air. Both factors permit a large amount of solar radiation to reach the ground in daytime and a similarly large amount of terrestrial radiation to be radiated to space at night. A further reason for the large diurnal temperature range is the barren or almost barren ground whose temperature becomes much higher during the day than that of soil which is covered by vegetation.

Desert and steppe types of climates are found in the regions of the subtropical high-pressure belts and farther poleward in the interior of the continents (see Frontispiece). Some of the most important regions with *B* climates in the Northern Hemisphere are the Sahara with its adjoining steppe regions, the Arabian Desert, and the dry regions of Iran, Afghanistan, and northwestern India, which continue into the interior of Asia, where they cover largely the southern part of the Asiatic Soviet Union, western China, and Mongolia. Considerable parts of the western United States and of the interior of Mexico also have dry climates. In the Southern Hemisphere the greater part of Africa beyond about 20° south latitude is taken up by dry climates. In South America the *B* climates extend along the west coast from about 5° south to 30° south latitude, while beyond about 25° south a dry region is found in the rain shadow of the Andes. At this latitude the *B* climates do not reach the Atlantic Ocean but beyond 40° south, where the South American continent becomes narrow, the *B* climate extends to the Atlantic coast. In Australia the largest part of the continent is taken up by dry climates.

The hot desert and steppe climates (*BWh and BSh*) owe their existence mainly to the subsiding motion in the subtropical anticyclones. This statement may at first seem at variance with the facts, for the map of the pressure distribution during July (Plate V) shows regions of low pressure over such typical dry regions as the Sahara and over the desert regions of the southwestern United States. However, these lows are of thermal origin and do not reach to any great heights, as is demonstrated by the map of the pressure distribution at 4000 m (Fig. 13), where a well-developed high-pressure area is found over Africa and where the isobars show an anticyclonic curvature over the southwestern United States.

The centers of the regions with *Bh* climates are found at around 20° to 25° north and south latitude and extend from approximately 15° to 30° north and south. On the west side of the continents the desert and steppe climates reach the water and even extend far over the adjoining oceans. As the coast is approached from the interior, however, the type of the climate changes somewhat. This modification is due to the presence of cool water offshore, which produces the characteristic *Bn* or *Bn'* climate with frequent fogs or at least with high relative humidity, to be discussed. On the east side of the continents the dry climates do not reach quite to the coast but are replaced by *A* types or, at higher latitudes, by *C* types because the trade winds with their easterly components bring moister maritime air over the continent. Furthermore, air that reaches the east side of the continent has moved in an east-to-west direction around the anticyclone. The plane in which this motion takes place ascends slightly toward the west, especially at somewhat higher levels above the ground, as pointed out by Bjerknes,[1] so that some lifting of the air arriving on the east coast of the continent has taken place. Owing to this lifting, the air may become conditionally unstable, and the possibility of rainfall is thus increased.

The annual mean temperature of the climatic types *BWh* and *BSh* is greater than 64°F according to Köppen's system. It has already been pointed out that the annual temperature range in deserts is large compared with the average for the same latitude. While the average annual range for 20° north or south

[1] Bjerknes, V., and collaborators, "Physikalische Hydrodynamik," p. 684, Verlag Julius Springer, Berlin, 1933.

latitude is only about 10°F according to Table 7 (page 24), temperature ranges of 30°F are quite common in the hot deserts and steppes. It is the high summer temperature which is responsible for the large annual variation, rather than the winter temperature, which is not particularly low. Of the same magnitude as the annual variation, or even greater, 40°F and more, is the daily temperature range. While the daily maxima during the warmer season are well above 100°F, the nights are relatively cool. In winter the temperature may occasionally fall even below freezing.

The cloudiness is very small. At In-Salah in the Sahara the average yearly cloudiness is only 1.3 tenths of the sky. At Yuma, Ariz., the cloudiness is 2.4 tenths in January, 0.8 tenth in June. The relative humidity is also low. At In-Salah the largest monthly mean occurs in January with 58 per cent, the smallest monthly mean in August with 19 per cent. At Yuma the maximum is 51 per cent in November, the minimum 35 per cent in May.

Precipitation is naturally very small. According to Köppen's definition the precipitation in a steppe climate with an annual mean temperature of 70°F is not more than 22 in., in a desert climate at most 11 in., provided that there is no annual period of precipitation (see page 115). In many deserts the precipitation is smaller, of course. The variability of the total annual rainfall from year to year is very great in dry climates. This variability of the precipitation from year to year is of economic importance in the marginal lands between the steppes and the humid climates. Here, the rainfall may be sufficient for agricultural needs during a number of years. But after some favorable years droughts occur, resulting in crop failure and showing that permanent cultivation of the region is impossible. Examples of such marginal lands are large tracts of the prairies of the United States although these regions have *BSk* rather than *BSh* climates.

While rain is rare and unreliable in the desert, occasionally very large amounts fall in one cloudburst. At Alexandria, Egypt, for example, the mean annual rainfall is 8 in., but 11 in. of rain fell during one single rainstorm. In such cases heavy floods occur and the dry river beds, called "wadis," become raging torrents. These sudden downpours, however, are of little benefit to vegetation since the flow of the water is unchecked. The water disappears again quickly, owing to the strong evaporation and the porousness of the soil.

Where precipitation falls in the hot steppe climates, it shows a pronounced seasonal variation similar to that of the adjoining humid climates. Thus, on the equatorial side of the *BSh* regions in North Africa, where the savanna climate (*Aw*) with summer rain is found, the scanty rains fall also in summer, while on the polar side, where the *BSh* climate changes into summer dry *Cs* climates, the winter is the rainy season. If an indication of the annual distribution of the precipitation is desired, the two types may be denoted, respectively, by *BShw* and *BShs*.

The regions of *Bh* climates are the principal source regions of tropical continental air. In winter, the Sahara is the only source region of tropical continental air in the Northern Hemisphere (Plate XVI) for only here are combined the conditions for its development, anticyclonic circulation and a warm continental area. In summer other land areas besides the hot deserts are warm enough to produce tropical continental air. At this time, consequently, the cool deserts and steppes, as, for example, in North America and Asia, also act as source regions for this air-mass type.

The climates of the coastal steppes and deserts (*Bn and Bn'*), which represent the westward extension of the *Bh* types, owe their peculiarities to the proximity of the ocean, and especially to the presence of a cool ocean current on the west coast of each continent, with the exception of Australia. This cool current reduces the temperature in the coastal region considerably below the normal for the continental regions in this area. The lower temperatures are particularly conspicuous during the summer, when the air at the ocean surface, in general, is cooler than the air over land. The temperature maps (Plates I and II) show this lower temperature during the summer months quite clearly, for example, along the west coasts of subtropical South America and South Africa during January and along the west coast of subtropical North Africa and North America during July. Since, during the summer months, the interior of the continents has a lower pressure than the oceans, the winds have a slight onshore component and the effect of the cold ocean currents on the temperature in the coastal zone is strengthened.

However, this effect does not reach far inland, and the temperatures increase rather rapidly with the distance from the coast. Thus, Windhoek in South Africa, 400 miles inland from Swakopmund (Table 34), has a yearly mean temperature which is 7°F warmer than that at Swakopmund even though it is more than 5000 ft higher. The most characteristic feature of the coastal deserts, however, is not their comparatively low temperature but their high humidity (*Bn'*), amounting in some regions even to frequent fogs (*Bn*). At Swakopmund, for example, there are on the average 150 days with fog per year. The relative humidity has here a monthly maximum of 84 per cent in February, a minimum of 70 per cent in June. At Windhoek, on the other hand, the maximum is 52 per cent during February, the minimum 24 per cent during September. The high humidities and fog frequencies are largely due to the cooling effect of the cold water, but there are a number of other factors involved that have been enumerated by Petterssen[1] in his discussion of the origin of the coastal fogs in California.

In some regions, especially on the coastal ranges of South America, the fog thickens into a drizzling mist during the morning hours. Nevertheless, the amount of precipitation is very small, values of around 1 in. per year being typical. Showers are very rare but may sometimes be heavy, as discussed on page 141. As in other desert and steppe regions the variability of the rainfall from year to year is very great, and the mean values of the annual precipitation have little meaning. At Iquique on the coast of northern Chile no rain fell during a period of 4 years. During the fifth year a single shower fell and gave 0.6 in. of precipitation. Thus the average annual rainfall for this period would be 0.12 in.

The cool and cold dry climates (*BSk*, *BWk*, *BSk'*, *BWk'*) are typical of the interior of the continents at higher latitudes. They owe their origin to their position in the interior of the continents, where any air arriving from the oceans has already become rather dry. Since the size of the continent must be large to ensure appreciable drying out of originally maritime air masses, *Bk* types of climates are found mainly in the heart of the large continental masses of the Northern Hemisphere. The only exception is the cold type *Bk'*, where the temperature even of the

[1] Petterssen, S., "Weather Analysis and Forecasting," pp. 130 ff., McGraw-Hill Book Company, Inc., New York, 1940.

warmest month is below 64°F. This type occurs only in Patagonia, and the conditions that produce it here are unique in so far as the Andes cause a rain shadow on the east side of South America that reaches to the Atlantic coast owing to the smallness of the continent at these latitudes. Moreover, the principal cyclone tracks lie farther to the south.

The presence of mountain ranges favors the aridity of the interior of the continents in other regions, also, and most of the actual desert regions in higher latitudes are surrounded by mountain barriers. Examples are the Gobi in Mongolia and the Great Basin in the United States. In winter a further factor contributing to the small precipitation is the anticyclonic circulation over the continents, which with its slowly subsiding motion of the air produces dry air masses. The cool dry climates with the exception of the cold type *Bk′* give way to *C* or *D* climates as the east coast is approached.

Owing to the continental position and the relatively high latitude at which these climatic types occur the annual range of the temperature is quite large. The diurnal temperature ranges are also large, as in the hot dry climates, and for the same reason. In many regions with these climates the winter temperature is well below freezing, with mean temperatures even below zero in some places, such as the Mongolian plateau. The summers are hot or at least warm, mean July temperatures of 80°F and above being not unusual.

Since the mean annual temperature according to Köppen's definition cannot be more than 64°F, the precipitation in the cool steppe climate cannot be more than 20 in. per year, or 25 in. if the precipitation maximum is in summer. In the cool deserts the annual precipitation can at the most be half as much. At higher latitudes precipitation occurs mainly in summer, the winter rains or snows being considerably smaller owing to the low temperature. Where cool dry climates reach into lower latitudes, the seasonal variation of the slight precipitation is the same as in the *Cs* climates, with stronger precipitation in winter owing to greater cyclone frequency. Since the cool steppe and desert types of climates extend well into the polar-front zone with its cyclone activity, the interdiurnal changes of the weather are greater than in the subtropical and tropical deserts. Even though the amount of precipitation is small snowstorms such

TABLE 34.—EXAMPLES OF *B* CLIMATES

Month	Jan.	Feb.	Mar.	Apr.	May	June	July	Aug.	Sept.	Oct.	Nov.	Dec.	Year	Range
Timbuktu, French Sudan (lat. 17°N; long. 3°W; el. 820 ft)														*BWhw*
Temp., °F	71.1	73.6	83.1	91.6	94.5	93.7	89.2	86.5	89.2	88.9	80.8	71.1	84.4	23.4
Precip., in.	0	0	0.1	0	0.3	0.9	3.5	2.8	1.1	0.4	0	0	9.0	
Yuma, Ariz. (lat. 33°N; long. 115°W; el. 141 ft)														*BWhs*
Temp., °F	54.0	58.5	64.0	69.3	75.9	84.6	90.5	90.1	83.5	73.0	61.9	54.7	68.4	36.5
Precip., in.	0.4	0.4	0.4	0.1	0.04	0.02	0.2	0.5	0.4	0.3	0.3	0.5	3.5	
Kayes, French Sudan (lat. 14°N; long. 12°W; el. 197 ft)														*BShw*
Temp., °F	77.2	80.8	88.7	94.1	96.4	90.5	83.7	81.7	82.2	84.5	83.1	77.2	84.9	19.2
Precip., in.	0	0	0	0	0.6	3.9	8.3	8.3	5.6	1.9	0.3	0.2	29.1	
Las Palmas, Canaries (lat. 28°N; long. 16°W; el. 30 ft)														*BShs*
Temp., °F	62.8	62.8	63.7	64.9	67.3	70.5	72.1	73.8	72.7	71.4	67.5	64.2	67.8	11.0
Precip., in.	1.4	0.9	1.0	0.7	0.3	0	0	0.1	0.2	1.1	2.8	2.6	11.1	
Fort Aleksandrovsk (Shevchenko), Russia (lat. 45°N; long. 50°E; el. 75 ft)														*BWkw*
Temp., °F	24.8	27.3	36.1	49.5	63.7	73.4	78.8	76.3	66.0	52.9	40.6	32.0	51.8	54.0
Precip., in.	0.3	0.2	0.3	0.2	0.3	0.7	1.0	1.3	1.3	0.6	0.4	0.2	6.8	
Havre, Mont. (lat. 49°N; long. 110°W; el. 2550 ft)														*BSkw*
Temp., °F	13.3	13.6	27.0	43.9	53.8	62.2	69.1	65.8	56.1	43.9	30.9	20.1	41.5	55.8
Precip., in.	0.7	0.6	0.5	1.0	2.0	2.9	1.8	1.2	1.3	0.7	0.6	0.6	13.9	
Santa Cruz, Argentina (lat. 50°S; long. 68°W; el. 39 ft)														*BWk'*
Temp., °F	58.6	57.6	54.7	47.7	40.8	35.2	35.2	38.3	43.5	48.7	52.9	56.3	47.5	23.4
Precip., in.	0.6	0.3	0.4	0.6	0.4	0.5	0.4	0.5	0.3	0.3	0.4	0.7	5.4	
Swakopmund, South-West Africa (lat. 23°S; long. 14°E; el. 20 ft)														*BWn*
Temp., °F	62.6	63.1	63.3	59.9	60.6	58.5	56.5	54.9	56.1	58.1	58.6	61.5	59.4	8.4
Precip., in.	0	0.1	0.2	0	0	0	0	0	0	0.1	0	0.2	0.6	

as the blizzards of the American prairies and the burans of Asiatic Russia are typical.

Owing to the large variations of the temperature between summer and winter the regions of *Bk* climates are in winter source regions of polar continental air, in summer source regions of tropical continental air (Plates XVI and XVII). Both air-mass types have a relatively low humidity content in common.

40. The Warm Temperate Rainy Climates (*C*). The *C* climates are distinguished from the dry types (*B*) by a sufficient amount of precipitation, from the tropical types by the existence of a cool season during which at least 1 month has a mean temperature below 64°F. On the other hand, no month has a mean temperature below 27°F since otherwise the climate would belong to the *D* types. Near the equatorial boundaries of the *C* types snow and frost are very rare; near the polar boundaries they are quite common. But the colder weather is often interrupted by a warmer spell so that a long period with a snow cover is not a regular feature.

In the tropical *A* climates the outstanding difference between the seasons is the amount of precipitation. In the *C*, *D*, and *E* climates the temperature variation is of equal or even greater importance. Another feature that becomes important as the climatic types of higher latitudes are considered is the variability of the weather. For in higher latitudes differences between air masses are much more pronounced than in the tropics, and hence frontal activity is much greater.

The climates of the *C* type are divided according to the seasonal variation of the precipitation and according to the temperature and the length of the summer season. Depending on whether the average amount of precipitation does not change much from season to season or whether there is a minimum in winter or summer, one distinguishes between the *Cf*, the *Cw*, and the *Cs* types. If the summer is hot, with a mean temperature of the warmest month of 72°F or more, the type is denoted by *Ca*. If the mean temperature of the warmest month remains below 72°F but if more than 4 months have temperatures higher than 50°F, the climate is of the *Cb* type. If the summer is short, with only 4 months or less with temperatures above 50°F, the type is *Cc*. The appropriate letter for the seasonal variation of precipitation is always inserted between the two letters indicating the main division and the summer temperature. Even though not all combinations of subdivisions according to precipitation and to summer temperature occur, there is a considerable variety of *C* climates.

The summer dry type (*Cs*) occurs on the subtropical margins of the middle latitudes. The coastal regions of the Mediterranean, the coast and part of the interior of California, central Chile, the southwestern tip of Africa, and parts of the southern and southwestern coast of Australia are under this climatic regime (see Frontispiece). The dry summers and wet winters of these regions are due to the seasonal shift of the polar-front zone and of the subtropical high-pressure belt. In summer these regions are little affected by frontal activity, being in the zone of the subtropical anticyclones. Therefore, the rainfall is light or absent. But in winter frontal passages are more frequent so that considerable precipitation occurs. Since the *Cs* type is found in regions whose latitude is sufficiently low so that they are dominated by the subtropical high-pressure belt in summer, the type *Csc* with a very short summer does not occur.

The type *Csa*, with hot summers, is characteristic for the Mediterranean coast, for the interior of southern California, and for a small area along the west coast of Australia. The other subdivision, *Csb*, occurs in more oceanic locations where the summer temperature is moderated by the sea.

The winter dry climates (*Cw*) as well as *Cf* are simply transitions from the corresponding *Aw* and *Af* climates of tropical regions to the snow-forest climates *Dw* and *Df*. Large parts of the regions with *Cw* climates are therefore counted amongst the tropical regions in view of the analogous annual variation of the rainfall. *Cw* climates are located in northern India, southern China, southern Brazil, Mexico, southern Queensland on the east coast of Australia, and on the plateaus of tropical and subtropical Africa. Many of these regions are considerably above sea level; and the lower winter temperatures, below 64°F, that distinguish these climates from the *A* types are in many instances an effect of the elevation rather than of the latitude. The otherwise tropical character of these high-level stations is also indicated by the small annual range of the temperature. In the tropical highlands, where the *Cw* type occurs, the dry winters and warm summers reflect the shift of the zone of the doldrums.

At higher latitudes this seasonal distribution of the rainfall is a result of the monsoonal variation of the wind, moist maritime air being more frequent in summer, dry continental air more frequent in winter.

As far as the temperature is concerned the *Cw* climates resemble the *Csa* or *Csb* types with hot or warm summers. Even stations that may be classified as *Cwb*, since the mean temperature of the warmest month remains below 72°F, are found in the tropics if their elevation is high enough.

The warm temperate rainy climates (*Cf*) *without dry season* cover a much larger part of the earth's surface than the *Cs* and *Cw* types. Even in the tropical highlands some stations show such a small monthly variation of the precipitation that they belong to this climatic group. Outside the tropics the *Cf* type is a typically oceanic climate covering almost 30 per cent of the whole ocean area (see Table 32). The *Cf* climates are found mainly in regions situated in the westerlies. Hence, the precipitation is principally caused by frontal lifting. In regions where the topography is favorable, orographic lifting is an important additional source of precipitation.

According to the temperature, three different types are distinguished. The type *Cfa* with hot summers is found in the eastern and central United States, approximately as far north as Boston, on the east coast of South America, especially in Uruguay, over the southern islands of Japan and farther south along the China coast, and on the east coast of Australia in New South Wales. The summers of the *Cfb* type of climate are cooler than those of the type *Cfa*, but still warm. This type is mainly found in locations under maritime influence, poleward from the regions with *Cfa* climates or, as on the west coast of North America, poleward of regions with *Cs* climates. This type represents the climate of the British Isles, large parts of France and Germany, the more northerly islands of Japan, the northern part of the west coast of North America, the southern part of the west coast of Australia, Tasmania, and New Zealand.

In continental locations the type *Cfa* or *Cfb* changes into *Dfa* or *Dfb* toward the pole, for the continental location with its absence of the moderating maritime influence on the winter temperatures brings about winter temperatures below 27°F. In regions under maritime influence, however, the type *Cfc* with cool and short summers—not more than 4 months with mean temperatures over 50°F—is found. Here the summers are short

TABLE 35.—EXAMPLES OF *C* CLIMATES

Month	Jan.	Feb.	Mar.	Apr.	May	June	July	Aug.	Sept.	Oct.	Nov.	Dec.	Year	Range
Quito, Ecuador (lat. 0°; long. 79°W; el. 9350 ft) *Cfbi*														
Temp., °F.	54.7	54.5	54.5	54.5	54.7	54.7	54.5	54.7	54.9	54.7	54.5	54.7	54.7	0.4
Precip., in. .	4.1	4.3	5.2	7.4	5.0	1.5	0.9	1.5	3.0	3.6	3.8	3.8	44.3	
LaPaz, Bolivia (lat. 16°S; long. 68°W; el. 12,000 ft) *Cwbi*														
Temp., °F.	50.4	50.4	50.0	48.4	47.7	44.8	43.5	46.2	49.3	50.0	51.8	51.4	48.7	8.3
Precip., in. .	3.8	4.9	2.5	1.5	0.5	0.1	0.1	1.0	0.8	1.2	1.5	4.2	22.1	
Athens, Greece (lat. 38°N; long. 24°E; el. 351 ft) *Csa*														
Temp., °F.	47.8	49.5	52.5	58.3	66.4	74.1	79.7	79.5	73.2	66.2	57.0	57.8	63.0	31.9
Precip., in. .	2.2	1.8	1.3	0.9	0.8	0.6	0.3	0.6	0.7	1.4	2.9	2.5	16.0	
Portland, Ore. (lat. 46°N; long. 123°W; el. 154 ft) *Csb*														
Temp., °F.	38.8	41.5	46.2	51.1	56.3	61.7	66.0	66.2	61.0	53.2	46.2	40.8	52.5	27.4
Precip., in. .	6.5	5.8	3.8	2.9	2.2	1.6	0.6	0.6	2.0	3.1	6.2	6.6	41.9	
Bandeirantes, Brazil (lat. 23°S; long. 44°W; el. 1748 ft) *Cwa*														
Temp., °F.	72.3	72.5	71.8	68.4	63.9	61.5	60.3	61.9	66.4	67.6	70.2	70.0	67.3	12.2
Precip., in. .	11.3	9.0	5.7	6.2	1.4	1.4	1.0	1.4	2.1	4.6	6.2	9.5	59.8	
Juliasdale, Southern Rhodesia (lat. 18°S; long. 33°E; el. 6070 ft) *Cwb*														
Temp., °F.	63.3	63.0	62.2	58.3	54.1	51.6	50.4	53.4	60.1	63.5	64.6	63.9	59.0	14.2
Precip., in. .	10.6	8.4	8.3	2.2	0.9	0.6	0.8	0.3	0.3	1.3	3.5	8.5	45.7	
Baltimore, Md. (lat. 39°N; long. 77°W; el. 125 ft) *Cfa*														
Temp., °F.	33.4	35.2	41.9	53.2	64.0	72.1	76.6	75.0	68.0	57.7	46.2	36.9	55.0	43.2
Precip., in. .	3.4	3.7	3.6	3.4	3.5	3.9	4.6	4.3	3.4	2.8	2.6	3.3	42.5	
Greenwich, England (lat. 51°N; long. 0°; el. 148 ft) *Cfb*														
Temp., °F.	38.7	39.7	42.3	47.3	53.4	59.2	62.8	62.1	57.6	50.0	44.1	40.5	49.8	24.1
Precip., in. .	1.7	1.7	1.7	1.5	1.7	2.0	2.2	2.2	1.8	2.5	2.3	2.2	23.5	
Trondheim, Norway (lat. 63°N; long. 10°E; el. 130 ft) *Cfc*														
Temp., °F.	27.3	27.9	30.6	38.3	45.9	53.6	57.2	55.4	48.9	40.5	33.1	27.8	40.5	29.9
Precip., in. .	3.4	2.9	2.2	1.8	1.5	1.7	2.2	3.0	3.3	3.4	3.1	2.6	31.1	

owing to the high latitude, but the winters are relatively mild owing to the tempering effect of the water. The west coast of Norway, the Faeroes and the Orkney Islands, the southern parts of Iceland, parts of the Alaskan coast, and the Aleutian Islands have a *Cfc* climate.

41. The Snow-forest Climates (*D*). These climates have a more severe winter than the *C* climates, with a more than occasional snow cover. On the other hand, they have a real summer, which is hot and long in more southerly regions with *D* climates, short and cool along the boundaries between the *D* climates and the polar *E* climates. The *D* types are separated from the *C* types by the isotherm 27°F of the coldest month and from the *E* climates by the isotherm 50°F of the warmest month. Thus, the annual range of temperature is at least 23°F, although the actual differences between the warmest and coldest months are much larger, at least twice as large or, in northern Siberia, even five times as large. With such great annual ranges of temperature the *D* climates are typical for regions in the interior of the large continental masses. Consequently, they are absent in the Southern Hemisphere. In the Northern Hemisphere they cover largely the continental areas from 40° to 50° latitude to the polar circle or even farther northward. Since westerly winds prevail in these latitudes, the west coast of North America and the western part of Europe have milder winters with *C* climates, which change gradually into the *D* type as the distance from the ocean increases. On the east coasts of the continents, on the other hand, the *D* climates extend to the coast. The variation of the temperature within the *D* climates is even greater than in the *C* types. For, in addition to the subdivisions *a*, *b*, *c* with hot, warm, and cool summers, respectively, there occurs in Siberia a fourth subdivision (*Dd*) with very cold winters.

Precipitation falls mainly during the summer, in the form of convective showers, which are frequently associated with weak cyclonic activity. The winter temperatures are so low that not much water vapor can exist in the atmosphere. Moreover, the presence over the interior of the continent in winter of a cold anticyclone with its subsiding air motion is not favorable to precipitation. In many regions with *D* types of climate, however, the difference between the precipitation during the wettest summer month and during the driest winter month is not sufficiently great to classify the climate as *Dw*. Only in eastern Asia is the amount of precipitation during the driest winter month less than one-tenth of that during the wettest summer month.

Therefore, only here is a winter dry region found that deserves the notation *Dw*; elsewhere, the notation *Df* is appropriate. The summer dryness of the *Cs* climates as a rule does not extend to these locations with cold winters. A *Ds* climate exists only in the interior of Oregon.

The climatic type Dw has the most typically continental characteristics in so far as the annual variation of the precipitation is of the monsoonal type, and the annual temperature range has high values. Owing to the monsoon winds from the ocean, the summer temperature is not nearly so much above as the winter temperature is below the average for the latitude.

The region of winter dryness in eastern Siberia does not extend to the coastal regions. Near the ocean the climate is of the *Df* type, as, for example, on Kamtchatka. In North America the type *Dw* does not occur, largely on account of the smaller size of the continent.

The Df type covers a much larger area than the *Dw* type. While the difference between *Df* and *Dw* is of meteorological interest, its effect on the vegetation is negligible, for even in the *Df* climates plant life has a rest period during the winter. Hot summers with a July temperature higher than 72°F are found in *Df* regions only in North America (*Dfa*), in *Dw* regions in northern China and Manchuria (*Dwa*).

In the *Dfb* climate the mean temperature even of the warmest month is below 72°F, but at least during 5 months it is greater than 50°F. This temperature is sufficient to permit the cultivation of many important crops and various kinds of fruits. The northern limit of the *Dfb* regions also represents, by and large, the northern limit of the fruit-growing districts. In North America the *Dfb* climate is found to the north of the region with the *Dfa* climate or, in the western part of the country, to the north of the cool dry climates (*Bk*). There are no *D* climates along the Pacific coast since the maritime influence prevents the winter temperature from becoming very cold. In Europe and Asia the *Dfb* climate occurs mainly in Poland, the Baltic States, and southwestern Russia. The region north of Vladivostok has the winter dry climate *Dwb*, except in a narrow coastal fringe, where the winter dryness is less pronounced.

North of the *Dfb* and *Dwb* climates the types with cool short summers (*Dfc* and *Dwc*) represent the transition to the polar climates. Again the winter dry type occurs only over eastern Asia. In the northernmost part of Asia the winter is so cold

TABLE 36.—EXAMPLES OF *D* CLIMATES

Month	Jan.	Feb.	Mar.	Apr.	May	June	July	Aug.	Sept.	Oct.	Nov.	Dec.	Year	Range
Des Moines, Iowa (lat. 42°N; long. 94°W; el. 860 ft)													*Dfa*	
Temp., deg F.......	19.9	23.4	35.8	50.0	61.3	70.5	75.4	73.0	65.5	53.1	38.3	26.1	49.3	55.5
Precip., in..........	1.1	1.2	1.7	3.0	4.5	4.8	3.4	3.4	3.7	2.5	1.5	1.2	32.0	
Portland, Maine (lat. 44°N; long. 70°W; el. 102 ft)													*Dfb*	
Temp., deg F.......	22.6	23.9	31.6	42.4	52.5	61.9	67.5	65.7	59.2	49.6	37.9	27.7	45.1	44.9
Precip., in..........	3.9	4.3	3.8	3.4	3.3	3.3	3.2	3.1	3.1	3.1	3.5	3.9	41.9	
Fairbanks, Alaska (lat. 65°N; long. 148°W; el. 492 ft)													*Dfc*	
Temp., deg F.......	−12.6	−0.4	10.8	28.6	47.1	58.5	60.3	55.2	43.2	26.4	2.7	−7.2	26.2	72.9
Precip., in..........	0.8	0.5	0.8	0.3	0.6	1.4	1.9	1.7	1.6	0.8	0.7	0.7	11.6	
Vilyuisk, Siberia (lat. 64°N; long. 122°E; el. 394 ft)													*Dfd*	
Temp., deg F.......	−38.9	−24.0	−4.7	17.6	39.4	58.5	65.3	57.4	41.2	16.9	−14.8	−35.1	14.9	104.2
Precip., in..........	0.4	0.3	0.3	0.3	0.9	1.2	1.3	1.6	1.1	0.7	0.5	0.4	9.0	
Peiping, China (lat. 40°N; long. 116°E; el. 131 ft)													*Dwa*	
Temp., deg F.......	23.2	27.7	38.8	55.4	67.6	75.6	77.5	76.1	66.7	54.5	38.3	27.0	52.3	54.3
Precip., in..........	0.1	0.2	0.2	0.6	1.4	3.0	9.4	6.3	2.6	0.6	0.3	0.1	24.9	
Minusinsk, Siberia (lat. 54°N; long. 92°E; el. 837 ft)													*Dwb*	
Temp., deg F.......	−3.5	−0.2	13.5	27.9	50.9	63.3	68.2	63.1	50.2	33.8	15.8	2.3	31.3	71.7
Precip., in..........	0.4	0.2	0.2	0.5	1.2	2.1	2.1	2.5	1.2	0.8	0.4	0.4	12.0	
Tunka, Siberia (lat. 52°N; long. 103°E; el. 2375 ft)													*Dwc*	
Temp., deg F.......	−14.4	7.4	13.3	33.8	46.8	58.5	62.6	58.5	46.0	30.0	6.8	−9.2	27.1	77.0
Precip., in..........	0.1	0.1	0.1	0.4	0.9	1.7	2.4	3.0	1.3	0.3	0.2	0.2	10.3	
Ust Maiskaya, Siberia (lat. 60°N; long. 134°E; el. 328 ft)													*Dwd*	
Temp., deg F.......	−43.8	−32.6	−8.3	17.1	42.1	57.9	65.7	57.9	44.1	18.9	−17.0	−35.3	13.8	109.5
Precip., in..........	0.1	0.1	0.2	0.3	0.4	0.8	1.2	1.4	0.8	0.3	0.2	0.2	6.0	

that the types *Dfd* and *Dwd* with mean temperatures below −36°F during the coldest month are introduced by Köppen.

In spite of the severe winters the summer temperature is still high enough to maintain coniferous forests, the taiga of Siberia. In the regions with *Dfd* climates and in the colder regions with *Dfc* climates the ground from a certain depth downward remains frozen throughout the year so that in summer the water at the surface cannot penetrate into the ground, giving rise to large swampy areas. At Yakutsk in Siberia, for example, the ground thaws only to the depth of 1 yd.

In winter the regions with *D* climates and to a lesser extent those with cooler *C* or *B* climates are the main sources of polar continental air. In summer only the colder regions with *D* climates (*Dc* and *Dd*) are sources of polar continental air, while tropical continental air is found in the remaining regions with *D* climates.

As in the regions with *C* climates the weather in the *D* climates is very variable owing to the frequent cyclone passages. During the winter these disturbances may give rise to severe snowstorms called "blizzards" in North America and "burans" in Asia. They occur also in the adjacent cool desert and steppe regions.

42. The Polar Climates (*E*). The boundary of the polar climates is determined by the 50°F isotherm of the warmest month. This line coincides sufficiently well with the limit of the distribution of trees poleward to warrant its choice as a climatic divide. Apart from the mountain regions the polar climates occupy the most northerly and southerly zones of the earth. The limits of the *E* climates are found over the northernmost parts of the Asiatic and North American continents, roughly at the Arctic Circle, but in some regions, especially over the ocean, they reach as much as 10° farther south. In the Southern Hemisphere the boundary between the polar and the tree climates lies between 45° and 55° south (see Frontispiece).

A special feature of the polar climates is the duration of the arctic night and day, which last half a year each at the pole and decrease in length toward the Arctic Circle. The long duration of the sunshine during the summer would make the incoming radiation at midsummer greatest at the pole if the atmosphere were absent (Fig. 2). However, owing to the obliqueness of the incident solar radiation and to the cloudiness in polar regions the maximum of the radiation reaching the ground at midsummer is at about 35° latitude (Table 3). A large part of the solar radiation arriving at the surface is furthermore reflected by the snow and ice, and a large percentage is expended in melting the snow and ice so that neither the temperature of the ground nor that of the adjacent air becomes really warm. But during the winter the temperature drops to such low values that the annual range of temperature is appreciable.

Because of the low temperatures the water-vapor content of the air is small, and precipitation is therefore slight by comparison with other regions. Over large portions of the polar land areas it is less than 10 in. per year, although in some locations, as, for example, along the Labrador coast, considerably larger amounts fall. Owing to the small evaporation and the slight amount of melting in the regions where the monthly mean temperature remains below freezing throughout the year (*EF* climate) even the sparse precipitation is sufficient to maintain large snow fields and glaciers.

Where the monthly mean temperature, at least during the warmest month, rises above freezing (but not over 50°F), a tundra vegetation can grow, and the subdivision *ET* (tundra climate) occurs. Those regions where the mean temperature even of the warmest month does not rise above freezing have a frost climate, *EF*. If either of these types is due not to high latitude but rather to the height of the land, this fact is indicated by addition of the letter *H*. This climatic type is found, for example, on the high plateaus of central Asia.

The tundra climates (*ET*) with their summer temperatures above freezing maintain a vegetation consisting largely of mosses, lichens, and grasses. In sheltered spots, flowering plants and, near the border of the tree climates, even some dwarf forms of trees are found. This scant vegetation is sufficient to support animal life, which in summer is quite plentiful.

In the Southern Hemisphere oceans cover the latitudes where the tundra climate occurs. Only on some of the most northerly fringes of the antarctic continent and over some small antarctic islands is the *ET* climate found on land. In the Northern Hemisphere, on the other hand, tundra climates extend along the north coasts of Canada, Alaska, Iceland, and Eurasia. The summer temperatures of various tundra regions do not differ greatly from each other, since the temperature of the warmest month must be below 50°F and above 32°F, according to the definition of the climatic type. Much greater variety is found in the winter temperatures.

The annual ranges of the temperature differ greatly according to the location. Where the maritime influence is strong, the temperature range is rather small. Cape Pembroke on the Falkland Islands in the South Atlantic, for example, has an

TABLE 37.—EXAMPLES OF *E* CLIMATES

Month	Jan.	Feb.	Mar.	Apr.	May	June	July	Aug.	Sept.	Oct.	Nov.	Dec.	Year	Range
Lake Harbour, Baffin Island (lat. 63°N; long. 70°W; el. 69 ft)													*ET*	
Temp., deg F.	−14.3	−12.8	−2.0	12.0	28.6	38.7	45.7	44.2	36.3	26.1	12.4	−3.8	17.6	60.0
Precip., in....	1.1	1.3	0.8	1.5	1.5	1.1	2.6	2.0	1.2	1.5	2.2	1.5	18.3	
Greenland, center of icecap.* (lat. 71°N; long. 41°W; el. 9940 ft)													*EF*	
Temp., deg F.	−56.9	−59.8	−38.7	−25.4	−7.1	−2.0	7.3	−0.8	−12.1	−31.5	−48.1	−42.7	−26.5	67.1
Precip., in....	No observations													
Mt. Washington, N. H. (lat. 44°N; long. 71°W; el. 6280 ft)													*ETH*	
Temp., deg F.	4.8	6.1	10.2	20.8	34.2	43.9	48.0	46.8	40.1	30.0	17.1	9.3	25.9	42.0
Precip., in....	4.2	4.5	5.8	5.9	6.0	9.2	10.1	8.4	8.5	7.3	6.7	5.2	82.1	
Cape Pembroke, Falkland Islands (lat. 52°S; long. 58°W; el. 69 ft)													*ET*	
Temp., deg F.	49.3	48.7	47.5	43.7	40.3	37.6	36.7	37.4	39.4	41.7	43.9	46.2	42.7	12.6
Precip., in....	2.8	2.6	2.2	2.4	2.5	2.1	2.0	2.0	1.1	1.6	2.2	2.8	26.3	

* One year, reduced.

annual range of only 13°F. At Sagastyr, on the arctic coast of Siberia, on the other hand, the difference between the mean temperature of the warmest and the coldest month is 77°F.

In the frost climates (*EF*) the mean temperature even of the warmest month remains below freezing. Data from these regions are scanty. One year's observations on the Greenland icecap (Table 37) have given an annual mean temperature of

$-26°F$, with a February temperature of $-60°F$ and a July temperature of 7°F.

On the antarctic plateau the temperatures may be even lower. The summer especially seems to be colder than over Greenland. Observations by the antarctic expeditions of Amundsen and Scott have shown temperatures of $-9°F$ during December and $-19°F$ during January. Precipitation is very scanty. It occurs, of course, almost exclusively as snow. Data on the amount of precipitation are very scarce, and the available figures are necessarily unreliable.

Polar climates owing to the elevation of the land (*EH*) may occur at any latitude provided that the altitude is high enough. Since the temperature changes much more quickly in the vertical than in the horizontal direction, transitions from one climatic type to another occur very rapidly with the ascent in mountain areas. In the tropics, such *E* climates show the characteristic small annual temperature range. Typical of mountain climates are the mountain and valley breezes, the foehn winds, and the increase of the precipitation with the altitude, although this increase takes place only up to a certain elevation (see page 72).

CHAPTER VIII

MICROCLIMATOLOGY

The data on which the discussions in the previous chapters were based have all been obtained by meteorological instruments with a "representative" exposure. That is, the instruments are placed sufficiently high, as a rule about 2 m above the earth's surface, so that they are beyond the zone where the local influences of the ground are felt most strongly. The meteorological and climatic conditions of the atmospheric layer below the 2-m level, however, are of great theoretical and practical interest. Their theoretical importance is due to the fact that it is the earth's surface which represents largely the direct source of heat and moisture for the atmosphere since the solar radiation passes to the ground without much absorption. Hence, a knowledge of the conditions in the lowest layers of the air through which these properties must be transported upward is of great value. Moreover, the friction at the ground affects the motion of the air decisively; consequently, the greatest changes of the wind velocity in the vertical direction are found in this layer.

The practical importance of the lowest atmospheric layer is obvious since most plants grow below the level at which the meteorological instruments are placed. Consequently, for agricultural purposes it is often not sufficient to have "representative" climatological data. Instead, it is necessary to know also how the particular soil properties and the topography of a given environment modify these values in the layer next to the surface.

The study of the typical climatic features of the lowest layer is referred to as "microclimatology." Besides the investigations of the surface layer, microclimatology deals with other local influences that modify the climate, as, for example, the climate of cities and forests. But these specialized aspects will not be discussed here.

Owing to the many local influences, it is impossible to give an exhaustive description of the numerous climatic peculiarities of the surface layer over the globe. Furthermore, microclimatological investigations have so far been undertaken only in temperate latitudes. But in typical cases one can study the deviations of these microclimates from the large-scale climates, or macroclimates. With the aid of such typical cases, conclusions can be drawn concerning the nature of the microclimate in other similar situations where only the ordinary climatic data are available. It is the purpose of the present chapter to discuss some of the more striking differences between the large-scale climate and the microclimate. For a more detailed study of the subject the reader is referred to special treatises[1] on microclimatology.

43. Distribution of the Meteorological Elements in the Microclimatic Region. *a. Temperature.* In the atmospheric layers next to the ground the turbulent mass exchange in the vertical direction is very small, for the proximity of a rigid boundary precludes large vertical motions. At somewhat higher levels the atmospheric turbulence tends to establish an adiabatic lapse rate; but next to the ground this regulating factor is absent, and the vertical temperature gradient is largely determined by the molecular conductivity of heat. The molecular conductivity of heat is about 10^5 times smaller than the conductivity due to turbulent mass exchange. When the earth's surface is heated by the sun in daytime, the temperature rise is transferred to the adjoining layers of air by molecular heat conduction only. Since the coefficient of molecular heat conduction is very small, the temperature rise of the air which is not in direct contact with the surface is much smaller than that of the ground. Hence, the temperature decreases rapidly through the first foot or so of elevation. With increasing height the turbulent mass exchange becomes more effective so that the vertical lapse rate becomes smaller.

An example of the rapid decrease of the temperature upward is shown in Fig. 26, after observations by Eaton, over a road with asphalt pavement. The observations were made at Riverside, Ill., on a clear day in August. During the early afternoon

[1] Geiger, R., "Das Klima der bodennahen Luftschicht," Friedrich Vieweg & Sohn, Brunswick, Germany, 1927; Mikroklima und Pflanzenklima in Köppen-Geiger, "Handbuch der Klimatologie," Vol. 1, D, Verlagsbuchhandlung Gebrüder Borntraeger, Berlin, 1930.

hours the temperature of the pavement is considerably higher than the air temperature. In Fig. 26 the temperature at 2 P.M. decreases 18°F through the first foot of elevation. This value is much higher than the autoconvective temperature gradient of 3.4°C/100 m at which the density does not change with altitude. Thus, in the case under consideration the density actually increases with the altitude. It may be mentioned in passing that it is this rapid decrease of the temperature with its resulting vertical increase of the density and hence of the optical index of

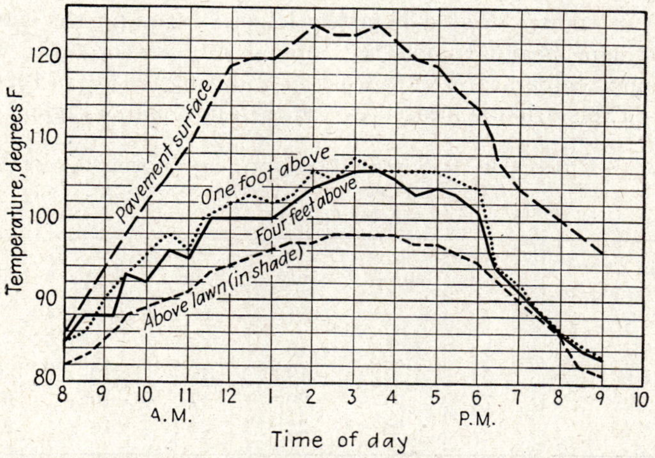

FIG. 26.—Diurnal temperature variation at different levels above asphalt pavement. (*After G. S. Eaton, Monthly Weather Rev.*, **47,** 801, 1919.)

the atmospheric refraction in the layer next to the ground which produces the mirages over paved roads, deserts, and similar locations with very high surface temperatures. Figure 26 represents an extreme case. The road surface is dry, while soil as a rule contains some moisture so that a part of the solar energy received at the ground is used, not to increase the surface temperature, but to evaporate some water. Nevertheless, the example is typical for the conditions that may prevail on clear summer days next to the ground.

Another factor on which the temperature in the lowest atmospheric layers depends is the coefficient of molecular heat conductivity of the soil. Where the heat conductivity is small, the surface of the ground is heated to a greater extent than where it is large. In the latter case, a considerable part of the heat energy received is transported to greater depths, preventing great rises of the surface temperature. The conditions are analogous to those which produce the difference between land and sea climate (see Sec. 7). Over the oceans it is the turbulent transport of heat downward into the water in addition to the evaporation from the water surface that prevents the temperature from rising as high as over land.

But even under less extreme conditions than those of Fig. 26 the great values of the maximum temperatures near the surface are very striking. Table 38, after Geiger,[1] shows for the station Schleissheim near Munich the annual number of days with maximum temperatures of at least 77°F during the years 1926 to 1929. The station is located in a garden in a marsh region.

TABLE 38.—NUMBER OF DAYS WITH MAXIMUM TEMPERATURES OF AT LEAST 77°F AT DIFFERENT ALTITUDES
(After Geiger)

Height above the ground, cm	May	June	July	Aug.	Sept.	Total
150	5	8	19	14	9	55
100	6	9	19	15	10	59
50	7	10	20	17	10	64
5	12	18	25	22	14	91

At night the lowest minimum temperatures occur at the ground, for the earth's surface loses its heat more rapidly by radiation than the air. As a result, pronounced inversions develop at night in the lowest layer, especially on clear, calm nights when the outgoing radiation is strong and when little or no heat transport from the upper layers by turbulent mass exchange takes place. As in the case of the daytime heating the intensity of the nocturnal cooling depends on the heat conductivity of the soil. If the heat conductivity is high, some of the heat loss at the surface will be replaced by the heat flow from greater depths so that the temperature does not sink so low as at the surface of the soil whose heat conductivity is smaller.

[1] *Ibid.*, p. D15.

Since the molecular conductivity of air is small, porous soil has a very low heat conductivity. Consequently, the nocturnal minimum temperatures are lower over ground prepared for sowing than over untilled soil. After a night during which hoarfrost has formed at the ground, such local differences in the minimum temperatures can often be observed directly in the distribution of the areas covered by frost and those free of frost.

Very bad conductors of heat are surfaces covered by a thick matting of dead grass, which forms an efficient insulator owing to the enclosed stagnant air. Over such a surface the daily maxima are very high and the minima very low, so that night frosts are considerably more frequent than over bare soil. A cover consisting of dead grass often forms in regions where a forest has been destroyed by some disaster. It is therefore important to begin with a proposed reforestation as quickly as possible before an insulating layer of dead grass has formed, since this affects the growth of young trees adversely.

Owing to the large nocturnal temperature fall next to the ground, the smaller the height above the earth's surface, the greater the number of days with minimum temperatures below freezing. Table 39, after Geiger,[1] shows the number of days with frost during the months of May to September, 1925, at some stations in southern Germany, at 5 cm and 150 cm above the ground. In all instances the frequency of frost increases toward the ground.

TABLE 39.—NUMBER OF FROSTS, MAY–SEPTEMBER, 1925
(After Geiger)

Station	Location	Number of frosts	
		At 150 cm	At 5 cm
Freienried	Orchard	2	6
Gräfrath	Forest clearing	4	10
Anzing	Forest clearing	12	23
Peiting	Marsh	5	22

b. Humidity. Like the temperature, the atmospheric moisture content shows rapid changes in the vertical direction through the lowest atmospheric layer. In order to understand the

[1] *Ibid.*, p. D14.

distribution of moisture in the lowest layers of the atmosphere, it is advisable to consider first the absolute humidity, *i.e.*, the density of the water vapor. This quantity permits one to recognize most directly the physical processes that regulate the moisture distribution. Since the source of atmospheric moisture is at the earth's surface, the maximum of the absolute humidity is almost always at the ground. Only when advection of moister air from a water surface or a marsh takes place or when dew forms at the ground will the greatest values of the absolute humidity be encountered at somewhat higher levels.

The daily variation of the absolute humidity next to the earth's surface indicates quite well the physical processes regulating the moisture content of the air at the ground. The absolute humidity shows a minimum in the early morning hours, when the temperature is low. With rising temperature the atmospheric moisture content increases rapidly until a maximum occurs sometime during the forenoon. Then the absolute humidity often decreases again, for two reasons. In the first place the top layer of the ground dries out since its water loss is only insufficiently replaced by transport of water from greater depths. On the other hand, the turbulent transport of water vapor from the surface upward increases toward noon. With the drying out of the uppermost layers of the ground the loss of water vapor to higher levels often can not be replaced by evaporation, so that the absolute humidity of the air next to the earth's surface reaches a minimum about the warmest time of the day. In very dry regions this afternoon minimum may even be stronger than the morning minimum. Toward the evening the absolute humidity rises again. At higher levels, for example, at 2 m above the ground, the diurnal variation of the absolute humidity is similar, but the absolute humidity is always smaller except occasionally when dew forms during the night.

As a rule, the relative humidity shows also a decrease with the altitude, even though its value depends not only on the atmospheric moisture content but also on the temperature. Differences of the relative humidity of 40 per cent between the air 5 cm above the ground and 200 cm above the ground are not uncommon. In general, the lapse rate of the relative humidity is greater at night than in daytime. At night the temperature of the air next to the ground is low, frequently lower than somewhat higher up. The existence of such a temperature inversion coupled with the decrease of the absolute humidity upward produces a rapid drop of the relative humidity in the vertical direction. In daytime, on the other hand, the temperature is very high next to the ground and decreases rather rapidly upward. Hence, the relative humidity next to the ground is not much larger than aloft, as during the night, for the high absolute humidity is counteracted by the high temperature next to the ground. During the noon hours, when the insolation is very strong, the surface layers as a rule become so warm that the relative humidity increases upward.

Actual measurements of the humidity distribution next to the ground are very scanty, but the few data available confirm the theoretical deductions of the preceding paragraphs.

c. Wind. Like temperature and humidity the wind changes rapidly in the surface layer. Next to the ground it is very small

TABLE 40.—PERCENTAGE FREQUENCY OF CALMS
(After Hellmann)

Height above ground, cm	Time of the day							
	0–3	3–6	6–9	9–12	12–15	15–18	18–21	21–24
5	41	36	25	6	5	16	46	43
25	31	26	10	1	0	7	33	34
50	22	18	4	1	0	3	22	25
100	18	13	2	1	0	2	16	19
200	13	12	5	0	0	1	12	15

owing to the effect of the surface friction. It increases with the altitude since the effect of the surface friction decreases with increasing distance from the ground. The vertical increase of the wind is greatest in the lowest atmospheric layer because the coefficient of turbulent mass exchange is here smallest.

The smaller wind speeds near the ground are shown in Table 40 which contains the percentage frequency of calms at various altitudes near the ground according to observations made by Hellmann[1] on 62 days. The frequency of hours with calms is always greatest next to the ground. During the late evening almost every second hour is calm. The number of calms

[1] HELLMANN, G., *Sitz.-Ber. preuss. Akad. Wiss. Berlin*, **22**, 404, 1919.

decreases upward, except during the time from 6 to 9 A.M., when the number of calms is greater at 2 m than at 1 m. This increase of the number of calms between 1 m and 2 m seems to be an irregularity that would disappear if a larger number of observations were used. Table 40 reflects also the diurnal variation of the wind velocity in the lowest layer next to the ground, with its maximum in daytime when the number of calms is small and its minimum at night when the number of calms is large.

44. Microclimate and Topography. In previous chapters the effect of mountains on the climate, especially on the temperature and on the precipitation, has been discussed. In a similar manner, as the large mountain ranges affect and modify the macroclimate, small elevations and differences in the slope of the ground create differences in the microclimate that are often surprisingly large.

The best known effect of topographical influences on the microclimate is the drainage of cold air downslope during the night and the resulting low temperatures in valleys and even in little dips in the ground. A characteristic example of such pockets of cold air has been published by Middleton and Millar.[1] These authors measured the temperature distribution in Toronto, Canada, on Feb. 22, 1936, along Yonge Street, one of the main thoroughfares, running in a general north-northwest direction from Lake Ontario to the northern limits of the city and beyond. The measurements were made by means of a resistance thermometer attached to an automobile at a height of about 27 in. above the ground. The results are reproduced in Fig. 27. The abscissa of this figure gives the distance in miles from the foot of Yonge Street at the lake front. The vertical profile of the terrain can be seen from the height scale on the right margin of the figure. The temperature run began 6 min past midnight at mile 8, reached mile 0 at 48 min past midnight, and returned to mile 8 at 18 min past 1 A.M. The results of the drive toward the lake and back are shown separately, the direction of each part being indicated by arrows on the temperature curves. The night was clear with a slight northerly breeze, which increased from 2 mph at the beginning to 4 mph at the end of the run. The

[1] MIDDLETON, W. E. K., and MILLAR, F. G., *J. Roy. Astron. Soc. Canada*, **30**, 265, 1936.

agreement of both curves is in general very satisfactory. Only the considerable drop in temperature at mile 5.8 that was observed during the run to the lake has vanished on the return trip. The warming of the air in this small valley is to be attributed to the increase in wind velocity during the period of observation. The slight increase of the air motion was sufficient to stir the air

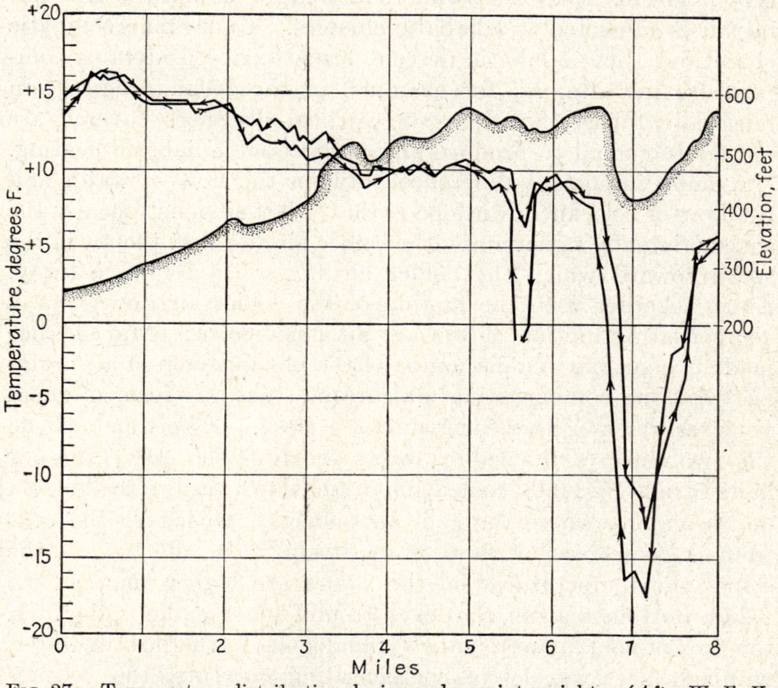

FIG. 27.—Temperature distribution during a clear winter night. (*After W. E. K. Middleton and G. F. Millar.*)

so that the temperature in the valley rose considerably. Even in the larger valley at 7 miles the increasing turbulence makes itself felt in a higher temperature toward the end of the observation period. It is also likely that the absence of large temperature drops in the smaller valleys closer to the lake, during the run southward, is already an effect of the increasing breeze.

The most noteworthy feature of Fig. 27 is the enormous temperature drop of 27°F in the valley at mile 7. Similar great temperature differences between valley bottoms and their crests are regularly occurring phenomena on clear, calm nights. During such nights the earth's surface cools rapidly by radiation, and the temperature of the air layer next to the ground drops considerably. The cold air along the slope flows downhill and collects at the valley bottom. The flow of the cold air is often compared to the flow of water downhill, although the analogy is by no means perfect. Owing to its compressibility the descending air is subjected to adiabatic changes. Under macroclimatic conditions such adiabatic descent leads to a considerable temperature increase, as, for example, in the foehn process. In microclimatology, however, the vertical distances covered are mostly too small to produce any appreciable adiabatic heating.

A more important difference between the flow of water and the flow of cold air downslope is that other air must occupy the space originally taken up by the colder air and that the air in the space toward which the colder air flows has to move away. Thus, together with the motion of the colder air downslope a compensating motion of warmer air must occur. Marvin[1] has made it plausible in a discussion of the phenomenon of air drainage that the compensating air currents take the form of small vortexes which bring warmer air above the valley bottom between the two sides of the valley toward each slope. Here this air in its turn is gradually cooled and drains downward to the bottom of the valley, where the cold air collects. Owing to the continuous compensating flow of relatively warm air toward the slopes the temperatures on the slopes are higher than at the valley bottom and on the level ground outside the valley. If the flow of cold air were entirely analogous to the flow of water, without the necessity of compensating currents, the warmer zones along the slopes could not develop.

The example in Fig. 27 representing the temperature distribution through Toronto in a north-southerly direction is a very striking case of cold-air drainage. It can easily be duplicated under similar topographical and meteorological conditions. But even very gentle slopes may produce analogous effects except that the temperature differences are not so spectacular. As an example, some observations by Geiger[2] may be quoted. He observed the minimum temperatures during the period of 1 year

[1] MARVIN, C. F., *Monthly Weather Rev.*, **42**, 583, 1914.

[2] GEIGER, R., in Köppen-Geiger, *op. cit.*, p. D30.

in an area of reforestation with a slope of only 1 in 200 near Munich in southern Germany. Where the slope changes into level ground, the flow of cold air was dammed up by older, fully grown trees. Table 41 gives the monthly minimum temperatures during May and June, when late frosts occur, and during October, when early frosts are to be expected. The observations were made at three different points along the slope, each at a height of 5 cm above the sloping ground. The first point of observation was at the bottom of the slope, the second 48 m from the bottom of the incline and 24 cm higher than the first point, while the third point was 80 m from the bottom of the incline and 40 cm higher than the first point. The mean monthly minimum temperatures are decidedly lower at the lower levels of the slope than higher up. The two last lines of the table show the number of nights during May and June when the temperature dropped

TABLE 41.—NOCTURNAL TEMPERATURE MINIMA, DEG F, AT 5 CM ABOVE THE GROUND ON A GENTLE SLOPE

	Relative height, cm		
	40	24	0
Mean minimum temperature during			
May, 1925	31.5	30.9	29.3
June, 1925	32.7	32.7	31.6
Oct., 1925	31.8	29.3	29.3
Number of nights with frost:			
May, 1925	17	20	22
June, 1925	12	12	15

below freezing. Here again the greater frost danger at lower levels is apparent. Although the actual differences in the minimum temperatures and in the number of nights with frost is not large, they are striking, especially if it is kept in mind that the slope which produces the cold-air drainage is so small that it is not perceptible without instruments.

The speed of the cold-air flow is small, probably rarely more than 1 m/sec. Owing to the stable stratification of the air the structure of the drainage current is laminar, not turbulent like the large-scale motions of the air.

At night only the inclination of the slope and its dimensions are important for the microclimate. In daytime, on the other hand, the orientation of the slope influences the temperature of the microclimatic space to a considerable extent. In order to discuss the differences in temperature due to different exposures of the slope it is simplest to consider a conical hill of circular symmetry. It will also be assumed that the inclination is not too steep so that the solar radiation received per unit area of the slope surface has a maximum at noon on that part of the slope which faces southward. When the slope is too steep, this will obviously no longer be true, especially in summer when the sun is high. The amount of solar radiation reaching the hill, on which the temperature largely depends, is symmetrical with respect to the meridian. Thus, a slope with an easterly exposure receives the same amount of solar radiation as a slope with a westerly exposure, at least when possible diurnal variations of cloudiness and of atmospheric turbidity are disregarded. Nevertheless, the temperature on the western slope will be higher than on the eastern slope. In the first place, during the afternoon, when the radiation on the western slope is particularly intense, the ground is already preheated by the air, whose temperature has increased since sunrise. The heating of an eastern slope, on the other hand, has to start from the lower temperatures of the early morning. These considerations apply no matter how small the inclination of the slope, even if it is so small that the slope with the eastern exposure receives direct solar radiation in the afternoon, also. For on the symmetrical hill that we are visualizing under these conditions the western slope receives the same amount of radiation in the forenoon as the eastern slope receives in the afternoon. Only if the inclination of the slope is different in different directions will the temperature be higher on the eastern than on the western slope. With irregularly shaped hills, of course, no general rule about the temperature distribution can be given.

Owing to the higher temperature of the western slopes compared with the eastern slopes on symmetrical hills the highest temperatures occur, not where the slope faces due south, but rather where it faces somewhat more toward the west. It must be added here that the rule concerning the higher temperature of westerly slopes, plausible as it appears, does not necessarily hold everywhere. For example, if there is a definite maximum of the cloudiness during the afternoon, as is often the case in continental climates during the summer (see Sec. 25), the direct solar radiation may be cut off to such an extent that the eastern slope has a higher temperature.

Another factor that favors higher maximum temperatures along the western slopes is evaporation. Especially in the early morning, but even later, during the forenoon, a considerable part of the incoming solar radiation is expended in evaporating water from the ground that is still moist from the preceding night. This heat energy is, of course, lost for the temperature increase. Later in the day the ground has already become drier so that the heat expended in evaporation is considerably less.

Besides the temperature, which has so far received most attention in microclimatic studies, the distributions of wind and precipitation show also characteristic deviations from conditions above the surface layer. On sufficiently high mountain ranges, *i.e.*, on those ranges which are high enough to produce macroclimatic effects, the windward side has always more precipitation than the lee side (see Sec. 19) because the air is forced to ascend on the windward side and the ascent causes adiabatic cooling, condensation, and precipitation. Under microclimatic conditions the elevations involved are not high enough to produce appreciable adiabatic cooling. The wind velocity is here the essential factor that determines the distribution of precipitation.

In discussing the wind distribution around a topographical obstacle a conical hill of circular symmetry will again be considered. The streamlines around such a hill are shown in a schematic fashion in Fig. 28. Observations made by Geiger[1] on an isolated small hill of conical shape bear out this schematic picture of the streamlines.

The streamlines are crowded together on the windward slope, especially along its upper portions and, as the upper sketch of Fig. 28 shows, on the flanks of the hill. A greater density of the streamlines implies a higher wind velocity since the condition of continuity must be satisfied. Hence the highest speeds occur on the windward side near the top of the hill. But since the air in part detours around the hill instead of flowing over it, two maxima will be found on the windward side near the hilltop between the central streamlines and the flanks of the hill. On the lee side the wind velocity decreases downward as shown in

[1] Geiger, R., in Köppen-Geiger, *op. cit.*, p. D27.

Fig. 28. The streamlines are here farther away from the hill, and a zone of vortexes is developed.

The distribution of precipitation corresponds to this distribution of the wind velocity. Where the wind velocity is high, little precipitation falls. Thus, the precipitation is higher on the lee side than on the windward side, contrary to what is the case under macroclimatic conditions.

That the amount of precipitation is greater behind the obstacle than in front of it can often be observed directly. For example, the accumulation of snow in the wind shadow of a rock, a tree,

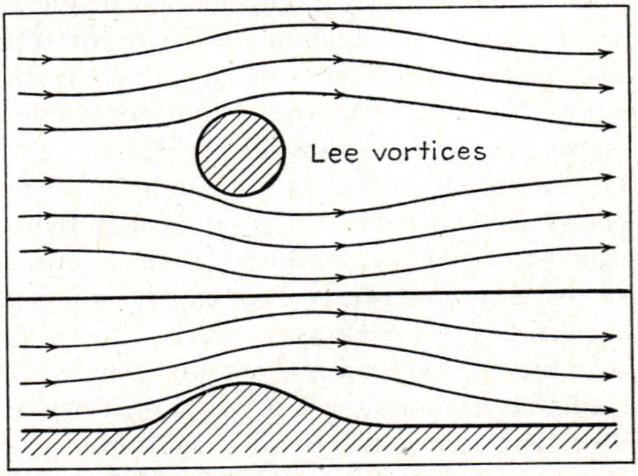

FIG. 28.—Wind distribution around a conical hill. (*After Geiger.*)

or a building is much larger than on the windward side. The action of snow fences is based on a similar principle. The wind when blowing through the interstices is retarded and small vortexes are set up behind the fence so that most of the snow carried by the air falls to the ground on the lee side behind the fence.

45. Microclimate and Vegetation. In the study of macroclimates references have already occasionally been made to the effects of vegetation on the climate. These effects become particularly conspicuous when the microclimate is considered because the plants grow in the microclimatic zone next to the earth's surface and hence their climatic influences are here greatest.

The existence of a vegetation cover makes the physical properties and the geometrical shape of the surface different from those of bare soil. That the geometric shape of the ground when covered by vegetation is different from that of bare soil is obvious. As far as the difference in physical properties is concerned, a vegetation cover absorbs larger amounts of the incoming short-wave solar radiation than bare soil. On the other hand, the absorptive and the emissive power in the long-wave region of the spectrum, where terrestrial emission to space takes place, is smaller for vegetal surfaces than for soil. Hence the radiative balance of a vegetation cover is more favorable for an equable climate than that of bare soil. Another favorable factor is that a vegetation cover, unlike bare soil, presents, not a single plane surface for absorption and emission but a great number of smaller surfaces in the form of its branches and leaves. Consequently, the higher plant surfaces afford at least partial protection to the lower ones against too large temperature extremes. Another factor that frequently acts favorably upon the heat balance of plants is the formation of dew, which may reduce the outgoing radiation as much as 20 to 30 per cent. However, evaporation is unfavorable since it is greater over plants than over bare ground. The presence of a plant cover increases the evaporating surface considerably. But on the whole the heat loss by evaporation is not sufficient to nullify the beneficial effects of a plant cover.

The existence of a vegetation cover affects the microclimate also in that it reduces the wind velocity within the vertical space in which the plants are growing. As a consequence, the turbulent mass exchange is also considerably smaller than at the same altitude over unplanted ground. Table 42 summarizes some measurements[1] made on two different days between and above heather vegetation in the north German plain. On both days the increase of the wind velocity with the altitude was very pronounced.

If the vegetation cover is higher, as in wooded country, the zone of small wind velocities extends higher, of course. Table 43 shows, after observations by Geiger,[2] the wind distribution in a stand of fully grown pines, 50 to 60 ft high. The records extend over 188 hr. The great reduction of the wind velocity

[1] Stocker, O., quoted from R. Geiger, "Das Klima der bodennahen Luftschicht," p. 135, Friedrich Vieweg & Sohn, Brunswick, Germany, 1927.
[2] *Ibid.*, p. 164.

between the trees is immediately apparent. A closer inspection shows that the wind velocity decreases rapidly downward to the tops of the tree crowns, a phenomenon similar to the rapid

TABLE 42.—WIND VELOCITY AND VEGETATION COVER

Vegetation cover	Elevation, cm	Wind velocity, m/sec
Jan. 11, 1921		
Between the heather	10	1.0
Between the tops of the heather	30	1.4
	50	3.7
Above the heather	180	9.3
Oct. 12, 1920		
Between the heather	2	<0.01
Between the tops of the heather	40	1.7
Above the heather	180	5.1

decrease of the wind velocity observed near the earth's surface over bare ground. Within the vertical space taken up by the trees the wind velocity is remarkably uniform. Only next to

TABLE 43.—WIND DISTRIBUTION IN A STAND OF PINES
(After Geiger)

Ht. of the anemometer, ft	Position among the trees	Average wind velocity, mph
55.3	Above the tree crowns	3.6
44.9	At the top of the crowns	2.0
34.6	Within the tree crowns	1.5
24.3	Upper part of the space among the trunks	1.5
13.9	Within the space among the trunks	1.5
3.6	Near the ground	1.3

the ground itself does a further decrease of the wind velocity take place.

Owing to the different conditions created by the vegetation in the microclimatic zone the temperature and humidity distribution differ from that at the same altitude over bare soil. Figure 29 may serve as an example of the modifications produced. The upper part of this figure represents the distribution of the noon temperatures in a grainfield near Munich during different times of the vegetation period, after Geiger. The height of the grain during the different stages of its growth is indicated by the vertically shaded area. The temperature distribution shows in all cases the effect of the absorption by the plant cover. Throughout the vertical space taken up by the plants the temperature changes rapidly owing to the greatly

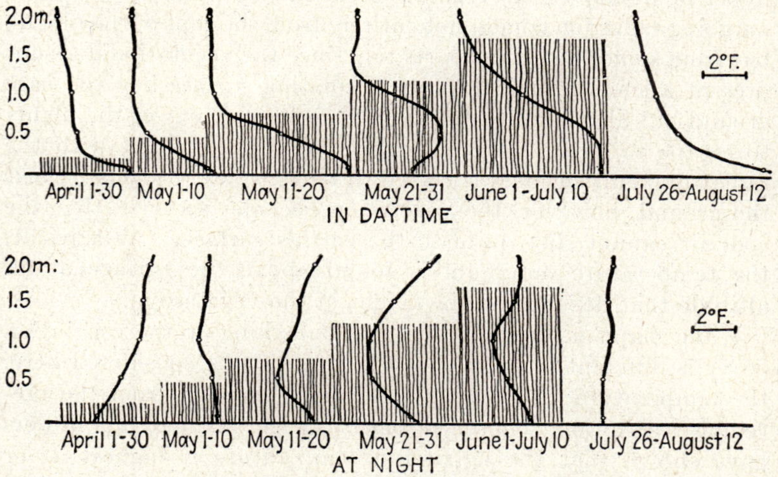

FIG. 29.—Temperature distribution in a grain field at noon and at night. (*After Geiger, R., Deutsches Meteorol. Jahrbuch, Bayern,* 1925, *Appendix B,* p. 5.)

reduced turbulent mass exchange, while above the plant surface the much slower normal decrease of the temperature with elevation begins. As the vegetation cover becomes higher and denser, less and less solar radiation penetrates to the ground, most of it being absorbed on the vegetation surface. Consequently, the temperature maximum moves upward from the ground to some level intermediate between the ground and the upper surface of the plant cover. These conditions are shown particularly clearly during the time from May 21 to 31. The grain was harvested on July 26, and hence the curve showing the mean temperature distribution during the period from July 26 to Aug. 12 represents conditions over bare soil. This temperature distribution is very similar to that during April, when the plants are still short and their lateral spread is small.

The temperature distributions at night are represented in the lower part of Fig. 29, where the minimum temperatures are plotted. The conditions are now opposite in most respects to what they are during the time of the daily maximum. With low vegetation cover the lowest temperatures are found near the ground, as over bare soil. As the plants become higher and grow denser, the minimum moves upward. However, for two reasons, the lowest temperatures are found somewhat below the upper boundary of the vegetation cover. In the first place, the outgoing radiation comes, not entirely from the tops of the plants, but from somewhat lower parts too, since the vegetation does not present a smooth, uninterrupted radiating surface like the bare ground but also permits radiation from lower parts of the plants to escape to space. Moreover, as the air in contact with the radiating plant surfaces is cooled, it sinks to lower levels. Near the ground, however, the vegetation becomes so thick that the cool air cannot sink quite to the earth's surface. As a result, the temperature minimum is found above the surface at an altitude that depends on the height of the vegetation.

If the shape of the individual plants making up the vegetation cover is different from that of the grain stalks considered here, the temperature distribution will vary somewhat from the preceding description. Observations by Geiger in a patch of flowers have shown that the maximum temperature is highest at or much nearer to the tops of the flowers than in the case of a grainfield. The leaves of the flowers form a much more continuous upper surface than the grain stalks, and hence the absorption is to a much larger extent emitted from the top of the vegetation cover than from a vertical space extending downward to a considerable depth.

The modification of the vertical temperature distribution produced by a forest differs from that caused by smaller plants because the crowns of the trees form here an almost continuous surface, which takes over to a large extent the function of the ground. Table 44 shows, after Obolensky,[1] the temperature distribution on clear days at 13:00 hr in an oak forest of 12-year-old trees whose height was about 1.5 m. Since the tree crowns

[1] Obolensky, V. N., quoted from R. Geiger, in Köppen-Geiger, *op. cit.*

form an almost continuous surface, the highest air temperatures during the daytime are found at the treetops, while the temperature decreases from here to the ground, which is in the shade. Upward from the treetops the temperature decreases, too. Similar to the conditions at the ground when vegetation is absent (see Sec. 43*a*) the vertical temperature lapse rate above the treetops is large owing to the small turbulent mass exchange next to a boundary surface.

At night the air in the forest is as a rule warmer than in the surrounding open country, for the ground is sufficiently shielded

TABLE 44.—VERTICAL TEMPERATURE DISTRIBUTION, DEG F, IN AN OAK FOREST (HEIGHT 1.5 M) ON CLEAR SUMMER DAYS AT 13:00 HR

Observation point	July	Aug.	Sept.
Ground	66.6	64.6	59.0
In the crown	68.2	64.9	61.0
On the treetops, 1.5 m	71.8	69.8	65.7
20 cm above the treetops, 1.7 m	70.9	68.7	64.2
At 2 m above the ground	69.4	67.6	62.6

by the crowns of the trees. On the other hand, in daytime the air in the forest is cooler owing to the shading effect of the trees. Occasionally, the temperature difference between the forest and its environment may even produce a feeble circulation analogous to the land and sea breezes, with the forest assuming the role of the water owing to its more moderate diurnal temperature variations.

Apart from the shade, the higher evaporation is another factor that moderates the daytime temperatures in a forest. Evaporation over any vegetation cover is higher than over bare ground. In the first place, plants are through their vegetation process always adding water vapor to the air; bare soil can become completely dry, and in this case evaporation ceases. Furthermore, as already pointed out, ground covered with vegetation presents a larger evaporation surface than bare soil. Owing to the continuous evaporation from plant surfaces the humidity between the plants is generally quite high.

PART B

Regional Climatology

CHAPTER IX

NORTH AND CENTRAL AMERICA

Extending from about 10° north to the interior of the Arctic Circle, this vast land mass is characterized by a wide variation in climate. However, as a result of the increase in width of the continent toward the north, the cold-climate types predominate, with a large area of tundra climate (*ET*) over the northern portion of Canada. Conversely, because of the tapering of the continent toward the south, the region of desert climate (*BW*) is confined to a comparatively small area. There is, however, a large expanse of steppe climate (*BS*).

The extensive land mass of Canada is covered with snow during the winter months and becomes a source region for cold continental air masses. The southeastward displacement of these air masses produces cold winters over the eastern United States. On the other hand, the narrowness of the continent in low latitudes is favorable for the production of warm humid maritime air masses. As a result of the northward movement of the moist air, precipitation is relatively abundant throughout the year over the southeastern portion of the continent. Consequently, the eastern United States experiences a moderate continental forest type of climate. Except for the mild Pacific coast climate, the remainder of the United States has an arid climate; this comprises the high western plateau.

The regions of tropical rain-forest climate (*Af* and *Am*) are confined solely to the narrow southern section of Central America. Farther north, the tropical savanna type of climate (*Aw*) predominates and extends along the Mexican coast. The plateau and northwestern section of Mexico are regions of *B* and *C* climatic types.

46. Physical Features. The most prominent characteristic of the topography is the extensive western highland, which extends from Alaska to Nicaragua. This region is essentially a vast plateau upon which is superimposed a series of mountain ranges, which can be subdivided into three main mountain chains, as follows:

1. The eastern range, or Rocky Mountains, where the main ridge is above 8000 ft. The broadest and most complex part of this range is located in the United States.

2. The western chain, which comprises the Alaskan ranges, the complex British Columbia ranges, the Cascade and Sierra Nevada ranges in the United States, and the Sierra Madre chain in Mexico.

3. The coastal ranges whose elevations rarely exceed 5000 ft, for example, the coastal regions of California, Oregon, British Columbia, and Alaska. The western chain is separated from the Rocky Mountains by a number of high plateaus and valleys. As shown by a topographical map, the entire highland system is widest about 40° north.

Smaller mountain ranges are observed in the eastern portion of the continent from the Labrador plateau to the Appalachian mountains. This mountain system is generally below 5000 ft and is considerably less extensive than the western highland region.

Between the two mountain systems is located the Great Central Plain, which extends from the arctic to the Gulf of Mexico. South of 30° north, with the tapering of the continent, the flat areas are confined mainly to the east and west coasts.

North America has numerous indentations. The most important are the Gulfs of Mexico and Alaska and Hudson Bay. Besides these water bodies the inland lakes of the Great Lakes system are also significant physical features. While Hudson Bay is ice-covered for a few months, the large lakes are rarely completely covered with ice.

The influence of the topography upon the wide-scale distribution of climate may be summarized briefly as follows:

1. The moderately warm and humid Pacific air masses are prevented, by the western highland chain, from invading the central and eastern portion of the continent without being subject to considerable modification. Especially north of 40° north,

this circulation results in heavy orographical precipitation along the west coast and lower precipitation amounts east of the mountains.

2. The Rocky Mountains have a smaller influence than the western chain upon the west-east distribution of precipitation. On the other hand, the Rocky Mountain system generally prevents any westward displacement of cold polar continental air, with the result that the winter surface temperatures are much lower near the east coast than near the west coast of the continent.

3. Because of the low elevation, the mountains near the east coast are not a very effective climatic barrier. However, during certain weather regimes, they have an important influence upon the weather in the adjacent regions.

4. The climate varies gradually from north to south. This is a result of the absence of a west-east mountain chain.

5. The inland water bodies have a controlling influence upon the climate. The vast ice-covered Hudson Bay region becomes a source region for deep polar air masses during winter. The large lakes of the Great Lakes network considerably modify any air masses that traverse this region. For example, during the winter cold air is heated rapidly by the underlying water surface; conversely, warm air is cooled during the summer as it crosses the lakes.

6. The Gulf of Mexico is an important source region of moisture for air masses that invade the continent from the south. The presence of these moist air masses accounts for the moderate precipitation of eastern North America.

Besides these large-scale climatic effects many minor topographical factors will be discussed in the succeeding paragraphs.

47. Mean Distribution of Pressure and Wind. The essential features of the sea-level circulation can be determined from the mean sea-level pressure maps for January and July. As a result of the rugged terrain in the western portion of the continent, the sea-level pressure gradients do not give a very accurate indication of the wind circulation. Here local orographical influences become quite important. On the other hand, near the coasts and over the Great Plains the mean wind circulation is indicated by Plates IV and V.

a. Winter. In summarizing the main aspects of the January pressure distribution, the following features are worthy of note:

1. Over the continent there are two main high-pressure regions, one in northwest Canada and a second between 30° north and 45° north in the United States. The predominance of high pressure over the land is consistent with the continental cooling during winter. However, the North American continental anticyclones are poorly developed as compared with the winter Asiatic anticyclone.

2. The western extension of the Icelandic low-pressure area results in a cyclonic circulation over northeastern America. It is important to note that, along the east coast, offshore winds prevail from the Arctic Circle to 35° north. Furthermore, as indicated by the pressure gradient, the mean wind speeds are highest over eastern Canada.

3. Partly because of the presence of the high coastal mountains the circulation of the Aleutian low-pressure system is not prolonged far inland. Even near the coast the sheltered inland region is characterized by low temperatures. Because of the fact that inland pressures are reduced to sea level, they appear quite high as compared with those over the immediate ocean; this sea-level correction, therefore, results in steep fictitious pressure gradients as shown in northern British Columbia and southern Alaska. North of 45° north the prevailing winds are mainly southerly. Because of the many fiords and the "banking" effect produced by the coastal mountains the mean surface winds do not conform with the general mean isobaric pattern. Southeast winds are frequently observed along the coast of British Columbia [see Prince Rupert (28), Table 45].

4. From Mexico southward there is a gradual decrease in pressure with a resulting high frequency of east and northeast trade winds along the east coast. Local influences cause variable surface winds inland and along the west coast. At Mazatlán (24, Table 45), for example, the persistence of northwest winds demonstrates the predominance of local influences over the general trade-wind circulation.

The daily variations of the pressure systems are frequently as important as the mean circulation. The moving cyclones and anticyclones produce considerable variations in the wind velocity. To illustrate the persistency of wind directions in the various regions, Table 45 has been constructed after excluding all calms. Wherever possible, stations with reasonably good exposures have been chosen, but it must be remembered that local factors may cause the surface wind to deviate appreciably from the gradient wind.

Because of the large number of migratory pressure centers, the region between 30° north and 50° north is characterized by

Table 45.—Percentage Frequency of Surface-wind Direction

Station	Lat., °N	Long., °W	January								July							
			N	NE	E	SE	S	SW	W	NW	N	NE	E	SE	S	SW	W	NW
East Coast																		
1. Colón	9	80	*61*	28	1	1	0	1	2	6	*22*	14	4	12	9	9	19	11
2. Progreso	21	90	11	31	*37*	11	5	0	1	4	7	32	*41*	17	2	0	0	1
3. Tampico	22	98	*22*	19	13	12	6	7	11	10	5	18	*25*	*25*	13	6	3	5
4. Key West	25	82	18	*33*	20	13	4	3	2	7	2	6	*43*	33	8	5	1	2
5. Savannah	32	81	10	17	9	6	12	11	*18*	17	5	9	7	11	21	*26*	18	3
6. Washington	39	77	12	11	5	6	19	7	10	*30*	15	6	7	8	18	*21*	8	17
7. Boston	42	71	12	4	4	4	6	18	*27*	25	4	7	14	4	8	*30*	22	11
8. Yarmouth	44	66	13	10	10	9	5	13	9	*31*	4	4	3	9	18	*35*	15	12
9. St. John's	48	53	7	7	2	9	6	24	13	*32*	7	10	8	10	14	*35*	10	6
10. Hebron	58	63	12	1	0	0	1	14	*68*	4	*32*	13	14	3	6	6	23	3
11. Lake Harbour	63	70	*54*	12	5	8	10	1	3	7	15	5	6	2	*41*	11	18	2
Inland																		
12. San José	10	84	4	*50*	42	4	0	0	0	0	4	34	*47*	9	1	0	1	4
13. Chimax	15	90	23	*48*	15	2	3	3	2	4	14	*60*	15	4	1	1	1	4
14. Memphis	35	90	12	12	8	11	15	15	9	*18*	10	10	8	10	16	*22*	9	15
15. Dallas	33	97	7	8	8	15	*23*	5	11	*23*	4	8	13	22	*41*	4	4	4
16. St. Louis	39	90	9	9	5	13	18	10	16	*20*	11	12	7	10	*24*	18	9	9
17. Chicago	42	88	7	7	4	8	14	18	*22*	20	7	*24*	9	12	10	20	12	6
18. Fort Hope	52	88	2	8	*40*	3	5	9	28	5	8	4	24	8	4	12	*31*	9
19. Winnipeg	50	97	10	3	3	18	12	11	15	*28*	13	9	5	18	14	11	11	*19*
20. Hay River	61	115	13	5	13	3	17	11	*29*	9	11	15	*29*	4	12	6	18	5
21. Chesterfield	64	92	7	4	4	3	3	2	5	*72*	12	10	4	9	10	6	10	*39*
22. McPherson	67	135	15	2	6	*43*	11	5	5	13	*29*	5	5	8	11	17	11	14
West Coast																		
23. Salina Cruz	16	95	*57*	30	1	2	4	5	0	1	*56*	31	1	1	3	4	1	3
24. Mazatlán	23	106	13	6	10	4	2	4	17	*44*	5	9	14	6	4	14	*35*	13
25. San Diego	33	117	16	21	8	6	5	8	13	*23*	6	1	0	1	5	28	29	*30*
26. San Francisco	38	122	*24*	7	5	17	7	9	11	20	0	1	0	1	5	*57*	36	0
27. Portland	46	123	8	8	10	21	*23*	9	3	18	20	2	2	8	7	6	4	*51*
28. Prince Rupert	54	130	9	7	28	*35*	10	3	5	3	2	2	6	16	16	22	*32*	4
29. Sitka	57	135	8	21	*41*	9	4	4	9	4	2	2	15	4	6	15	*36*	20

large variations in the wind direction. In tropical latitudes the trade-wind circulation produces steady east or northeast winds as demonstrated by San José (12). North of 50° north the deep cyclones are usually located over the neighboring oceanic regions. Therefore, as a result of the absence of frequent migratory lows, the land stations show relatively minor variations in the surface-wind direction, for example, Hebron (10), Lake Harbour (11), McPherson (22), and Sitka (29).

b. Summer. Plate V illustrates the mean sea-level pressure distribution during July. Because of the pronounced thermal heating, low-pressure areas are observed over the continent instead of the winter thermal high-pressure areas. Furthermore, the Aleutian Low is almost entirely absent while the Icelandic Low is very weak. The Atlantic subtropical anticyclone extends over the land in summer causing a predominance of southwest winds at stations like Savannah (5), Boston (7), and Yarmouth (8). The corresponding high-pressure cell in the Pacific is very pronounced and produces steady northwesterly winds along the west coast of the United States. A remnant of the winter polar anticyclone is located over extreme northern Canada. South of the Tropic of Cancer the summer pressure distribution differs only slightly from the winter distribution. As illustrated by Progreso (2) and San José (12) the winds are persistently from an easterly direction.

Over the continent as a whole, middle and late summer is the calmest period of the year. With the diminished pressure gradients, the surface winds are affected by local factors, and consequently the data in Table 45 are not entirely representative of the mean air motion during the summer season. Consistent with the northward progress of the mean position of the polar front, the region of maximum occurrence of migratory cyclones and anticyclones is located nearer Canada during July. In spite of the possible nonrepresentativeness of the wind observations, it is evident from Table 45 that these moving pressure systems cause greater wind variations at most stations north of 50° north in summer than in winter. On the other hand, between 30° north and 45° north there are, in general, less frequent wind variations in summer than during the cold season.

c. Upper Air. The January upper-air circulation at 4 km is given in Fig. 12. The principal features of the mean upper-air motion may be summarized as follows:

1. Cyclonic motion predominates north of 30° north, with a closed low-pressure center over northeastern Canada. The axis of the Atlantic cyclone is displaced westward with elevation, consistent with the location of the coldest air over the Hudson Bay region.

2. The subtropical anticyclone is displaced toward the equator with elevation. This displacement arises from the southward increase in temperature.

3. As a result of these pressure systems, the upper winds north of 30° north are from a general westerly direction. The mean wind speeds increase with elevation. In low latitudes the winds are easterly, and the speeds decrease with elevation; finally, they become westerly.

The mean July upper-air circulation differs from the January pattern in the following respects:

1. The wind speeds in middle and high latitudes are considerably weaker in summer than in winter.

2. Because of the continental heating the low-pressure center is not present over northeastern Canada while the southwestern surface cyclone is replaced by an upper-level anticyclone.

3. The subtropical belt of high pressure is located farther north in summer than in winter. Moreover, because of the strong heating north of the equator, the axis of the anticyclone remains, even at high altitudes, north of 20° north. Consequently, between 0° and 20° north, east winds prevail to high elevations. Furthermore, unlike the conditions in winter, the wind speeds increase with elevation.

48. Ocean Currents. Plate VI shows the main ocean currents in the vicinity of North and Central America. The large-scale currents conform to the mean circulation of the atmosphere. Thus, there are the northward-flowing Gulf Stream system along the east coast and the southward-moving California Current along the west coast. The motion of all currents is restricted by the shape of the coast line, with the result that the water circulation in restricted areas does not conform to the mean air motion, for example, the currents in the Gulf of Mexico, Baffin Bay, and Bering Sea regions.

The ocean currents have an important influence upon the distribution of temperature since they modify the air masses with which they come in contact. The February and August sea-surface isotherms are given in Plates VII and VIII. The outstanding features can be summarized as follows:

1. In all seasons, from 20° north to 35° north, the warm Florida Current produces warmer water along the east coast than is found along the west coast, where the cold California Current pre-

dominates. The maximum temperature difference occurs in summer when, between 30° north and 37° north, the Atlantic coastal water is about 12°C warmer than the Pacific coastal water.

2. The pronounced summer temperature minimum north of San Francisco arises from the prevailing northerly winds, which produce an upwelling of cold water. Because of the steepening of the pressure gradient together with the northward movement of the Pacific anticyclone, the summer winds are usually fresh.

3. The West Greenland Current reverses its direction in Davis Strait and joins the southward-moving Labrador Current. Near the coast the cold water moves as far south as Cape Hatteras. In contrast, because of an eastward extension of the North Pacific Drift (Alaska Current), relatively warm water is found along the northwest coast of the continent. Consequently, north of 45° north the sea-surface isotherms are higher along the west coast than along the east coast.

4. In tropical latitudes the temperature difference between the east and west coasts is always very slight. The isotherms demonstrate the marked homogeneity of the temperature distribution.

5. The decided temperature minimum near New Orleans in February is produced by the outflow of cold water from the Mississippi River.

The effect of the ocean currents upon the climate of North America will be evident later when the mean air temperature, precipitation, cloudiness, and fog distributions are discussed in detail.

49. Air Masses. The general source regions of the air masses, which affect the weather and climate of the continent, are illustrated in Plates XVI and XVII. The seasonal properties of the various air masses will now be discussed briefly.

a. Polar Continental Air (*cP*). During the winter portion of the year the continent is covered with snow north of 45° north. Cold, dry *cP* air masses are produced when air is allowed to stagnate over this region during the formation of a cold anticyclone. As a result of the radiational cooling from below and the upper-level subsidence, *cPWs* air is always very stable in its source region. Clear skies are quite common except in those regions where excessive radiation produces ice-crystal fogs.

Warmer maritime air masses are prevented from entering the source region by the shielding effect of the western mountain chain. However, at upper levels, warmer air sometimes invades the western section. Consequently, in the west, real *cP* air is frequently confined to the lowest 2 or 3 km. The deepest and coldest air masses are generally located over central and eastern Canada, where polar maritime invasions are rare. Furthermore, the formation of deep air masses is aided by the low elevation of the surface in the vicinity of Hudson Bay.

When *cP* air moves away from its source region, the air-mass properties are modified by the absorption of heat and moisture from the underlying surface. The original stability of the air is affected not only by the turbulent transfer of heat and moisture along the vertical but also by the presence or absence of marked subsidence. For example, pronounced instability results when the air flows southward over the Great Lakes in the rear of a deep cyclone. This flow generally produces stratocumulus clouds and rain showers or snow flurries on the lee side of the lakes. If the air moves southward over a snow-covered surface, under anticyclonic control, the original stability is only slightly affected.

In the summertime, because of continental heating in middle latitudes, the source region of *cP* air masses is restricted to the far north. During this season the properties vary considerably. However, the humidity and temperature are always relatively low. The pronounced wintertime stability is not present during summer. These air masses are usually shallow and are rapidly modified as they move southward over the continent, finally becoming convectively unstable.

b. Tropical Maritime Air (*mT*). Source regions of tropical maritime air masses are located over the ocean where the subtropical anticyclones dominate the atmospheric circulation. Because of the uniformity of the underlying surface, *mT* air is decidedly homogeneous. It is everywhere warm and moist in the lower layers. On the other hand, the upper-air properties vary considerably from the western to the eastern end of the subtropical anticyclone. In the central and eastern portions of the anticyclone, for example, off the Pacific coast, subsidence produces a stable *mTs* air mass, while in the western portion of the high-pressure cell, for example, off the southeastern United

States, the absence of pronounced subsidence favors a less stable type of air mass.

During the winter, when *mT* air invades the eastern section of the United States, it is cooled from below. This cooling increases the stability in the lowest layer of the atmosphere. On the other hand, the subsidence inversion is usually weakened, especially when the air moves northward under cyclonic control. In this case, convergence often results in widespread cloudiness and precipitation. During the summer months, the continental heating steepens the lapse rate so that, in general, the air mass is definitely unstable. In the absence of pronounced subsidence inversions, widespread convective cloudiness and instability showers are quite common during the summertime.

Because the circulation around the subtropical anticyclone discourages *mT* invasions over the western portion of the United States, this air mass is present only in the warm sector of cyclones. The air is never so warm as Atlantic *mT* air because of the cooler ocean surface. Furthermore, as a result of turbulent mixing, it usually reaches California as a convectively indifferent air mass. With the intensified summer anticyclonic circulation, *mT* air rarely invades the western section of the continent.

South of the Tropic of Cancer, tropical maritime air is very warm and moist. The stability of the air mass depends upon the circulation. If the air is under decided anticyclonic control, subsidence increases the stability (*mTs*), while if the air converges toward a low-pressure trough it becomes unstable (*mTu*). In the latter case the release of the ever-present convective instability results in convective phenomena, for example, line squalls and thunderstorms. This *mTu* air mass mainly affects the climate of Central America. Especially during late summer, when the equatorial low-pressure belt is farthest north, *mTu* air produces heavy convective showers north of Panama.

c. Polar Maritime Air (*mP*). Over the oceans, north of 35° north, cold arctic and continental air masses as well as warm tropical air masses are modified by the underlying sea surface into polar maritime air. Polar maritime air is moist and cool although, in winter, very warm in the lower layers as compared with *cP* air. The stability of the air depends upon its trajectory over the ocean. If it is of continental origin and has had a relatively short cyclonic path over the ocean, it arrives along the northwest coast as a relatively unstable air mass (*mPKu*) with low temperatures aloft. On the other hand, if it has moved from warmer regions, it arrives as a stable air mass (*mPW*). Modified *cP* air that has had a long trajectory over the ocean usually shows an indifferent state of convective stability.

Polar maritime invasions are very frequent along the west coast during winter because of the dominating influence of the Aleutian cyclone. The air is modified by forced ascent over the mountains so that it reaches the Rocky Mountains plateau region and the eastern portion of the continent with considerably lower specific humidities. Over the Great Plains, modified *mP* air is warmer than *cP* air. Frequently, because of the adiabatic heating, relatively high temperatures are observed on the lee side of the Rocky Mountains (foehn effect). When *mP* air is allowed to stagnate over the snow-covered mountains, radiational cooling tends to produce an air mass whose characteristics resemble those of a *cP* air mass.

Pacific polar maritime invasions of the continent are restricted to high latitudes during the summertime since the dominant circulation south of 55° north is approximately parallel to the coast line.

The polar maritime air that invades the east coast from the Atlantic Ocean has generally had a very short trajectory over the ocean. The modifications of the original *cP* air have therefore been confined to the lowest kilometer. During the winter season the air is relatively warm and moist at the surface, while at upper levels the characteristics are similar to those of *cP* air over the eastern section of the continent. North of 40° north, invasions of Atlantic polar maritime air frequently take place during the summer months. The cold coastal waters produce a cooling from below; this results in widespread coastal fog.

d. Tropical Continental Air (*cT*). As a result of the tapering of the continent toward the south, the source region of tropical continental air is confined to the southwest section of the United States and northern Mexico. The formation of this air mass is restricted principally to the summer months. It is a very warm, dry, and convectively unstable air mass. Convective clouds and precipitation arise when there is an invasion of moisture or, more frequently, when the air moves northward over a moister underlying surface. However, within the source region, the low moisture content prevents appreciable amounts of precipitation.

50. Fronts, Cyclones, and Anticyclones. *a. Frontal Zones.* The principal frontal zones for the North American area are illustrated in Plates XVI and XVII. The Atlantic polar front, during the winter season, continually oscillates between the Canadian border in the north and the West Indies in the south. The frontogenetical wind field is the field of deformation between the Atlantic subtropical anticyclone and the cold continental anticyclone, whereas the source of the temperature contrast is the difference in temperature between the cold land and the warm ocean. The most intense fronts are observed when the line of frontogenesis coincides with the eastern coast line, *i.e.*, when the polar front is quasi-stationary in the region of maximum temperature gradient.

Over the Pacific Ocean there are two frontal zones that have an important indirect influence upon the weather of the continent. In the wintertime, on daily weather maps, two high-pressure cells are frequently observed in the Pacific, with the result that two main polar fronts may be distinguished. The western front is generally the more intense because of the marked air-mass contrast between the cold continental Siberian air and the warm Pacific air. The eastern Pacific polar front is located in the field of deformation between the two subtropical anticyclones. As in the case of the Atlantic polar front, these frontal regions continually oscillate in accord with the displacements of the large anticyclones.

A third important frontal region is located along the northwest coast. The frontogenetical wind field is the field of deformation that is present in the extensive mean-pressure trough that extends from the Aleutian low-pressure region. The temperature contrast across this front arises from the difference in temperature between the cold continental air and the warmer Pacific polar maritime air.

Because of the weaker temperature gradients and the less favorable wind fields, fronts are considerably weaker during the summer months. Moreover, the mean frontal regions are located farther north than in the winter. The Arctic front separates the warm Pacific air from the colder arctic air. Near the Canadian border, in the east, a second main frontal region separates the warm, humid, modified *mT* air from the cooler *cP* air.

During summer and winter the intertropical front is located in the equatorial low-pressure belt that moves northward with the approach of summer. Here the temperature gradient is very weak and hence unfavorable for frontogenesis. However,

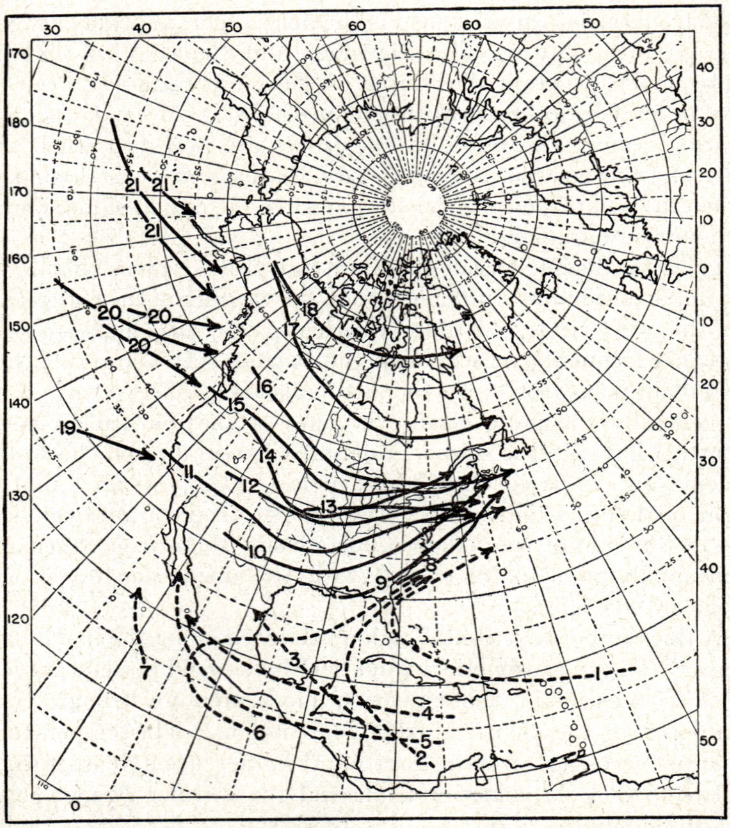

FIG. 30.—Cyclone tracks of North and Central America. Solid lines are extratropical cyclones; broken lines are tropical cyclones.

the converging trade-wind systems are decidedly frontogenetical. In view of the slight temperature contrast and of the low Coriolis force, the intertropical front is quite different from those of middle and high latitudes. Essentially it is a zone of convergence that, because of the moist, unstable air masses, is accompanied by widespread convective phenomena.

b. Extratropical Cyclones and Anticyclones. The migratory pressure systems which are connected with the displacements of the frontal zones, in middle and high latitudes, will now be discussed in detail. The mean cyclone tracks in Fig. 30 are based upon the following references: type 1 from Tannehill,[1] types 2 to 6 from Page,[2] and types 8 to 16 from Bowie and Weightman.[3] Types 8 to 16 represent the average January tracks. Instead of depicting the mean July tracks for the United States the seasonal variations of types 8 to 16 will be discussed in the following paragraphs.

In connection with all mean cyclone or anticyclone tracks it is important to realize that they give only a very general picture of the movement of pressure systems. The movement of an individual cyclone depends upon a great number of factors, for example, the movement of adjacent pressure systems, deepening, filling, and the upper-air conditions. Consequently, many cyclones follow paths entirely different from the mean path. However, the average tracks serve to indicate the more general displacements of pressure centers.

1. Cyclone Frequency over the United States. The seasonal variation of the cyclone tracks across the United States is summarized in Table 46. The values are based upon 100 cyclones a year of types 8 to 16. Besides the variations illustrated by Table 46 it is important to note the seasonal oscillation of the mean tracks. In general, the region of maximum cyclonic activity is displaced northward with the approach of summer on account of the similar displacement of the mean frontal zones. Hence the mean paths are farther north in summer than in winter. For example, while the mean January paths of types 11 to 13 pass south of Chicago, the mean July paths cross northern Lake Michigan. A similar longitudinal displacement is observed in types 8, 9, 10, 15, and 16. Finally, it is important to note that types 8 and 9 are rare during the summer season.

2. Winter Cyclones and Anticyclones North of Mexico. Over the Pacific area three general cyclone tracks may be distinguished:

[1] Tannehill, I. R., "Hurricanes," Princeton University Press, Princeton, N. J., 1942.

[2] Page, J. L., *Monthly Weather Rev.*, suppl. No. 33, Washington, 1930.

[3] Bowie, E. H., and Weightman, R. H., *Monthly Weather Rev.*, suppl. No. 1, Washington, 1914.

i. One group of cyclones (type 19) is observed to develop in comparatively low latitudes and move toward the coast of southern California. The warm sector of the cyclones comprises *mT* air, but frequently these systems are occluded by the time they reach the coast. If they arrive as open warm-sector lows, the occlusion process is accelerated by the mountains along the west coast. After passing inland they may either fill or continue to move across the southwestern states (type 11) and perhaps finally redevelop east of the plateau. This group of cyclones produces heavy frontal and orographical rain over California.

TABLE 46.—CYCLONE FREQUENCY OVER THE UNITED STATES
(Mean velocities of centers, in mph, included in parentheses)
(After Bowie and Weightman)

Type	Dec.–Feb.	Mar.–May	June–Aug.	Sept.–Nov.	Year
8. South Atlantic	0.7 (28)	1.0 (21)	0.5 (21)	1.2 (24)	3.4 (23)
9. East Gulf	0.8 (30)	0.8 (22)	0.3 (17)	0.9 (24)	2.8 (24)
10. Texas	4.5 (31)	2.4 (26)	0.7 (20)	1.7 (21)	9.3 (27)
11. South Pacific	2.2 (29)	2.6 (23)	0.8 (22)	1.4 (22)	7.0 (25)
12. Colorado	3.2 (29)	3.7 (23)	2.5 (22)	2.8 (25)	12.2 (25)
13. Central	1.3 (29)	2.0 (25)	2.4 (21)	1.6 (23)	7.3 (23)
14. Northern Rocky	0.7 (32)	1.2 (24)	1.7 (21)	1.2 (23)	4.8 (23)
15. North Pacific	6.0 (30)	4.0 (23)	1.6 (19)	3.8 (24)	15.4 (26)
16. Alberta	9.9 (30)	8.0 (24)	9.9 (21)	10.0 (26)	37.8 (25)
All types	29.3 (30)	25.7 (24)	20.4 (21)	24.6 (25)	100.0 (25)

ii. From the eastern Pacific polar front, cyclones develop and move toward the northwest coast of the continent (type 20). In general, they arrive as deep occluded pressure systems, which subsequently fill on moving inland. Again, however, a new deepening sometimes occurs on the eastern side of the Great Divide, giving rise to a new cyclone of type 14, 15, or 16. Failing this development, the Pacific occluded front may frequently cross the mountains and be displaced eastward as an upper front. Because of the frontal activity and orographical lifting, type 20 cyclones produce heavy precipitation along the Pacific coast.

iii. The cyclones (type 21) that develop on the western Pacific polar front most frequently move toward the mean low-pressure region near the Aleutian Islands, arriving as deep occluded systems. Subsequently, a weak cyclone may move across northern Canada (type 17 or 18). Since the deep Aleutian cyclones produce southerly winds along the northwest coast, steady orographical precipitation is common.

Four main cyclone paths may be recognized over the continent, *viz.*, one across northern Canada; a second across the northern United States, Great Lakes, and St. Lawrence valley region; a third from the Colorado area across the southern Lakes territory; and finally a track across the southern states and then parallel to the Atlantic coast. The variations from these four main cyclone paths are illustrated in Fig. 30.

i. The cyclones that form over the southern United States (types 8 to 11) develop either upon the southern extremity of the Atlantic polar front (types 8 and 9) or as a rejuvenation of old cyclones that have moved inland from the southern Pacific (types 10 and 11). From California to western Texas, type 11 cyclones produce relatively light precipitation. When the cyclones are east of 100° west, *mT* air is generally in juxtaposition to modified *mP* or *cP* air. As a result of the large temperature contrast and the high moisture content of the warm mass, there is an abundant supply of energy for the subsequent development of a deep cyclone. The cyclones that develop over Texas and the southern states region usually deepen only slightly while gradually moving eastward. Later, they may continue to move off the southern Atlantic states without being subject to appreciable deepening until they curve northeastward over the Atlantic Ocean. More frequently, these cyclones curve northward about 90° west and commence to deepen rapidly. Subsequently, the occluding cyclones follow a northeasterly path. On occasion new cyclones develop near the Virginia capes region, very often as secondary developments of the primary cyclones when these have followed a more westerly track than usual. Here the temperature contrast is very pronounced because of the presence of cold *cP* air over the land to the north, together with very warm, moist, and unstable *mT* air to the southeast. Cyclones that develop in this region usually deepen and occlude very rapidly while moving northeastward. All cyclones that originally form or deepen over the southeastern states are associated with widespread cloudiness and moderate to heavy precipitation as a result of frontal lifting and convergence within the *mT* air mass, which comprises the warm sector of these lows.

ii. The region around Colorado, east of the Rockies, is a favored area for the development of the cyclones (types 12 and 13). In most cases, they are preceded by the eastward movement of a frontal trough from the Pacific coast. Cyclones that develop in this region sometimes move southward and later follow a path like the Texas cyclones. Others move in a general easterly direction, gradually curving toward the St. Lawrence valley. Those lows which develop in the Montana region (type 14) usually move southward before recurving northeastward. Because of the frequent absence of moist, unstable *mT* air, cyclones of this group (types 12 to 14) do not deepen to the same extent as those which form near the Gulf of Mexico or the Atlantic coast. Furthermore, precipitation amounts are less than in the case of types 8 to 10.

iii. The Alberta cyclones (type 16) develop on the southern extremity of the Pacific Arctic front and follow a general southeasterly path across the northern portion of the United States, finally recurving eastward in the vicinity of the Great Lakes. Type 15 cyclones follow a similar path. These lows (types 15 and 16) generally move quite rapidly across the eastern portion of the country. They do not usually develop into deep cyclones because of the absence of pronounced air-mass contrasts and because of the low specific humidities. Likewise, the accompanying precipitation amounts are only light to moderate. It is apparent from Table 46 that this group is of very frequent occurrence.

iv. Type 17 and 18 cyclones move across central and northern Canada. Since the cold temperatures of this region favor anticyclogenesis, this group is less frequent than types 15 or 16. Furthermore, the low humidities of the continental air masses account for the light precipitation that accompanies each cyclone.

The anticyclone movements are closely connected with the displacement and deepening of cyclones. For example, a cold anticyclone is displaced southward over the continent in the rear of a deep cyclone moving northeastward. The extent of the southward displacement of a *cP* anticyclone depends on the latitude of formation of the deep cyclone and on the location of the warm anticyclones. When deep lows develop over the southeast, a cold anticyclone frequently moves toward the Gulf of Mexico. On the other hand, the anticyclones in the rear of the Alberta type cyclones do not generally move very far south. Because of the mountain barrier to the west, the paths of cold *cP* highs are usually directed toward the southeast. Nevertheless, cold *cP* air occasionally invades the west coast; these invasions are usually the result of the northeast winds to the rear of a deep cyclone located over the southern portion of the high western plateau.

Another family of anticyclones is observed to move inland from the Pacific or develop over the high plateau about 40° north. These anticyclones comprise modified *mP* air, and with their eastward progress the winter temperatures over the Great Plains are higher than during the periods of *cP* air-mass invasions. As would be expected, the eastward velocities of the anticyclones conform closely to the eastward velocities of cyclones in the corresponding regions.

3. Mexico and Central America (Winter). During the winter season the most important secondary circulations, south of the United States, are (*a*) the migratory cyclones over northwestern Mexico and (*b*) the occasional strong north winds that mainly affect the east coasts of Mexico and Central America.

With the southward displacement of the westerlies during the cold months, extratropical cyclones are observed in relatively low latitudes. Figure 30 shows a cyclone track (type 11) near the Mexico–United States border. These lows, especially when they are displaced farther south than usual, produce weather similar to that observed in the vicinity of cyclones over the southwestern United States.

As a cold anticyclone is displaced southward across the central United States, the cold continental air may cross the western portion of the Gulf of Mexico and finally arrive along the east coast of Mexico as a cold, humid air current. Such winds are called *northers* and represent an extreme southerly trajectory of *cP* air. Naturally, because of the warm Gulf water, the *cP* air is rapidly modified as it moves southward, so that the norther reaches Yucatán, Tabasco, or Vera Cruz as a cool, unstable air mass. Heavy rains and temperature falls of almost 10°F occur along the Gulf coast. As a result of the orographical lifting and downslope adiabatic heating the air reaches the Pacific coast as a warm, dry air current. On occasion, the norther extends its influence through Central America to Panama, but in these cases it produces only small temperature changes since the long trajectory has produced an air temperature comparable to the sea-surface temperature.

4. Summer Cyclones and Anticyclones North of Mexico. Consistent with the northward displacement of the main frontal zones in summer, extratropical cyclones follow a more northerly path than during the winter season. Furthermore, summer is a period of weak temperature gradients and light winds, with the result that cyclones are generally slow-moving and poorly developed. Table 46 summarizes the essential characteristics of these summer cyclones.

Only a few cyclones develop over Texas, the Gulf states, and the South Atlantic states. They move slowly, mainly northeastward, and rarely develop into deep cyclones as in winter. A more favored cyclone path is found across Colorado (types 11 to 13) and the Great Lakes area. However the majority of the summer cyclones develop over the northwestern United States or western Canada (type 16). The low-pressure systems that form near the Canadian border follow a more northerly path across the central portion of the continent than during the wintertime. Cyclones are observed to develop relatively frequently on the arctic front in the vicinity of Alaska (types 17 and 18). These lows follow a general southeasterly path before recurving eastward over the northeast section of the continent.

While the winter cyclones are frequently associated with widespread frontal precipitation, the summer cyclones are not accompanied by large areas of rainfall. In general, summer rainfall results from local convective showers. However, the weak fronts that are connected with summer cyclones often cause the release of the convective instability, with the result that showers and thunderstorms are frequently observed in the vicinity of the summer low-pressure systems.

As in the winter season, the anticyclonic movements are closely associated with the cyclonic movements. The high-pressure area over the southeastern United States is quasi-permanent, with the result that this region is only occasionally invaded by well-developed migratory anticyclones. The permanency of this subtropical anticyclone is consistent with the rare occurrence of deep cyclones. Likewise, the Pacific anticyclone is relatively stationary. On the other hand, high-pressure systems migrate eastward from the northwest portion of the continent, as in the winter season. These anticyclones follow the relatively frequent cyclones that traverse the northern section of the country (type 16). Consistent with the weakly developed cyclones, the migratory summer anticyclones are also poorly developed; in general, the summer circulation is weak as compared with the more intense circulation around winter high-pressure areas.

Figure 30 shows that cyclones rarely traverse the Hudson Bay region. Here comparatively cold anticyclones are commonly observed. These high-pressure areas are displaced southward in the rear of lows that traverse the eastern region of southern Canada. The east winds south of the high-pressure center frequently persist along the North Atlantic coast, sometimes as far south as 40° north. On these occasions, a well-marked front separates the cold air to the north from the warm tropical maritime air south of the front. These cool eastern anticyclones may be displaced southward in the rear of comparatively deep lows off the Atlantic coast. The cool air mass is rapidly modified into a warm, humid air mass after it is displaced south of 40° north.

c. Tropical Cyclones. South of the United States border the secondary circulations are markedly different from those of middle and high latitudes. In general, this region lies between the subtropical belt of high pressure and the band of equatorial low pressure. In connection with the prevailing easterly winds, migratory low-pressure areas move in a general east to west direction. The violent cyclones that affect the climate of Mexico and Central America are tropical hurricanes; the most frequent tracks of these tropical storms are given in Fig. 30. It is apparent that two main classes may be distinguished, *viz.*, hurricanes which develop over the warm waters of the Caribbean Sea and the neighboring Atlantic Ocean and those which develop or rejuvenate over the eastern Pacific near the coast of Central America. The subsequent path of any tropical cyclone depends on the prevailing surface and upper-air circulation as well as on the location and movement of adjacent pressure systems. During the earlier part of the hurricane season (June–August) the West Indies storms frequently move on land or recurve farther west than during the latter half of the season (September-November). This difference is due mainly to the subtropical anticyclone, which is well developed during midsummer and prevents the recurvature of a storm until it reaches a region of southerly winds. With the approach of autumn the Atlantic High is less permanent, and a hurricane is more likely to recurve northward into a trough of low pressure over the western Atlantic. Schiaffino[1] has subdivided all tropical storms that affect Mexico into the following five main classes. It should be remembered that these tracks represent only average conditions and that, in some cases, because of the neighboring pressure systems hurricanes have very erratic trajectories.

1. The Antillean hurricanes recurve in the eastern portion of the Gulf of Mexico (type 2). They are most frequent during the months of August, September, and October. Besides affecting the weather of the Yucatán peninsula they produce marked weather changes over the southern United States. During August, hurricanes recurve farther north than during October.

2. Another commonly observed hurricane path (type 3) extends across the Caribbean Sea, the Yucatán peninsula, and then inland along the northeast coast of Mexico or the coast of Texas.

3. Sometimes the storms, after crossing the Yucatán peninsula, follow a trajectory across central Mexico (type 4) and reappear over the Pacific, where they move in a general northwesterly direction.

4. Similar to the above category, another cyclone track may be detected across Central America, then running northwest parallel to the Pacific coast, and finally penetrating the Gulf of California (type 5). Schiaffino includes in this group those hurricanes which develop south of the Gulf of Tehuantepec.

5. The final group of cyclones (type 6) form over the southeast Pacific and move on to the western coast of Mexico. If these storms are of sufficient intensity, they may possibly cross the land area and finally move northeastward across the Gulf of Mexico.

Besides the storms mentioned in the foregoing the cyclones of type 7 are worthy of note. Apparently these tropical cyclones develop south of the Revillagigedo Islands.

In general, August, September, and October are the months of the most frequent tropical cyclones. The most important

[1] Schiaffino, P. V., "Atlas climatológico de la República mexicana," Servicio meteorológico mexicano.

weather characteristics are the very strong wind velocities and the torrential rainfall that accompany the hurricane. Because of the increased friction and the absence of an immediate moisture source, tropical cyclones generally fill when they move inland. However, they may continue to move inland as shallow low-pressure areas and finally cross the country. On reaching the coast again, the shallow cyclone may deepen and redevelop into a tropical hurricane. This sequence of filling and deepening applies to those tracks in Fig. 30 which indicate a trajectory across Mexico or Central America.

51. Climatic Regions. Köppen's classification of the climates of North and Central America is given in the Frontispiece, while the mean distributions of the various meteorological elements are illustrated by the various plates. On the basis of these diagrams and the data presented in the previous paragraphs a brief discussion will be presented of the main climatic regions of the continent.

a. Central America. From Panama to the southern border of Mexico a tropical rainy climate (*Af*) predominates except in the mountainous region, where the temperatures are lower than along either coast line. The prevailing wind circulation is the easterly or northeasterly trades, which persist along the entire Atlantic coast. The northeast winds predominate along the Pacific coast (Papagayos) only during the winter months. The southeast trades of the Southern Hemisphere are deflected to become southwest winds along the Pacific coast of southern Central America. The latter are intensified by the summer circulation toward the semipermanent low-pressure area east of Panama, with the result that southwest winds predominate over the entire west coast during summer. The land area is under the influence of tropical maritime air throughout the year, and consequently the temperature is uniformly high. Likewise, the precipitation is heavy and occurs chiefly in the form of convective showers with frequent mild thunderstorms.

The principal features of the climate are illustrated in Table 49 (page 208) by the mean temperatures and rainfall amounts at selected stations. From about December to April the *mT* air is not so markedly unstable as during the summer and early autumn. This increased stability results in diminished amounts of precipitation along the Atlantic coast and very light amounts along the west coast, the latter because of the prevailing easterly winds and the shielding effect of the central mountain chain. The January dryness at San Salvador (70), San José (49), and Salamá (50) demonstrates the effect of the downslope adiabatic heating upon the rainfall distribution. Because of the prevalence of unstable mT air over the entire country, the rainy season occurs during summer and early autumn. The amounts are excessive in some regions, as evidenced by the average monthly rainfall of 27 in. between July and November at Greytown, Nicaragua. Summer is also the cloudiest period. Consistent with the slight precipitation, the average cloudiness over the west coast is low during the winter months. During autumn the northers produce extended periods of overcast skies and steady rain.

b. Mexico. The principal climatic zones depend on the prevailing winds, mainly the northeast trades, and on the large-scale topographical features. As a result, there are four major climatic regions: (1) the east coast from the Yucatán peninsula to the foothills of the Eastern Sierra Madre; (2) the plateau of Mexico, generally above 6000 ft, extending from the United States border to the border of Central America; (3) the south and west coast to 25° north; (4) the desert region of northwestern Mexico.

Since the prevailing winds are usually onshore, the eastern section of Mexico is mainly under the influence of mT air from the Gulf of Mexico. The Gulf waters are always warm, and consequently the temperatures over the adjacent land area are usually high, especially during the summer and early autumn. January is generally the coldest month because of the greater tendency toward offshore winds, which are relatively cool as a result of the continental cooling during winter. Over the northern section, coastal observations show minimum temperatures of less than 40°F. Farther south, toward the Yucatán peninsula, there is no pronounced source of cool continental air so that, even during the winter months, warm weather persists. The rainfall intensity is everywhere less than along the east coast of Central America, probably as the result of more stable mT air and occasional winter invasions of relatively dry continental air masses. Over the northern Gulf region the precipitation is confined mainly to summer and autumn convective showers within the *mT* air mass. The occasional heavy rainfall, which accompanies tropical hurricanes, also contributes to the summer maximum. Farther south, as represented by Puerto México (33, Table 49), there is a region of heavy rainfall throughout the year. Besides heavy summer convective precipitation, the unstable early winter northers produce appreciable amounts of precipitation. The Yucatán peninsula, illustrated by Mérida (32, Table 49), is another region of summer convective rains. The entire eastern section of Mexico, however, has more winter rainfall than the west coast with the exception of northwestern Lower California. Compare, for example, Ensenada (74) and Tampico (34). In general the mean monthly cloudiness is in accord with the mean monthly precipitation, *viz.*, a summer maximum of cloudiness.

Because of its elevation the Mexican plateau is comparatively cool for the latitude, as shown by the mean monthly temperature at Puebla (52). In the north there is a continental temperature distribution. For example, in the state of Chihuahua, annual extremes of 100°F in summer to 18°F in winter are not unusual. Farther south the range is less pronounced, but a temperature of 32°F has been recorded at Puebla. Throughout the entire plateau region, north of 18° north, a summer-rainfall regime predominates, with frequent thunderstorms due to the continental heating. Near the United States border appreciable winter precipitation arises from extratropical cyclonic activity. However, here the annual total is small as compared with that of southern Mexico. In general, the plateau region is characterized by warm summers with frequent convective showers and cool, dry winters.

The southwestern portion of Mexico resembles the northwest part of Central America. The annual temperature range is small as demonstrated in Table 49 by Acapulco (71) and Mazatlán (72). Since the prevailing winter circulation produces offshore winds, the downslope adiabatic heating is unfavorable for extensive cloudiness and precipitation. While the winters are very dry, heavy summer precipitation is quite general. The convective precipitation is occasionally augmented by the downpours that accompany tropical hurricanes. In general, the precipitation intensity decreases north of about 20° north—compare Mazatlán (72) and Acapulco (71). Fogs are not infrequent in the inland valley regions during the winter months, for example, Guadalajara. They arise from the radiational cooling of maritime air.

From 25° north the rainfall amounts are very light, with the result that there is an extensive desert regime, which extends toward Arizona. It includes the major portion of Lower California. During the summer, northwest winds persist along the coast under the influence of the Pacific subtropical anticyclone. Over the land, warm, dry air masses prevail, and therefore precipitation is unlikely in summer. Moreover, throughout the year the subtropical belt of high pressure is unfavorable for precipitation from cyclonic activity. The rainfall that is observed occurs mainly in the late summer in the form of convective showers. Because of the aridity, the summers are very warm. On the other hand, because of cold-air invasions from the United States low winter temperatures are not infrequent. For example, at Ures (elevation 1410 ft) the average annual temperature extremes are 109°F and 33°F.

In the northern part of Lower California a winter-rainfall regime predominates as illustrated by Ensenada (74). This rainfall occurs chiefly as the result of migratory cyclones from the Pacific (type 19). Summer is a dry season because of the northward displacement of the subtropical belt of high pressure. With the prevailing westerly circulation, winter is not a very cold season. Likewise, the coastal location of the major part of this area ensures moderately cool summer temperatures. The entire west coast of Lower California is subject to frequent summer fogginess and low stratus clouds (*Bn* climate; see Sec. 39). This summer fog is confined to the coastal land stations and is the southern California type of low stratus described in numerous meteorological texts.

c. Alaska. From a climatic standpoint, Alaska may be subdivided into three main areas.

1. The southern coastal area from Ketchikan to Attu has a mild and humid climate (*Cfc*). Because of the high frequency of cyclones located over the neighboring ocean, this section has abundant precipitation, with an autumn and winter maximum. With the prevailing onshore winds, orographical lifting contributes to the heavy rainfall. On the adjacent mountain slopes very heavy snowfalls are quite frequent. Cordova (80, Table

49) demonstrates the influence of the relatively warm *mP* air masses on the mean temperature distribution. Throughout the entire coastal region, winter temperatures are unusually warm for these high latitudes. On the other hand, as a result of occasional polar continental outbreaks, minimum temperatures of 5°F to −5°F sometimes occur along the continental coast line. Cordova (80) also illustrates the relatively low monthly variation of mean temperature. In accordance with the frequent precipitation the average cloudiness is very high and fogs are quite frequent.

2. The mean temperatures at Dillingham (81) and Nome (82) demonstrate the rapid northward decrease of temperature along the west coast of Alaska. The low-lying coastal region from 60° north to Mackenzie Bay on the Arctic Ocean is a region of cold arctic winters with mild summers. Unlike conditions on the south coast of Alaska, the prevailing easterly winds result in frequent invasions of cold *cP* air across the west coast, which produce low temperatures. Over the northern section of Alaska, cold air masses are generally present throughout the year. The total yearly precipitation is considerably diminished toward the north, for example, Kodiak, 61.3 in.; Dillingham, 27.1 in.; Nome, 18.1 in.; and Barrow, 5.3 in. Furthermore, the precipitation type changes from an autumn and winter maximum at Kodiak to a decided summer maximum at Barrow. The latter can be attributed partly to the increase in moisture content of the air with the breaking up of the ice and partly to the migratory summer cyclones. The winter minimum near Barrow is consistent with the presence of a cold anticyclone. Summer cloudiness and fog are quite common along the north coast.

3. The climate of the interior of the country is characterized by extremely cold winters. The high coastal mountains generally prevent the invasion of *mP* air masses at ground level. Radiational cooling during the long winter nights produces very low surface temperatures while at upper levels warmer temperatures prevail. The continental heating results in warm summer temperatures, which sometimes exceed 88°F. The rainfall maximum occurs during summer in the form of convective showers and weak frontal precipitation.

d. Pacific Coastal Region. The climate of the west coast varies from north to south principally as a result of the seasonal displacement of the subtropical anticyclone and the migratory cyclones that reach the coast, generally from the southwest. Like southern Alaska the Pacific coast of Canada experiences heavy rainfall throughout the year. This is the result of the frontal and orographical lifting associated with the deep cyclones of these latitudes. There is a decided autumn and winter maximum when the cyclonic activity is most effective. Toward the south the yearly precipitation total diminishes because of the smaller number of deep cyclones and the northward displacement and intensification of the semipermanent anticyclone in summer. This summer anticyclone produces winds parallel to the coast, and consequently there is little or no precipitation from orographical lifting. As would be expected, the presence of this summer circulation results in increased summer dryness and a higher percentage of winter precipitation toward the south. The general distribution of precipitation is illustrated by Table 47, which shows the total amount of precipitation as well as the percentage that occurs in the 4 months representative of winter, spring, summer, and autumn. Thunderstorms rarely occur along the coast but are more common inland as a result of the increased summer heating; the accompanying rainfall is usually light.

TABLE 47.—PRECIPITATION AT SELECTED WEST COAST STATIONS

Station	Lat., °N	Long., °W	Yearly precipitation total, in.	Monthly percentage of yearly total			
				Jan.	Apr.	July	Oct.
Cordova	61	146	141.3	6	7	6	15
Prince Rupert	54	130	98.1	10	8	5	12
Tatoosh Island	48	125	83.6	14	7	2	9
North Head	46	124	59.1	15	7	2	8
Eureka	41	124	39.8	18	8	<0.5	6
San Francisco	38	122	21.3	22	6	<0.5	4
Los Angeles	34	118	14.9	22	7	<0.5	4
San Diego	33	117	10.3	20	8	<0.5	7

The temperature varies from north to south and from the coast to the neighboring inland valleys. Because of the prevailing maritime air currents, coastal areas are relatively warm throughout the winter as compared with inland stations, where radiational cooling lowers the temperature appreciably. For example, the mean January temperature at Eureka (77, Table 49) is 47°F as compared with 17°F at Sioux City. Extreme minimum temperatures occur when cold *cP* air crosses the mountains and encompasses the Pacific coast. However, *cP* air rarely invades the west coast, and consequently very cold winter temperatures are quite infrequent. During the summer months the persistent northerly winds produce low coastal temperatures. With this prevailing wind circulation, *mP* air scarcely invades the continent, and there is a marked difference in temperature between the coastal stations and the inland valley regions. For example, mean July temperatures of 81°F at Fresno and 59°F at Prince George may be compared with the corresponding temperatures of 70°F at Los Angeles (75) and 56°F at Prince Rupert (79).

The fog regime may be subdivided into winter radiation fogs over the inland valleys and summer fogs along the coast. When moist *mP* air stagnates in the interior of the continent, radiational cooling may result in long periods of foggy weather. For example, in December there are, on an average, 9 days of dense fog at Fresno and 4 days at Portland, Ore. The summer coastal fog is very persistent and arises from advectional cooling of maritime air over the cold coastal water. South of 35° north low stratus clouds predominate, with frequent early-morning radiational fog. The persistence of this surface fog is evidenced by the following average number of days with dense fog in August: Prince Rupert (79), 4, Tatoosh Island, 15; Eureka (77), 5; Los Angeles (75), 2. Despite the relatively low fog frequency in southern California there is a very high frequency of low stratus clouds—the California stratus.

e. Plateau Provinces. The climate of the wide mountainous western region from Alaska to Mexico depends to a considerable degree on the topography. The high elevation results in comparatively cool summer temperatures, except in the south, while winter temperatures may reach very low minima. Because of the irregular topography, local conditions have a marked influence on the cold season temperatures. The high mountains generally prevent the westward progress of *cP* air from the northern plains so that major cold waves with high winds are quite rare. Because of the continental nature of this region, summer heating is very effective and causes higher temperatures than at Pacific coastal stations; for example, compare Lander (61) and Eureka (77) in Table 49.

Pacific cyclones and *mP* air that reach the high plateau region produce comparatively small amounts of precipitation. The air masses have been desiccated by the forced ascent up the west slope of the mountainous region, and the entire area is in a rain shadow as compared with the heavier precipitation along the west coast. The winter rainfall also decreases from north to south. During the summer, the intensive heating produces frequent afternoon thundershowers, but the total rainfall is again generally light. Over the southern portion of this region the subtropical belt of high pressure results in a desert climate, as demonstrated by Yuma (57) and Phoenix (58) in Table 49.

Consistent with the diminished precipitation the mean cloudiness is less than along the Pacific coast. Nevertheless, in the northern portion the average cloudiness exceeds five tenths. Fog occurs in the inland valleys during the winter months, especially north of 40° north, as the result of radiational cooling of modified *mP* air.

f. The Great Plains. The continental location has an important influence upon the climate of the plain states. The temperature varies considerably from summer to winter because of the frequent presence of cold *cP* air during the winter months and warm continental air in the summertime. The mean January and July temperatures at Chesterfield (67), Hay River (66), and Chicago (60) in Table 49 clearly demonstrate the yearly range of the mean temperature. Near the Arctic Ocean the weather is cold throughout the year. With the absence of an east-west mountain barrier there is a consistent latitudinal variation of mean temperature. During winter, cold *cP* air outbreaks have an important influence upon the temperature south of the Canadian border. The severe cold waves (blizzards) are accompanied by snow showers and strong winds. They are preceded by a rapidly moving cold front in the rear of a deep cyclone moving northeastward. When the cyclones (Alberta type) cross the middle portion of the continent from west to east, the real *cP* air does not penetrate very far south. Consequently, on these occasions, comparatively mild temperatures are observed over the southern plains. During the summer, warm continental air persists in the south, while farther north there are frequent cool outbreaks in the rear of eastward-moving cyclones. These cool invasions result in a mean July temperature considerably less than the average extreme maximum temperatures; for example, Bismarck has a mean of 70°F as compared with a yearly extreme maximum of 100°F. During the warmer months of the year the diurnal temperature variation is everywhere very pronounced.

On occasions, when polar maritime air crosses the mountains warm chinook winds produce comparatively high temperatures over the middle and northeastern plain states. This weather phenomenon is most striking during the colder season, when the temperature is usually very low under the influence of *cP* air.

TABLE 48.—PRECIPITATION AT SELECTED GREAT PLAINS STATIONS

Station	Lat., °N	Long., °W	Yearly precipitation total, in.	Monthly percentage of yearly total			
				Jan.	Apr.	July	Oct.
Smith	60	112	11.3	5	3	14	8
Le Pas	54	101	14.6	4	5	16	7
Winnipeg	50	97	19.8	4	5	15	7
Sioux City	42	96	26.8	3	10	13	6
Kansas City	39	95	37.1	3	9	11	8
Fort Smith	35	94	38.8	6	10	9	8

Over the northern prairies summer is definitely the season of maximum precipitation as a result of convective and frontal rainfall. This is a continental type of annual variation of precipitation (see Sec. 20). During the winter months the presence of cold anticyclones prevents appreciable amounts of precipitation. Furthermore, the high mountains to the west result in very small amounts of rainfall in connection with frontal systems from the Pacific coast. South of the Canadian border the yearly precipitation total increases slightly with decreasing latitude. On the other hand, the latitudinal variation is small as compared with the precipitation increase from west to east. The western minimum arises from the shielding effect of the Rockies and the very infrequent presence of *mT* air over their eastern slopes. The mean cyclone tracks demonstrate that deep cyclones almost always follow a northeasterly path from the southern states. Hence frontal precipitation is most pronounced over the eastern section of the Great Plains. As in the northern states, the precipitation in the south also shows a summer maximum and occurs in the form of convective showers and frequent thunderstorms. Table 48 shows the general features of the rainfall distribution.

The mean cloudiness follows a pattern similar to the mean precipitation amounts. When the prevailing winds are easterly, the upslope motion produces low clouds, fog, and light orographical precipitation on the eastern slopes of the Rockies. This weather regime usually occurs in connection with the east winds in advance of a cold anticyclone that is moving southward. Farther west the clouds result from either occasional frontal lifting or radiational cooling in winter and from convection during summer. Except for infrequent occurrences of frontal fog and early-morning radiational ground fog, there is a low fog frequency over the Great Plains.

g. Gulf States. Because of the latitude and the close proximity of a large, warm water surface the temperature remains reasonably high throughout the year. During the winter this is a region of frequent frontal activity, and thus there are periods of onshore and offshore winds. High temperatures prevail when *mT* air moves northward, while north winds produce low temperatures. As mentioned in Sec. 50, cold anticyclones occasionally result in an extreme southward displacement of very cold polar continental air. These cold waves are accompanied by temperatures well below freezing throughout the Gulf states. For example, the average annual minimum at New Orleans is 26°F. Because of the presence of the subtropical anticyclone the winds transport air from the Gulf of Mexico over the southern states throughout the summer season. However, as a result of the high moisture content and the convective activity, the summer temperature maxima rarely exceed 95°F. At the same time, the relative humidities are generally high.

During the winter season, precipitation over the Gulf states arises either from frontal lifting or from horizontal convergence. With the high moisture content of *mT* air, heavy precipitation occurs quite frequently. This winter precipitation is usually connected with cyclones that develop over the southern states or over the Gulf of Mexico. The summer precipitation is almost entirely of convective origin—afternoon showers and thunderstorms are common, especially inland from the Gulf coast. This convective rainfall results in a summer precipitation maximum.

Consistent with the frequent precipitation the mean cloudiness is high as compared with that of the western plains. With the cooling of mT air as it moves northward, light fog and low stratus clouds are sometimes observed during the colder months. Similarly, poor visibility conditions exist during periods of continuous light warm-front precipitation. The drainage from the Mississippi River causes cold coastal water to be present around New Orleans during the late winter. The advective cooling of mT air over this colder water produces fog and low stratus clouds around this section of the Gulf coast.

The state of Texas has a wide variation of climate. For example, the annual precipitation at Galveston, on the Gulf coast, is 46.1 in. as compared with a yearly total of 8.8 in. at El Paso, in extreme western Texas. The climate of eastern Texas resembles that of the other Gulf states, while the middle and northern region has a climate similar to that of the southern states of the Great Plains. The climate of western Texas verges on the dry climate of New Mexico and Arizona.

h. Eastern Section of North America. The winter climate of the eastern portion of the continent can be explained by the migratory cyclones and polar anticyclones that traverse the east coast. With the prevailing offshore circulation the mean winter temperatures are low as compared with those of the west coast stations; compare Norfolk (39) and San Francisco (76) in Table 49. Over the southeast, frequent invasions of mT air ahead of an eastward-moving cyclone result in short periods of high temperatures. Florida, because of its insular characteristics, has a relatively high mean winter temperature. The southern tip of Florida even has a tropical type of climate. Outbreaks of cold air in the rear of deep cyclones produce periods of extremely cold weather throughout the eastern section of the continent. Over eastern Canada the offshore circulation is more persistent than over the United States, and consequently cold weather prevails.

Throughout the summer, high temperatures predominate south of about 40° north. North of this region frontal movements result in periodic invasions of cool air, either from the Hudson

TABLE 49.—METEOROLOGICAL MEANS AT SELECTED STATIONS
(Temperature in degrees Fahrenheit, precipitation in inches, cloudiness in tenths)

Station	Lat., °N	Long., °W	Temperature				Precipitation				Cloudiness				Elevation, ft
			Jan.	Apr.	July	Oct.	Jan.	Apr.	July	Oct.	Jan.	Apr.	July	Oct.	
East Coast															
30. Colón	9	80	80	81	80	79	3.5	4.5	15.5	15.2	6	6	7	7	0
31. Belize	18	88	75	80	83	79	5.1	1.5	9.6	11.0	5	4	6	6	0
32. Mérida	21	90	73	82	82	79	1.3	0.8	5.2	4.1	5	4	6	6	70
33. Puerto México	18	94	70	79	80	78	7.0	1.8	11.7	19.4	7	5	7	7	50
34. Tampico	22	98	66	76	82	77	1.5	1.5	4.9	5.0	6	5	4	5	40
35. Corpus Christi	28	97	56	71	83	73	1.5	1.8	1.5	2.4	6	6	5	4	20
36. New Orleans	30	90	54	68	81	70	4.3	5.3	6.3	3.2	6	5	6	4	50
37. Key West	25	82	68	75	83	78	1.7	1.3	3.0	5.7	4	4	5	5	20
38. Jacksonville	30	82	55	68	81	71	2.8	2.4	6.6	4.4	5	4	6	5	40
39. Norfolk	37	76	40	57	78	62	3.1	3.3	5.6	3.0	6	5	5	4	90
40. Baltimore	39	77	33	53	77	58	3.4	3.4	4.6	2.8	6	6	5	5	120
41. Boston	42	71	29	47	72	55	3.2	3.5	4.0	2.6	6	6	6	5	120
42. Portland	44	70	23	42	67	50	3.9	3.4	3.2	3.1	..	..	..	..	100
43. Eastport	45	67	21	38	60	47	3.9	2.9	3.0	3.5	7	6	6	6	80
44. St. John	45	66	19	39	61	47	4.1	3.2	3.3	4.1	6	6	6	6	120
45. Chatham	47	65	13	37	66	45	3.5	3.0	4.0	3.8	5	6	6	5	100
46. Belle Isle	52	55	9	27	49	37	1.9	2.6	4.8	5.4	6	6	8	7	440
47. Hebron	58	63	−6	20	47	31	0.9	1.2	2.3	1.7	5	7	7	7	50
48. Pangnirtung	66	66	−18	8	47	26	0.6	0.8	1.3	2.3	..	..	..	..	50
Inland															
49. San José	10	84	66	69	68	67	0.2	1.1	8.3	13.3	5	6	8	8	3770
50. Salamá	15	90	68	77	74	71	0.0	1.9	4.0	3.6	..	..	..	..	3050
51. Chimax	15	90	61	65	66	66	5.2	3.4	11.4	12.9	7	6	8	8	4280
52. Puebla	19	98	54	65	63	60	0.2	0.8	5.9	2.7	2	4	6	5	7120
53. Zacatecas	23	103	51	61	61	58	0.5	0.2	4.1	1.6	3	3	5	4	8570
54. Chihuahua	29	106	50	66	77	65	0.2	0.2	3.6	0.9	5	5	7	4	4670

TABLE 49.—METEOROLOGICAL MEANS AT SELECTED STATIONS.—(*Continued*)

Station	Lat., °N	Long., °W	Temperature				Precipitation				Cloudiness				Elevation, ft
			Jan.	Apr.	July	Oct.	Jan.	Apr.	July	Oct.	Jan.	Apr.	July	Oct.	
55. Amarillo	35	102	35	56	76	57	0.5	1.9	2.8	1.6	4	4	4	4	3670
56. Little Rock	35	92	41	62	80	63	4.6	5.3	3.4	2.7	6	5	5	4	360
57. Yuma	33	115	54	69	90	73	0.4	0.1	0.2	0.3	2	1	2	1	140
58. Phoenix	33	112	51	67	89	70	0.8	0.4	1.5	0.6	4	3	3	2	1110
59. Des Moines	42	94	20	50	75	53	1.1	3.0	3.4	2.5	6	6	5	5	860
60. Chicago	42	88	25	48	74	55	1.8	2.8	3.2	2.6	6	6	4	5	670
61. Lander	43	109	18	43	68	43	0.6	2.1	0.7	1.3	4	5	4	4	5370
62. Helena	47	112	21	44	68	45	0.8	1.0	1.1	0.9	7	6	4	6	4110
63. Havre	49	110	13	44	69	44	0.7	1.0	1.9	0.7	6	5	3	5	2510
64. Cochrane	49	81	0	32	63	39	1.5	1.6	3.3	2.6	5	5	5	6	930
65. Churchill	59	94	−21	17	54	27	0.6	1.0	1.8	1.3	4	5	5	7	40
66. Hay River	61	115	−15	23	60	32	0.6	0.6	1.3	0.9	5	4	4	7	530
67. Chesterfield	64	92	−26	1	48	24	0.2	0.9	1.8	1.1	..	..	..	..	10
68. Fairbanks	65	148	−13	29	60	26	0.7	0.3	1.9	0.7	..	..	..	..	500
69. Fort Yukon	67	145	−22	20	61	21	0.5	0.4	1.1	0.7	..	..	..	..	400
West Coast															
70. San Salvador	14	89	72	76	74	73	0.1	1.6	12.3	10.3	2	6	7	6	2150
71. Acapulco	17	100	78	80	83	81	0.6	0.0	6.0	5.9	2	2	6	5	10
72. Mazatlán	23	106	68	71	82	80	0.6	0.0	5.1	2.4	4	3	6	3	260
73. Guaymas	28	111	64	72	87	81	0.2	0.2	1.7	0.4	2	2	4	2	20
74. Ensenada	32	117	55	60	70	63	2.6	0.9	0.1	0.6	..	..	..	..	20
75. Los Angeles	34	118	56	60	70	66	3.3	1.0	0.0	0.6	4	4	3	3	340
76. San Francisco	38	122	49	54	57	59	4.4	1.7	0.0	1.1	5	4	4	4	150
77. Eureka	41	124	47	49	54	55	7.0	3.4	0.1	2.3	7	6	6	6	60
78. Portland	46	123	39	51	66	53	6.5	2.9	0.6	3.1	8	6	4	6	150
79. Prince Rupert	54	130	35	43	56	47	9.6	7.1	4.7	12.0	7	7	7	8	170
80. Cordova	61	146	27	37	55	42	8.5	9.7	8.5	21.6	..	..	..	..	30
81. Dillingham	59	158	16	29	56	36	1.9	1.2	2.7	2.6	..	..	..	..	80
82. Nome	65	165	1	17	50	29	1.0	0.6	3.0	1.4	4	5	7	6	20

Bay region or from the Atlantic Ocean. In the latter case the air does not usually penetrate very far inland, with the result that inland stations are consistently warmer than coastal stations. For example, the mean July temperature is 71°F at Rochester, N.Y., as compared with 67°F at Portland, Maine. North of 60° north the summer temperatures are very low as shown by Pangnirtung (48, Table 49).

The monthly distribution of precipitation shows that there is no pronounced seasonal maximum or minimum. With the warm, moist, unstable *mT* air the yearly precipitation amounts are high over the southeast. There is a gradual decrease toward the north. Winter precipitation can be attributed to the cyclonic activity; the heaviest falls occur with low-pressure systems that develop in the south and subsequently follow a northeasterly path. Eastward-moving cyclones from the Alberta region also contribute to the precipitation over the northeastern United States and Quebec. In summer, south of 40° north, the rainfall is mainly the result of convective showers and thunderstorms. Occasional hurricanes intensify the rainfall when they move onshore or near the east coast. It has been seen that summer cyclonic activity is confined to relatively high latitudes. Hence, over eastern Canada, convective showers frequently result from the passage of weak cold fronts. In the far north the precipitation is less because of the absence of moist air masses and active cyclones.

One of the outstanding features of the mean-cloudiness maps is the high winter frequency of clouds over the Great Lakes region. The prevailing northwest winds on crossing the open lakes rapidly gain heat and moisture from the underlying surface. The turbulent and convective mixing results in persistent cloudiness; these clouds extend to the Appalachian Mountains. The cold water around Newfoundland—the Labrador Current—is instrumental in the production of fog and low clouds. Advectional cooling of warm air from the neighboring oceanic regions gives rise to the fog. During summer warm air from the continent also produces fog over the cold water. With the wind circulation more favorable for the transport of warm air over the Labrador Current and cold coastal waters and also for the transport of the fog inland, the summer season is a period of fog maximum along the entire northeastern section of the continent.

For example, in July, the average number of days with fog at the following stations is Belle Isle, 19; Halifax, 10; Yarmouth, 14; Eastport, 12; Nantucket, 14. Summer fog and low clouds are also prevalent near the Hudson Bay area because of the cooling of warm continental air by the colder water surface. Inland winter advection and radiational fogs are observed, in *mT* air, over the southeastern United States. During the cold season of the year, frontal fogs also contribute to occasional poor visibility over the entire eastern section of the country. However, these fogs are considerably less frequent than those which occur along the northeast coast; for example, Nantucket averages 85 days of dense fog per year as compared with 19 days at Atlanta and 14 days at Louisville.

Thunderstorms occur more frequently during the summer season as a result of the solar heating in the south and weak cold frontal lifting in the north. Winter frontal thunderstorms are also observed in the south. As would be expected from the general location of unstable *mT* air, the thunderstorm frequency decreases northward; for example, there is an average of 60 days with thunderstorms each year at Atlanta as compared with 16 days at Eastport. Plates XIV and XV illustrate the mean distribution of thunderstorms.

CHAPTER X

SOUTH AMERICA

The continent of South America extends from approximately 10° north to 55° south. Over this range of 4500 miles there is a considerable variation in climate, not, however, so pronounced as in the case of North America. The essential differences in climate between the two continents can be attributed to the poleward tapering of the land mass of South America. Because its greater part is located within the tropics, South America is principally a warm-climate continent; except in the high mountainous regions, the snow-forest climates are entirely absent. The mildness of the climate in the far-southern temperate latitudes results partly from the small longitudinal extent of the continent and partly from the presence of an open water surface between Cape Horn and Antarctica.

Unlike North America and Eurasia, South America is never invaded by extremely cold polar continental air masses. Likewise, south of 40° south, summer heating is not very pronounced. Consequently, South America lacks the continental variety of temperate climate that characterizes the land masses of the Northern Hemisphere. The widest region of South America is located about 5° south, where the distance from the west to the east coast exceeds 3000 miles. This vast land mass contains the greatest expanse of tropical rain-forest (*Af*, *Am*) and tropical savanna (*Aw*) climates of all the continents of the earth.

Besides having the greatest area of typical equatorial climate, South America has also the greatest area in tropical latitudes of tundra (*ET*) climate. The presence of regions of temperate and tundra climates can be explained by the high elevation of the country in the vicinity of the lofty Andes to the west and the highlands of Brazil to the east. The highest levels of the Andes are perpetually covered with snow even at the equator; consequently, a small region of frost climate (*EF*) is also found within tropical latitudes.

South of the Tropic of Capricorn various classes of temperate climates (*C* types) are observed. As a result of the close proximity of the oceans the cold snow-forest climates (*D* types) are almost entirely absent. However, as in the case of South Africa and Australia, a region of desert climate is observed south of the tropics. Its counterpart in North America is the desert area of Arizona and Nevada.

The sections of this chapter that deal with the more detailed aspects of the weather, such as air masses and cyclone paths, are based to a great extent upon the publications of Serra and Ratisbonna of the Servico de Meteorologia, Rio de Janeiro.

52. Physical Features. The most prominent topographical feature of South America is the continuous chain of lofty mountains that extends from Venezuela to Cape Horn. This mountainous region is widest near 20° south, where the land is over 10,000 ft for a distance of 300 miles—from within 70 miles of the Pacific coast to the Bolivian plateau. Except for several isolated high peaks, the mountains south of 40° south are below 10,000 ft. The Andes have a significant influence on the climate of the entire continent. In tropical latitudes they prevent the passage of Pacific air masses across the vast lowlands of Brazil. South of 35° south, where the mountains become a less formidable barrier, it is possible for the Pacific air to appear at low levels over Argentina. However, the moisture content of these air masses is decidedly modified by the forced ascent over the mountains.

Over southeastern Brazil there is another region of highlands. However, few of the mountains are over 5000 ft. Despite the lower elevation, these highlands have an important effect upon the weather of Brazil. When the winds have an easterly component, the air is forced to ascend along the Atlantic coast and this may result in orographic precipitation. Likewise, as in the case of the Andes, the interior of the continent is sheltered, to a certain extent, from invasions of moist Atlantic air masses.

Smaller areas of highlands are observed in northern Brazil, Guiana, and southern Venezuela. They have an important influence on the climate of the regions in the immediate vicinity. On the other hand, they do not form a formidable barrier to maritime air masses from the northeast, and consequently invasions of North Atlantic maritime air can be expected to extend into the interior of Brazil.

Between these three main regions of hills and mountains extend the lowlands of Argentina, Uruguay, Paraguay, and Brazil. The land that surrounds the vast network of tributaries of the Plata and Amazon rivers is generally less than 1000 feet above sea level. For example, the central Amazon basin is below 500 ft. These water surfaces form a source of moisture for the air masses that invade the central portion of the continent. South America is singularly free of inland lakes and prominent indentations. In this respect, the continent differs from North America.

The influence of the topography upon the weather and climate of South America is illustrated by the preferred trajectories of cold polar maritime air when it is displaced from south to north.[1] The cold air outbreaks occur with northward-moving anticyclones and high-pressure wedges. In general, the cold air follows one or more of three favored paths, as follows:

a. West of the Andes. As a result of the high elevation of the Cordillera, especially north of 30° south, the cold air mass remains west of the mountains and finally reaches the trade-wind region of the Pacific. In Peru, with the change in direction of the coast line, the cold air is prevented from penetrating inland by the same mountain chain. On rare occasions in the wintertime strong cold outbreaks may cross the Andes.

This trajectory, west of the Andes, is quite common in summer. The cold air invasions are called *suradas*. On the other hand, in winter, the cold outbreaks occur more frequently to the east of the Cordillera.

b. Atlantic Ocean and East of the Brazilian Highlands. This oceanic trajectory of polar anticyclones usually occurs in the rear of a cyclone located somewhere over the Atlantic Ocean. The cold air mass is preceded by a cold front, which generally becomes very indistinct as it approaches the equator. On these occasions the weather of Brazil depends upon whether the cold mass is deep enough to cross the Serra do Mar chain. With deep invasions heavy frontal and orographic rains occur in southeastern Brazil as the cold air crosses the mountains toward the interior of the continent. If the cold air mass is shallow, it will not traverse the Serra do Mar chain. The cold front then tends

[1] Serra, A., and Ratisbonna, L., "As Ondas de Frio da Bacia Amazonica," Ministério da Agricultura, Serviço de Meteorologia, Rio de Janeiro, 1941.

to parallel the coast line and become quasi-stationary, giving rise to steady rain of 4 or 5 days' duration.

c. Interior of the Continent. The cold air sometimes follows a continental path between the Andes and the Brazilian highlands and, in the wintertime, may reach the equator. Exceptionally cold outbreaks over central Brazil are called *friagens*, because of the abnormally large temperature falls that accompany the cold air invasion. The air mass spreads across the Amazon valley as it approaches the equator and is quite shallow as compared with the deeper mass about 20° south.

In the wintertime, with exceptionally strong polar outbreaks, the air may follow all three trajectories simultaneously. Nevertheless, it is quite obvious that the topography has a decided influence on the direction in which the air may move. In turn, this affects the distribution and variation of the meteorological elements.

Numerous other examples could be cited showing the influence of topography upon the weather and climate of South America. However, these will be presented in later paragraphs in connection with a more detailed description of the sectional climatology.

53. Mean Distribution of Pressure and Wind. *a. Sea Level.* Plates IV and V show the principal features of the sea-level pressure and surface-wind distribution. The mean-pressure charts illustrate the following:

1. Throughout the year there are two semipermanent anticyclones, over the Pacific and Atlantic oceans, which touch the coast lines of South America. On the mean-pressure maps the anticyclones are found to vary in intensity and to oscillate with latitude as shown in Table 50.

TABLE 50.—ATLANTIC AND PACIFIC SUBTROPICAL ANTICYCLONES

Month	Central pressure, mb		Latitude of center	
	Atlantic	Pacific	Atlantic	Pacific
Jan	1021	1021	28°S	32°S
July	1024	1024	23°S	26°S

2. A circumpolar zone of low pressure extends to about 40° south, with the mean-pressure trough slightly to the east of the

continent. The deep semipermanent cyclones of the Weddell and Belgique seas are located south of Cape Horn. In contrast to the mean circulation of the Northern Hemisphere it is noticeable that the lows are deeper in summer than in winter.

3. As a result of the continental heating, a thermal low is located over Paraguay in January. During July this area is under the influence of a weak anticyclone.

4. The northern coast of the continent is located on the outer extremities of the Northern Hemisphere Atlantic anticyclone.

5. A trough of low pressure is located near the equator in all seasons. This doldrum belt oscillates with the season along the coastal regions as shown in Table 51. The range of the oscillation is greater over the central portion of the continent than over the oceanic regions.

TABLE 51.—APPROXIMATE MEAN LOCATION OF THE DOLDRUM BELT OF LOW PRESSURE

Month	Atlantic coast	Pacific coast
Jan.	1°S	0°
July	5°N	7°N

The mean-pressure maps have not been included for April and October. They would probably show that the April chart resembles the January map more closely than the July map, and vice versa for the October pressure map.

The mean wind distribution can be interpreted from the pressure charts. General westerly winds are observed in all seasons south of 40° south. Along the coast of Chile and Peru the winds are from a southerly direction; about 30° south they are decidedly stronger in summer than in winter because of the intensification of the pressure gradient between the continental thermal cyclone and the Pacific subtropical anticyclone. On the east coast of the continent northeast winds are observed from 10° south to 35° south in January. The circulation in this vicinity is weaker in July because of the feeble continental anticyclone. In all seasons east and southeast winds prevail along the coast of Brazil from 10° south to the equator. The northeast trade winds of the Northern Hemisphere are observed to the north; they are stronger and more persistent in the latter half of summer when the doldrum low-pressure belt is farthest south.

Over the continent the surface winds are generally light and variable except during the Southern Hemisphere summer when the thermal low pressure is well developed. Furthermore, as a result of this low pressure, the northeast winds across the north coast of the continent are more pronounced in January than in July.

Table 52.—Percentage Frequency of Surface-wind Direction

Station	Lat.	Long.	January								July							
			N	NE	E	SF	S	SW	W	NW	N	NE	E	SE	S	SW	W	NW
Atlantic Coast																		
1. Port-of-Spain...	11°N	61°W	11	*54*	9	14	2	3	2	5	7	*43*	12	20	2	6	2	8
2. Paramaribo....	6°N	55°W	14	*62*	23	1	0	0	0	0	12	*54*	28	3	1	2	0	0
3. Fernando de Noronha.......	4°S	32°W	0	0	32	*55*	13	0	0	0	0	0	32	*62*	6	0	0	0
4. Curitiba.......	25°S	49°W	9	21	*25*	15	4	5	9	12	19	*26*	16	7	3	6	9	14
5. Santa Cruz.....	50°S	69°W	7	1	7	8	12	*27*	25	13	10	4	4	3	18	18	*29*	14
6. Staten Island...	54°S	64°W	12	2	2	8	17	16	16	*27*	9	7	1	11	14	18	15	*25*
Pacific Coast																		
7. Iquique........	20°S	70°W	0	1	1	2	5	*89*	1	1	5	10	1	7	3	*61*	1	12
8. Caldera........	27°S	71°W	7	3	1	1	17	*62*	5	4	18	8	5	4	*29*	24	2	10
9. Valparaiso.....	33°S	72°W	5	6	4	7	11	*52*	11	4	19	14	11	10	17	*21*	5	3
10. Punta Galera...	40°S	74°W	22	1	0	2	*54*	11	4	6	*35*	5	6	6	10	6	10	22
11. Evanjelistas....	52°S	75°W	8	1	0	1	5	14	30	*41*	8	2	8	12	16	*21*	13	20
Inland																		
12. Manáos........	3°S	60°W	23	15	*35*	5	17	0	5	0	19	12	*28*	6	27	1	7	0
13. Cuiabá........	16°S	56°W	*48*	7	5	3	7	4	7	19	*34*	6	5	6	31	5	5	8
14. Corumbá......	19°S	58°W	2	10	17	22	*31*	8	6	4	2	3	26	26	*32*	10	0	1
15. Tucumán......	27°S	65°W	12	16	3	14	12	*36*	3	4	21	18	5	8	12	*30*	3	3
16. Córdoba.......	31°S	64°W	17	*29*	8	14	22	4	3	3	18	*25*	7	15	24	5	2	4
17. Punta Arenas...	53°S	71°W	8	8	3	2	6	21	*32*	20	12	14	6	3	3	*21*	*21*	20

The day-to-day variations from the mean circulation become more pronounced toward Cape Horn. The circumpolar zone of low pressure is constantly traversed by cyclones, which move from northwest to southeast. In the rear of these cyclones there are cold air outbreaks in the form of cold anticyclones of varying intensity. These move in a northerly direction as described in Sec. 52; their most frequent direction is from southwest to northeast. On many occasions, over northern Argentina, the anticyclones are preceded by the deepening of a cyclone that subsequently moves southeastward over the Atlantic Ocean. Wind changes accompany these migratory pressure systems in accordance with the passage of centers, troughs, and wedges.

Near the equator the wind systems are more stable along the coast lines; the principal variations are seasonal. In the interior of Brazil the winds are generally so light that the variations are insignificant. The frequency of wind direction at a series of stations is shown in Table 52. With respect to this table it should be remembered that the surface winds may be local, for example, land and sea breezes, valley and mountain winds. Even with these reservations Table 52 supports the previous general remarks concerning the persistent circulation over northern Chile and northern Brazil.

b. Upper Air. The upper-air circulation is described by Serra and Ratisbonna.[1] The principal features of the January circulation may be summarized as follows:

1. The thermal low over Paraguay extends to 1 km, but by 3 km there is a well-developed upper-level anticyclone.

2. The circumpolar zone of west winds extends northward with elevation. The wind speeds increase with elevation.

3. The latitude of the subtropical high-pressure belt varies only slightly with elevation. However, there is an increase in the west-wind speeds to the south and a small increase in the east winds to the north. This is consistent with the thermal wind relationship when it is remembered that the region of highest temperature lies between 20° south and 25° south.

The main features of the winter upper-air circulation may be summarized as follows:

1. Unlike the condition in summer, the northeast circulation along the northern portion of the continent is very shallow. The northeast winds extend to at least 2000 m in January and to only 500 m in July; above the 500-m level they are practically due east.

2. The circumpolar zone of west winds extends rapidly northward with elevation.

3. The axis of the subtropical high-pressure belt is displaced toward the equator with elevation. For example, the latitude of the axis of high pressure at sea level is about 29° south, at 5 km

[1] SERRA, A., and RATISBONNA, L., "As Massas de Ar da America do Sul," Ministério da Agricultura, Serviço de Meteorologia, Rio de Janeiro, 1942.

about 15° south, and at 18 km about 5° south. Consequently, in winter, while the west winds of high latitudes increase in speed with elevation the east winds of low latitudes decrease with elevation and finally undergo a reversal of direction.

As in other regions of the world, the upper-air winds are definitely more persistent in direction than the surface winds. Shallow surface cyclones appear as minor troughs in the upper westerly circulation. Deep cyclones appear as small closed centers at 3000 m and 5000 m. Consequently, above 2000 m and between 30° south and 50° south, changes in wind direction are relatively small. In low latitudes the wind velocities depend on the location of the subtropical anticyclones.

54. Ocean Currents. The principal ocean currents are illustrated by Plate VI, while the sea-surface isotherms are given in Plates VII and VIII. Corresponding to the circulation around the oceanic subtropical anticyclones there is a warm southward-moving current (Brazilian Current) along the east coast and an extensive northward-moving cold current (Peru Current) along the west coast. The Atlantic Equatorial Current and the Pacific Equatorial Countercurrent dominate the oceanic circulation in low latitudes. About 55° south the Southern Hemisphere is completely surrounded by the West Wind Drift Current. The Peru Current may be considered as a branch of this cold drift, and the Falkland Current along the southeast coast of the continent is a further branch. It is significant to note that near the east coast the drift ice extends to about 40° south and west of Cape Horn only to about 50° south.

In high latitudes the Falkland Current is somewhat colder than the Peru Current. This temperature difference can be explained partly by the prevailing wind circulation, which is southwest in the rear of the Weddell Sea Low, and partly by the close proximity of Antarctica on the east side of South America. These factors account for the transport of extremely cold water and drift ice from Graham Land toward the east coast of the South American continent.

The mean isotherms show the influence of the currents on the surface temperature distribution in the vicinity of South America. The essential features can be tabulated as follows:

1. South of 35° south the Atlantic is colder than the Pacific because of the Falkland Current.

2. North of 35° south the Atlantic is warm because of the Brazilian Current, while the Pacific is cool because of the Peru Current. At 10° south the mean temperature difference is about 6°C.

3. North of the equator the two oceans have about the same temperature.

4. The bending of the sea-surface isotherms toward the north along the coast of Chile and Peru arises from the upwelling of cold water. This is caused by the southerly winds and is most noticeable in summer when the south winds are stronger and more persistent. The cold coastal water favors fog and low stratus cloud formation.

55. Air Masses. South America comes under the influence of numerous air masses, and Plates XVI and XVII show their approximate source regions. The principal properties of these air masses are given below.

a. Tropical Maritime Air (*mT*). All types of *mT* air are characterized by high surface humidities and moderately high surface temperatures. Although there are temperature differences between the air over the Pacific and Atlantic oceans south of the equator, the *mT* air masses are homogeneous in the lowest layers. The principal variations arise from the presence or absence of upper-level stability.

Within the subtropical high-pressure cells, subsidence produces an air mass that is stable aloft. This *mTs* air is always present along the west coast of Chile. Here the subsidence is decidedly more pronounced than off the east coast of the continent. Over the latter region, *mT* air is convectively unstable and may produce convective showers when lifted to saturation.

Unstable tropical maritime air (*mTu*) is observed away from the circulation of the subtropical anticyclones and on their west side. The absence of anticyclonic subsidence and the convergent flow toward a low-pressure trough result in the formation of a deep unstable air mass. This air mass invades eastern Brazil throughout the year so that instability precipitation is frequent—particularly in winter when there is a deep current of maritime air. Tropical maritime air from the Northern Hemisphere reaches the north coast of South America as a pronouncedly unstable air mass because of its long trajectory. The invasions are deeper in January than in July. Similarly, very stable *mTs* air off the Chilean coast becomes less stable farther north and reaches Panama as an *mTu* mass. Heavy convective showers are quite common within these *mTu* air masses.

Over the low-lying land of Brazil another form of *mT* air generally prevails. Besides the convergence toward the doldrum belt, continental heating accentuates the instability so that an *mTKu* air mass is formed. Thunderstorms are quite common within this mass. In summer (January) *mTKu* air occupies a larger region than in winter (July). Often during the summer the air mass invades southeastern Brazil.

b. Tropical Continental Air (*cT*). During the summer, as the result of strong heating, a warm, dry air mass is located over the continent from 30° south to 40° south. This is the region of a thermal low-pressure area. Tropical continental air is convectively unstable, but because of the low humidity little convective precipitation is observed. Tropical continental air is rarely present in winter.

c. Polar Maritime Air (*mP*). From 40° south to Antarctica, the southern oceans are under the influence of polar maritime air masses. Actually it is a region where cold antarctic and warm tropical air masses are transformed by the circulation around migratory cyclones and anticyclones. The properties of an *mP* air mass depend upon the air trajectory. For example, low-level instability occurs when the air moves northward. Conversely, southward-moving air masses become increasingly stable in the lower layers.

Polar maritime air frequently invades southern Chile and Argentina, and on occasions there are outbreaks toward lower latitudes. The trajectories of these northward invasions of *mP* air were described in Sec. 52.

56. Fronts, Cyclones, and Anticyclones. The mean location of fronts has been illustrated by Plates XVI and XVII. The positions of these fronts are subject to continual variations, the more important of which are discussed in this section.

a. Perturbations of the Antarctic Front. Apparently, during the winter season, there is a successive passage of cyclones from the southwest to the northeast in the Belgique Sea region. After each family of lows there are polar continental outbreaks from Antarctica. The cold anticyclones sometimes invade the Patagonia region, but they are more frequently displaced toward the Atlantic. In summer, the front is displaced well to the south.

b. Perturbations of the Pacific Polar Front. During winter, cyclones form in the mid-Pacific on the front that separates tropical Pacific and polar Pacific air masses. As these lows occlude, they move southeastward and generally reach the continent as occluded cyclones. Occasionally open warm-sector depressions reach the Chilean coast; but, because of the Andes barrier, they are forced to occlude rapidly. The frontal precipitation over southern Chile is generally of the warm-front occlusion type, with polar Pacific air in the lower levels and tropical Pacific air aloft. With the passage of the cyclones across the Andes, orographical lifting intensifies the precipitation along the west coast. Because of the downslope motion only slight precipitation occurs in Patagonia (foehn effect).

Sometimes cyclonic centers pass to the south of Cape Horn, with the result that precipitation is diminished over southern Chile. Finally, the cyclones are displaced toward the Weddell Sea, where they form the very deep, semipermanent low that is located in this general region.

In the rear of these cyclones cold polar anticyclones move northward and may take one of the three trajectories that have been described previously. This sequence of lows and highs takes place in summer as well as in winter. However, the cold anticyclones frequently move along the west coast of Chile in summer, whereas in winter they are usually displaced toward the Atlantic.

The occluded fronts that cross the Andes generally appear as upper fronts over Patagonia. Because of the upper front, a trough appears at the ground, and frontogenesis may result in a new surface front, which later intensifies as it approaches the axis of dilatation of the Atlantic polar front. This is frequently followed by the deepening of a new cyclone in the Plata River area. Especially in autumn, deep lows form in this region. On other occasions, the upper occluded front traverses the continent without the formation of a new surface front.

c. Perturbations of the Atlantic Polar Front. In advance of a cold outbreak, the cold front may become stationary in middle latitudes or else move toward the tropics.

If the Atlantic subtropical anticyclone is well developed or if the cold outbreak is weak, the northward motion of the cold front is retarded and finally the front becomes stationary. As a result of the forced ascent of tropical Atlantic air, continuous rains occur along Argentina and southern Brazil, especially when the front moves southward as a warm front.

With strong cold air outbreaks the front moves into low latitudes, where it is subject to frontolysis. A new polar Atlantic front is later created, to the south of the moving cold anticyclone, between modified polar air and the cold polar westerlies. Because of the influence of the Brazilian mountains, the northward-moving cold front and anticyclone may undergo the following main changes:

1. When the cold mass is deep enough to cross the Serra do Mar, frontal and orographical rain occurs. Heavy showers and thunderstorms are also observed. If the cold front continues to move north or northwest, the cold air invades Brazil as south winds and produces the phenomenon called the *friagem.* On the other hand, when the winds in the cold air are southwest, the front moves along the Brazilian coast and produces frontal phenomena as far as Pernambuco. At this latitude the front dissolves so that the polar maritime air finally mixes with the trade-wind air and aids in the intensification of the subtropical high.

2. When the cold mass is shallow, the air cannot traverse the Serra do Mar chain. The western portion of the front is rapidly subjected to frontolysis, whereas the eastern portion becomes parallel to the Brazilian coast line. The latter front acquires the characteristics of a stationary front with continuous rain and low ceilings. Finally, it dissolves, and the cold air becomes part of the subtropical anticyclone. This weather sequence is typical of the spring and summer seasons.

The cyclonic storms that develop on the Atlantic polar front form frequently in the Plata River region. However, they also develop farther north when the front becomes quasi-stationary. The deepening lows are usually displaced toward the southeast.

d. Perturbations in the Equatorial Zone. The variations from the mean circulation pattern become less pronounced near the equator. On the other hand, they are definitely significant as regards the day-to-day changes in the weather. For convenience, the ensuing discussion will be divided according to two main seasons (Southern Hemisphere seasons).

1. Winter and Spring. During this period, tropical Atlantic air encompasses the hilly section of southern Brazil and extends to about the latitude of São Salvador. In general, this is a region of clear skies and small amounts of precipitation. Along the coast from São Salvador to Belém, an *mTu* air mass prevails and is associated with a showery regime, particularly south of Natal. The north coast is under the domination of *mTu* air from the Northern Hemisphere, while Pacific *mTu* air surrounds the Pacific coast. In the interior of northern Brazil, *mTKu* air is accompanied by very frequent rain showers and thunderstorms.

The latitudinal displacement and the small disturbances of the intertropical front produce important weather changes. During the winter the front is in the Northern Hemisphere, where, on the Atlantic side, it separates Northern and Southern Hemisphere air masses. Except when the front is farthest north (August and September), no deep cyclones develop; small cyclones, which move slowly westward, are not uncommon, and they are accompanied by continuous heavy rains from the Atlantic Ocean to the Andes. When the front is displaced north or south, the rain belt is similarly displaced. One of the most pronounced causes of a rapid northward movement of the front is the invasion of a deep polar mass from the south. Because this is a feature of the variation of the polar Atlantic front, it is now convenient to consider the influence of the latter upon the weather of these equatorial regions.

We have already seen that, in its maritime trajectory, the cold front may proceed as far north as Natal. The principal weather changes are a marked prefrontal heating, a temperature decrease after the front passage, and an intensification of the winter rains, together with extended periods of cloudy skies over the northeast coast of Brazil. North of Natal the modified cold air is comparatively unstable; and, because of the easterly circulation, it has a tendency to advance toward the northwest, producing rain along the coast of Brazil as far as Belém. It is apparent, however, that the interior of the continent is unaffected by the cold mass when it follows this maritime trajectory.

Sometimes the cold air advances toward the interior of Brazil, where it causes profound weather changes. After crossing

Bolivia and Mato Grosso the cold air penetrates into the interior of Brazil, where it deviates toward the west, crossing Acre, the upper Amazon, and Colombia. Along the cold front, cumulonimbus clouds and showers are generally observed. When this modified cold mass is in the Amazon valley, there is a dry period of a few days' duration. The showers and thunderstorms of the *mTKu* air mass are then observed farther north. Finally, owing to the surface heating and evaporation, the polar air is transformed into a new unstable *mTKu* air mass. Throughout the world, invasions of markedly cold air into equatorial regions are usually rare, and consequently the *friagem* is a very interesting meteorological phenomenon. The *friagem* generally persists for 3 to 5 days and is preceded by a prefrontal heating, a decrease in relative humidity, and a fall in pressure. It occurs about three times a year at Cuiabá (16° south) and about two and a half times a year at Senna Madureira (9° south). The depth of the cold air mass varies generally from 3000 m in Mato Grosso to 500 or 1000 m in the Amazon valley. The effect of the *friagem* can be illustrated by the severe cold outbreak that occurred during June, 1933. At Cuiabá (16° south, elevation less than 1000 ft) the temperature fell from 91°F on June 18 to a minimum of 34°F on June 21 while the atmospheric pressure rose nearly 20 mb. At the same time, the temperature decreased to 41°F at Cruzero do Sul (7° south), where the elevation of the station is again less than 1000 ft.

The polar masses practically never invade the central zone of Brazil (Pará, Goyaz, Piauhy, Ceará, Bahia, and Minas). Hence, in the lower Amazon, the trades are never affected by cold invasions and the air masses in this region remain more stable. Consequently, the precipitation is less than in the upper Amazon.

2. Summer and Autumn. The intertropical front, or the doldrum belt, reaches northeast Brazil as far south as the interior of Goyaz during summer and early autumn. The unstable *mTKu* air of the vast Amazon valley, which is associated with daily rain showers, is separated from *mT* air by the mountains of Goyaz and Bahia. The area covered by the Atlantic trades (Ceará, Rio Grande do Norte, Parahiba, Pernambuco) is characterized by comparatively slight rainfall except along the coast from Alagoas to Natal, where the air is more unstable.

As in the winter season, there are frequent heavy showers in the immediate vicinity of the intertropical front. However, no deep tropical cyclones are observed. Important weather changes accompany the displacement of the front since the latter is associated with a rain belt. For example, when the subtropical anticyclone of the South Atlantic extends toward the interior, the intertropical front is displaced northwestward so that Atlantic maritime air masses occupy a region usually under the influence of *mTKu* air. The heavy convective thunderstorms cease with the advection of the more stable air mass. Again, with cold polar outbreaks from the Northern Hemisphere, the northeast winds of South America are intensified. This results in a southeastward displacement of the intertropical front so that the northeast section of the country is invaded by *mTKu* air. In these cases rain showers and local thunderstorms sweep the usually arid regions of Minas, Bahia, and Ceará.

Perturbations of the polar Atlantic front also cause short-period changes in the low-latitude weather. When the fronts move northward along the coast, they cause a considerable prefrontal warming as a result of the turning of the winds toward the northwest and the advection of *mTKu* air. With the cold-front passage, rain and thunder frequently occur and are followed by a slight decrease in temperature. As a result of the displacement of *mTKu* air, the rain along the northeast coast of Brazil (south of Natal) ceases for 2 or 3 days. Finally, the advection of this modified cold mass intensifies the weak intertropical front in the vicinity of Belém. If the polar masses advance toward the interior of the continent, a slight *friagem* is produced. The cold air moves farther east than in winter. These weak invasions produce little change in the occurrence of precipitation; heavy convective showers persist. Consequently, the weather changes in the central tropical regions of South America are less dependent on the movements of the polar front in summer than in winter.

57. Climatic Regions. The mean distributions of the meteorological elements are shown in Plates I-V and IX-XVII, and the Köppen classification of climate is given in the Frontispiece. A brief discussion of these will now be undertaken.

a. Coast of Chile and West Slope of the Andes (30° *South to* 55° *South*). The climate of this region is influenced by the action of the polar Pacific and tropical Pacific air masses in connection with the cyclones that originate on the Pacific polar front. South of 36° south, rainfall due to orographical and frontal lifting is observed in all seasons, with a maximum at 53° south. Because of the northward displacement of the polar front in winter and the southward displacement of the subtropical anticyclone in summer, the rainfall north of 40° south occurs mainly in the winter half of the year (*Cs* type of climate). The annual variation of cloudiness follows the annual variation of rainfall since the causes are the same. Because the prevailing winds are from the ocean, the annual temperature range is comparatively small. The strongest surface winds generally blow from the southwest; winds in excess of Beaufort force 7 are quite common near Cape Horn.

From 36° south to 30° south there is a gradual transition toward desert conditions. Here, frontal and orographical precipitation occurs during the winter season, and the amounts diminish rapidly toward the north. The rainfall regime is illustrated by Coquimbo (54), Contulmo (55), Morro Lobos (56), and Evanjelistas (57) in Table 54.

b. West Slope of Andes from 30° *South to* 5° *South.* This region is under the continuous influence of the Pacific anticyclone. The winds in the lower layers are more or less parallel to the coast line. Consequently, as a result of the absence of frontal or orographical lifting and of the presence of a stable air mass, this is a region of no precipitation. In the higher levels, orographical lifting of upper-air currents produces rain and snow along the high mountain chain as shown by La Quiaca (46), Ollagüe (45), La Paz (44), and Puno (43).

The sea-surface isotherms show a relatively uniform water temperature along the entire coast; moreover, the water is quite cold for these latitudes. This results in comparatively uniform cool air temperatures near the coast. Because of the elevation, the Andes are a region of low temperatures.

Land-station observations about 15° south indicate a maximum of low cloudiness and of surface fog during winter and spring [see Iquique (52) and Mollendo (51)]. The fogs appear to be confined to a narrow coastal belt but extend to 5° south. On the other hand, coastal ship observations, about 15° south, show a fog maximum in summer and a low cloud maximum in the autumn. The land fog is probably the result of nighttime radiational cooling of the maritime air, which never penetrates far inland, whereas the sea fogs are partly caused by the advection of warm air over the cold coastal water. Nighttime radiational cooling from the sharp upper-level inversion of the subtropical anticyclone also aids in the production of surface fogs and low stratus clouds; compare the California summer stratus. Farther north, low stratus clouds persist over the eastern Pacific, where diverging ocean currents produce cold surface water near the Galápagos Islands. The clouds cover a wide area west of the continent and arise from the surface cooling of northward-moving maritime air, especially during the winter half year.

The surface winds are persistently southerly; as mentioned previously, they are stronger in summer than winter. On rare occasions the dry Peruvian coast comes under the influence of the equatorial calms as the precipitation belt is carried southward.

c. East of the Andes, South of the Tropic of Capricorn. The polar westerlies and the Pacific cyclones deposit their moisture on the windward side of the Andes, with the result that dry air masses arrive over the Patagonian plateau. Precipitation is therefore considerably reduced over the high plateau and over the east coast. The lowering of the mountains, south of 50° south, allows reasonably humid air masses to reach the eastern plateau so that, as compared with conditions farther north, a slight increase in precipitation is observed from disturbances moving toward the east. Invasions of modified antarctic air masses during the winter also contribute to the dryness of the east coast of Patagonia. Except during midwinter, when antarctic air masses make frequent invasions over the east coast, the Pacific coast is colder than the Atlantic coast of Patagonia. Over the southern portion of the continent the annual variation of temperature is greatest over the eastern plateau region because of the winter continental cooling and summer continental heating. It will be noticed from Table 53 that this continental influence is small as compared with that of the Northern Hemisphere continents.

From Patagonia to the tropics the plains are dominated in summer by tropical continental masses; it is the zone of greatest surface heating, with slight evaporation. The slight summer rains in this region must be attributed principally to the invasions of *mTKu* air from the north. The low specific humidity in the polar air masses results in very slight winter rainfall. Likewise, because of the presence of these air masses, the winter temperatures are low, especially during periods of south winds or calms. Higher winter temperatures occur during periods of west winds, because of adiabatic heating (foehn effect). Consistent with the surface dryness, cloudiness is small throughout the year.

The eastern zone of the continent between 40° south and the tropics shows rainfall well distributed throughout the year—see Buenos Aires (25), Rio Grande (24), and Rio de Janeiro (23) in Table 54. Three rainfall regimes can be distinguished:

1. Tropical rainfall dependent on the convective precipitation of *mTKu* air masses; minimum in winter, maximum in summer especially toward the northeast.
2. Rainfall from the Atlantic polar front, with a maximum in autumn.
3. Orographical precipitation with the polar invasions of winter and spring.

Both cloudiness and precipitation increase toward the Brazilian plateau as a result of frontal, orographical, and convective lifting. Thunderstorms are more frequent in summer, when the tropical air masses become quite unstable. Frontal and advective fogs are frequently observed over the land in winter, whereas advective fogs occur over the ocean about 40° south in all seasons of the year. The latter are produced by the cooling of warm maritime air over the cold waters of the Falkland Current. As with the North American temperature distribution, the annual variation of temperature is greater along the east coast than along the west coast.

In order to compare the South American and North American temperate regions, Table 53 shows meteorological means for a few selected stations. The summer and winter means include the month before and after January and July.

Besides illustrating some of the climatic aspects of South America, Table 53 shows the essential differences in climate between the two continents. For example, in temperate latitudes, the continental nature of North America results in lower winter temperatures and higher summer temperatures over its central and eastern portions. Furthermore, because of the frequent cyclones that move northeastward, east central North

America experiences greater precipitation than the corresponding section of South America.

TABLE 53.—DISTRIBUTION OF METEOROLOGICAL MEANS IN SOUTH AND NORTH AMERICA
(Values represent the monthly means taken over 3 months)

Item	South America			North America		
	West coast	East central	East coast	West coast	East central	East coast
	About 40° lat.					
	Punta Galera	Choele Choel	Mar del Plata	Eureka	Spring-field, Ill.	Atlantic City
Precipitation, in.:						
Summer	3.0	0.7	2.6	0.4	3.3	3.8
Winter	12.0	0.5	2.1	6.7	2.2	3.6
Temperature, deg F:						
Summer	57	73	66	54	74	70
Winter	49	46	47	47	29	34
	About 35° lat.					
	Punta Car-ranza	Victoria	Monte-video	Los Angeles	Mem-phis	Wil-mington
Precipitation, in.:						
Summer	0.2	2.8	3.1	0.0	3.3	6.1
Winter	5.1	0.4	3.1	3.0	4.6	3.1
Temperature, deg F:						
Summer	59	73	72	69	79	77
Winter	51	46	51	55	43	47

d. Central and Eastern Zones North of the Tropics. The interior of the continent, from the eastern slopes of the Andes to the western slopes of the Brazilian highlands, is under the influence of modified tropical maritime air. The rainfall is chiefly of a convective nature except along the mountain slopes, where orographical lifting promotes precipitation. The summer maximum and winter minimum arise not only because of the pronounced summer instability of *mTKu* air but also because of the extension of more stable Atlantic air over the continent during the winter season. As is typical of low latitudes, the temperature is always high and the annual range is small, as shown by Corumbá (38) and Cuiabá (37) in Table 54. The convective instability of *mTKu* air is demonstrated by the tall, towering cumulonimbus clouds during the late afternoons. Except in winter, thunderstorms are very frequent. Occasional cold-air outbreaks, mainly during winter, cause the short-period weather variations that were discussed in Sec. 56.

About 20° south the high plateau of the Andes has its greatest width, and there is a desert or steppe regime. This is produced by the warm, dry upper-level air of the trades that invades the high levels of Bolivia. As we proceed toward the north, this zone is more confined to the coast and west slope of the Andes, while the rainy season (summer) becomes continually longer. The La Quiaca (46), Ollagüe (45), La Paz (44), Puno (43), and Cajamarca (42) means in Table 54 illustrate these general features of the rainfall distribution. The precipitation arises from the unstable *mTKu* air, which is displaced westward by the east winds.

The eastern coastal zone between the tropics and 15° south [Rio de Janeiro (23), Ilhéos (22)] experiences, in summer, the rain of the *mTKu* air mass and, in autumn, the rain from the northward-moving modified polar masses. During the winter and spring the region is comparatively dry. Likewise, cloudiness is at a minimum during July along the coast line, while the interior, under the influence of stable Atlantic air, is characterized by clear skies. Thunderstorms occur inland because of the invasions of *mTKu* air with the approach of summer.

The east coast of Brazil, north of 15° south, has a regime of winter rainfall produced by the orographical lifting of modified polar air. The rainfall distribution is shown by Recife (21) in Table 54. In Bahia the rainy period commences in April. The northern portion of this region is affected by the autumn rainfall regime on the southern edge of the doldrum belt, while the southern portion is subject to the summer precipitation of the continental *mTKu* masses. Over this eastern section of Brazil, the cloudiness is mainly of a convective nature, but the convection is not sufficient to produce widespread thunderstorms. The temperature is always high, but lowest in winter as a result of the invasions of modified polar air masses.

The interior section of northeast Brazil is a comparatively dry region [see Morro do Chapéo (30), Quixeramobim (29), and Barra do Corda (28)]. The winter rainfall of the east coast does not extend far inland, and the summer precipitation of *mTKu* air affects this region only weakly because the northeast trades of late summer oppose the eastward displacement of this unstable mass. However, rainfall occurs in the autumn, when the intertropical front is farthest south. As a result of the dryness in early summer and winter the annual variation of temperature is large for this latitude.

e. Coastal Band to the North of the Continent (*East of* 60° *West*) *and the Lower Amazon.* The rainfall of this area is produced by the southward displacement of the North Atlantic air masses with a maximum in autumn (Southern Hemisphere) when the doldrums, or the intertropical frontal zone, are in their extreme southern position. The rainfall is of a convective nature, and the maximum occurs along the coastal regions owing to the added effect of orographical lifting. Around the mouth of the Amazon the continental heating produces an *mTKu* air mass. According to the extent of the southeast trades over this region, at the approach of spring the rainfall gradually diminishes and the dry season commences. These climatic features are illustrated by Belém (20, Table 54).

Over the coastal regions of the Guianas, the rainfall regimes have two maxima and two minima, especially to the west [see Georgetown (19)]. During March and April, when the doldrums are far to the south, the North Atlantic air masses are relatively stable in these latitudes and there is a comparatively dry period. Likewise, another relatively dry period occurs in September and October, when the intertropical front is farthest north and the coast is under the influence of the southeast trades. The two rainy seasons, June-July and December-January, arise from the convective precipitation of the doldrums. This precipitation, which occurs during night or day, results from the convergence toward the weak low-pressure trough of the intertropical front. The temperatures are always high, with the warmest periods during the dry seasons. The winds oscillate from southeast to northeast in accordance with the dominant trade-wind system.

f. Upper Amazon. Here rainfall is very abundant during the entire year, with a maximum in the Southern Hemisphere summer. The rainfall is convective and results from the instability of the *mTKu* air mass, which is present in all seasons. The relative humidity and temperature are uniformly high throughout the year.

g. North of the Equator and West of 60° *West.* In the continental region there is a regime of tropical rains with a summer maximum and a winter minimum (Northern Hemisphere seasons) depending on the presence or absence of unstable maritime air. During winter (December-January) the maritime air of the Northern Hemisphere causes only light rains [see Calabozo (34, Table 54)]. The temperature is uniformly high, and the annual amplitude is small.

The precipitation over the northwest coast of the continent and the northern Andes depends upon the displacement of the trades and the equatorial calms. When the doldrums have their extreme southerly position, the rain extends to 5° south. South of the equator there are distinct dry and rainy periods, the latter coinciding with the southern location of the intertropical front (February). The duration of the rainy season diminishes southward to 6° south, where the desert commences, as illustrated by Guayaquil (49) and Trujillo (50). The rainfall and cloudiness are entirely of a convective nature. The doldrums have their extreme northern position in the Northern Hemisphere autumn, and convective precipitation dominates Panama. As in the case of the Guianas, two distinct rainy periods and two dry periods occur near the Gulf of Panama.

Over the Cordillera of the Andes the precipitation follows the same pattern as along the coast line and is illustrated by Bogotá (40) and Medellín (48). In every case, the rainfall is of convective origin, and thunderstorms are quite frequent.

The stations in Table 54 have been chosen to illustrate the various climatic regions. Unfortunately some of the means are based upon very few observations, but the table does illustrate the vast climatic variations over South America.

TABLE 54.—METEOROLOGICAL MEANS AT SELECTED STATIONS
(Temperature in degrees Fahrenheit, precipitation in inches, cloudiness in tenths)

Station	Lat.	Long.	Temperature				Precipitation				Cloudiness				Elevation, ft
			Jan.	Apr.	July	Oct.	Jan.	Apr.	July	Oct.	Jan.	Apr.	July	Oct.	
East Coast															
18. Caracas	11°N	67°W	64	68	68	68	0.9	1.6	4.1	3.9	6	7	8	7	3,420
19. Georgetown	7°N	58°W	79	81	80	82	7.3	5.0	9.0	3.0	6	6	6	6	10
20. Belém	1°S	48°W	78	78	79	80	7.6	17.8	2.3	0.5	6	7	4	3	30
21. Recife	8°S	35°W	82	82	77	81	2.0	8.6	10.3	1.0	5	5	6	4	100
22. Ilhéos	15°S	39°W	79	78	71	77	9.6	10.6	5.0	3.6	..	..	..	..	10
23. Rio de Janeiro	23°S	43°W	79	75	69	72	4.5	4.3	1.7	3.2	7	6	5	7	200
24. Rio Grande	32°S	52°W	73	68	55	62	3.5	3.0	5.2	3.3	5	6	7	6	10
25. Buenos Aires	35°S	58°W	74	61	49	60	3.1	4.8	2.1	3.3	4	4	5	5	80
26. Santa Cruz	50°S	69°W	59	48	35	49	0.6	0.6	0.4	0.3	6	6	5	6	40
27. Staten Island	54°S	64°W	49	43	37	41	5.4	6.2	4.8	4.2	7	8	8	7	40
East Inland															
28. Barra do Corda	6°S	45°W	78	77	76	80	4.7	3.9	0.3	1.7	7	7	3	6	270
29. Quixeramobim	5°S	39°W	83	80	80	83	3.7	7.0	1.3	0.1	6	6	4	3	680
30. Morro do Chapéo	12°S	41°W	69	68	62	68	3.9	3.5	1.9	2.0	..	..	..	..	3,540
31. Uberaba	20°S	48°W	73	71	65	73	9.9	5.9	0.5	4.1	8	6	3	6	2,490
32. Bandeirantes	23°S	44°W	72	68	60	68	11.3	6.2	1.0	4.6	7	6	4	7	1,750
33. Santa Maria	30°S	54°W	77	68	57	65	6.3	5.2	5.4	3.9	4	5	5	5	460
Central															
34. Calabozo	9°N	67°W	80	84	79	80	0.0	2.4	8.9	5.0	..	..	..	..	330
35. Iquitos	4°S	73°W	78	77	74	77	10.0	6.6	6.5	7.1	..	..	..	..	350

36. Manaus	3°S	60°W	80	80	81	83	9.2	8.5	2.2	4.1	6	7	6	6	150
37. Cuiabá	16°S	56°W	81	80	75	83	9.6	4.1	0.2	4.4	8	6	4	6	660
38. Corumbá	19°S	58°W	80	79	71	78	6.4	5.0	0.3	3.9	6	4	4	5	380
39. Córdoba	31°S	64°W	74	62	51	63	3.7	2.1	0.6	2.2	5	5	5	5	1,390
West Inland															
40. Bogotá	5°N	74°W	58	59	57	58	2.3	5.7	2.0	6.3	6	9	9	8	8,730
41. Quito	0°	79°W	55	54	54	55	4.2	7.4	0.9	3.7	6	7	4	6	9,350
42. Cajamarca	7°S	78°W	58	59	56	59	5.2	3.7	0.4	4.7	..	..	..	..	9,220
43. Puno	16°S	70°W	48	48	42	49	7.6	3.5	0.4	5.0	8	5	2	6	12,540
44. La Paz	16°S	68°W	50	48	44	50	3.8	1.5	0.2	1.3	6	5	2	6	12,000
45. Ollagüe	21°S	68°W	50	44	35	48	0.8	0.0	0.0	0.0	4	2	2	2	12,120
46. La Quiaca	22°S	66°W	54	51	37	51	3.2	0.3	0.0	0.2	..	..	..	..	11,360
47. San Juan	32°S	69°W	77	61	47	65	0.8	0.1	0.3	0.2	3	3	3	3	2,180
West Coast															
48. Medellín	6°N	76°W	71	71	71	69	2.7	6.6	4.1	6.9	..	..	..	..	4,950
49. Guayaquil	2°S	80°W	79	80	75	77	9.7	5.3	0.4	0.4	7	7	6	8	40
50. Trujillo	8°S	79°W	77	72	64	67	0.2	0.0	0.1	0.0	..	..	..	..	200
51. Mollendo	17°S	72°W	70	67	60	62	0.0	0.0	0.0	0.1	6	6	8	8	80
52. Iquique	20°S	70°W	70	65	60	63	0.0	0.0	0.0	0.0	5	4	8	7	30
53. Antofagasta	24°S	70°W	70	63	57	61	0.0	0.0	0.0	0.0	3	4	4	5	310
54. Coquimbo	30°S	71°W	64	59	54	57	0.0	0.0	1.1	0.0	5	6	6	6	90
55. Contulmo	38°S	73°W	62	55	48	54	1.3	6.6	11.6	2.6	3	5	6	5	120
56. Morro Lobos	42°S	73°W	56	50	44	48	4.6	11.8	11.3	4.1	5	6	8	6	230
57. Evanjelistas	52°S	75°W	47	44	40	42	11.5	11.6	9.2	8.7	9	9	8	9	180

CHAPTER XI

EUROPE (EXCLUDING RUSSIA)

Although western Russia is included in Europe politically, for climatological purposes it is desirable to discuss the climate of Russia together with the climate of Asia. Consequently, this chapter will be confined to those countries west of 25° east. Despite the meridional extent of Europe, 35° north to 70° north, there is no wide range in climatic types as observed in North America. Except for the mountainous region of Scandinavia, Europe is a region of mild temperate climate. The major portion of Norway and Sweden comprises the principal area of snow-forest climate.

The mildness of the European climate arises from the oceanic exposure to the north, west, and south. The proximity of warm water, even as far north as the Arctic Circle, prevents long periods of cold weather as observed in Russia and Asia. In the latter regions the extensive land masses are covered with snow during the winter months; and, as the result of the continental exposure, very cold air masses predominate. Consequently, across Europe there is a gradual transition from the western temperate climate to the continental forest climate in the east. The most abrupt change occurs north and south of the mountain ranges of southern Europe, where the Mediterranean type of climate (*Cs*) is separated from the humid temperate climate (*Cf*) to the north.

Because of the presence of the vast African continent south of 35° north there is no immediate source region of tropical maritime air. In view of the latitude of the continent and of the absence of very warm moist air masses, tropical rainy climates are entirely absent. Also, because of the frequent presence of maritime air masses over Europe, there are no regions of desert climate. There are, however, isolated areas that have a steppe climate.

58. Physical Features. In contrast to the American continents there is no extensive south to north range of mountains. Instead, the principal mountain systems are arranged in a general west-east direction. The main highland regions may be classified into four sections:

1. The Scandinavian mountains are orientated in a southwest to northeast direction. The highest peaks rarely exceed 7000 ft and are located nearer to the west coast than the Gulf of Bothnia. Consequently, the major portion of Sweden is below 3000 ft in elevation.

2. The Pyrenees chain separating France from Spain extends above 9000 ft and forms one of the main mountain systems of Europe. Farther south there are smaller mountain ranges extending above the plateau region of Spain.

3. The Alps are the highest European mountain chain and extend from southern France to southeastern Austria. This mountain system branches southward in the form of the Dinaric Alps and the mountainous regions of Albania, Bulgaria, and Greece. Farther east there is a semicircular mountainous region that includes the Transylvanian Alps and the Carpathian Mountains. These smaller mountain systems are considerably lower than the main ridge of the Alps, where there are many peaks with elevations greater than 10,000 ft.

4. Stretching the entire length of Italy, the Apennines are one of the major mountain chains of southern Europe.

Besides the above, numerous smaller mountainous regions are located in southern Europe. The outstanding feature of these ranges is their general east-west orientation.

In direct contrast to South America, Europe has many large indentations, for example, the Black, Aegean, Adriatic, North, and Baltic seas, the Gulfs of Bothnia and Finland, and the Bay of Biscay.

In addition to the coastal areas around the inland seas, the low-level regions consist of the northern sections of France, Belgium, Holland, Germany, and the countries along the Baltic Sea. The British Isles and Hungary are also regions where large areas of land are located below 1000 ft.

The influence of the topography on the general distribution of climatic types can be summarized briefly as follows:

1. Because of the absence of a north-south mountain barrier, the relatively warm Atlantic maritime air masses can invade the entire land area. Therefore, there are no abrupt changes in climatic zones from western to eastern Europe.

2. The several mountain chains, extending west-east across southern Europe, separate the warm southern region from the cooler lands to the north. Similarly, they tend to form a natural barrier to the northward progress of storms that frequent the Mediterranean, principally during the winter half of the year. Consequently, the main cyclone tracks of the Mediterranean region show a due eastward displacement of low-pressure centers.

3. The mountains of Scandinavia provide an effective barrier to the prevailing air circulation of high latitudes, with the result that the west coast is a region of heavy orographical precipitation. Likewise, the climate of Sweden and Finland is influenced by the Scandinavian mountains. The low yearly rainfall can be attributed to the sheltering effect of the high Norwegian plateau.

4. The numerous seas and gulfs that indent the continent provide additional moisture sources for many of the air masses that invade Europe.

5. The flatness of the country of eastern Europe and Russia has an important effect upon the winter weather of the entire continent. Because of the absence of a mountain barrier, cold continental air is free to spread westward and produce extended periods of cold winter weather.

59. Mean Distribution of Pressure and Wind. The mean-pressure maps for the months of January and July demonstrate the general features of the sea-level circulation. In some regions the pressure gradients may be fictitious and may not represent the true mean air circulation, for example, adjacent to the Scandinavian mountain chain and adjacent to the Alps. Moreover, over the high plateau regions the true air circulation is better illustrated by the mean upper-level pressure charts.

a. Winter. To consider first the mean winter circulation, Plate IV shows the following characteristics:

1. Except for the Mediterranean countries the main air flow over Europe is from the southwest. The trough of low pressure from Iceland to Novaya Zemlya is a northeastward extension of the deep semipermanent low centered south of Iceland. Similarly, the belt of high pressure across southeastern Europe represents an extension of the Siberian anticyclone. The predominance of high pressure across the Alps is consistent with the wintertime cooling. The high-pressure wedge over southern

Spain may be considered as an eastward extension of the Atlantic subtropical anticyclone.

In its broadest features the wind regime over western Europe resembles the air flow along the Pacific coast of North America—a pronounced southwesterly circulation. The main distinction between the two continents lies in the topographical differences. The western chain of mountains and the vast Alaskan land area shelter the American continent from the oceanic low-pressure circulation. Instead of a cyclonic circulation over Canada there is a thermal belt of high pressure. In contrast, the northeastward extension of a comparatively warm oceanic area and the absence of a western mountain barrier allow the cyclonic circulation to extend across northern Europe from about 45° north to beyond 70° north.

2. Over the warm Mediterranean Sea there is a belt of low pressure that extends from Spain to Arabia. On the northern side the predominant circulation is easterly, *i.e.*, across Greece and northern Italy. A weak west-wind regime predominates over northern Africa between the low-pressure region and the subtropical belt of high pressure. Over the western Mediterranean, northwest winds persist in the rear of the mean position of the cyclonic center.

Since large sections of the land masses surrounding the Mediterranean Sea have irregular terrain, local conditions have an important influence upon the surface winds. For example, the sloping land regions and river valleys contribute to the strength of the north winds in the rear of cyclones centered over the Gulf of Genoa. This unusually strong cold winter wind is called the *mistral*. A similar wind regime exists at the head of the Adriatic Sea and along the Dalmatian coast where it is called the *bora*. South of the Mediterranean basin a hot, dry wind, called the *sirocco*, frequently blows across the African coast. The sirocco occurs in advance of eastward-moving cyclones. The southward extension of the mistral over Africa is called the *tramontana*. The mistral, bora, sirocco, and tramontana blow in response to migratory pressure systems; therefore, they cannot be considered solely as local wind regimes. Nevertheless, the local terrain frequently contributes to their intensity and influences their direction. Consequently, they demonstrate some of the characteristics of local circulations.

Because of migratory pressure systems the winds in any particular region are subject to daily variations. The persistency of the winds at various European stations is given in Table 55. Here it should be remembered that local conditions contribute to the wind variability.

TABLE 55.—PERCENTAGE FREQUENCY OF SURFACE-WIND DIRECTION

Station	Lat.	Long.	January								July							
			N	NE	E	SE	S	SW	W	NW	N	NE	E	SE	S	SW	W	NW
West Coast																		
1. Skömvar	67°N	12°E	7	6	8	8	18	*25*	16	12	15	*28*	9	2	7	21	11	7
2. Malin Head	55°N	8°W	8	4	3	9	*27*	25	13	11	15	4	9	8	12	15	*19*	18
3. St. Mathieu	48°N	5°W	11	14	15	10	8	*16*	14	12	19	10	7	1	6	15	17	*25*
4. Porto	41°N	8°W	7	6	13	*19*	16	16	10	13	16	6	6	5	3	16	*29*	19
Mediterranean Coast																		
5. Murcia	38°N	1°W	3	13	3	6	1	*29*	16	*29*	0	13	*42*	28	6	3	1	7
6. Barcelona	41°N	2°E	17	10	3	0	2	10	24	*34*	2	14	10	4	*23*	*23*	17	7
7. Marseille	43°N	5°E	2	20	18	9	2	4	9	*36*	1	10	4	7	6	17	24	*31*
8. Livorno	44°N	10°E	5	*54*	18	3	5	5	8	2	6	12	4	1	9	17	*36*	15
9. Palermo	38°N	13°E	6	7	4	3	23	*36*	12	9	6	*37*	21	3	4	6	7	16
10. Venedig	45°N	12°E	25	*35*	4	5	6	11	3	11	8	17	12	*42*	11	7	1	2
11. Athens	38°N	24°E	17	*26*	7	6	18	9	6	11	22	*32*	5	2	12	15	6	6
Inland																		
12. Helsingfors	60°N	25°E	10	11	8	7	11	*25*	15	13	8	7	12	8	10	*33*	10	12
13. Göteborg	58°N	12°E	3	8	*19*	17	12	*19*	16	6	1	5	10	7	5	20	*42*	10
14. Warsaw	52°N	21°E	6	5	6	13	19	16	*20*	15	15	6	4	7	9	13	*24*	22
15. Leipzig	51°N	13°E	5	7	9	9	22	*24*	16	8	9	4	6	6	13	18	*24*	20
16. Paris	49°N	2°E	15	15	7	9	*23*	19	6	6	17	12	4	5	15	*24*	11	12
17. Madrid	40°N	4°W	9	*25*	11	10	9	20	8	8	7	16	7	9	7	*24*	15	15
18. Florence	44°N	11°E	17	*18*	10	13	16	11	7	8	8	10	7	6	8	*28*	25	8
19. Sarajevo	44°N	18°E	6	6	*32*	6	7	7	19	17	2	5	*30*	11	5	5	26	16
20. Sulina	45°N	30°E	*23*	12	3	5	8	19	7	*23*	*16*	14	6	13	*16*	14	5	*16*

Besides verifying the general discussion regarding the mean circulation, Table 55 demonstrates that the wind is everywhere subject to large variations in direction. These changes can be attributed chiefly to the wind shifts that accompany the passage of low- and high-pressure systems. Local conditions also definitely add to the wind variability.

b. Summer. The principal variation from the winter to the summer circulation is the intensification and northward displacement of the Atlantic anticyclone. In summer the Icelandic Low is less intense, but it still persists with a pressure trough extending northeastward. The cyclonic circulation is considerably less intense. Over Norway and Sweden, for example, the mean pressure gradients are very weak, and there is even a second minor cyclonic center east of the mountains. This change from winter to summer is illustrated by the prevailing northeast winds at Skömvar (1) during July.

While the cyclonic circulation is diminished, the Atlantic anticyclone is very well developed in summer and extends its influence far inland. The anticyclonic circulation over Europe is also intensified by the circulation around the pronounced thermal cyclone of Asia. Plate V further demonstrates the difference between the circulation over Europe and over the western portions of North America. In the latter case the thermal heating occurs near the coast, and consequently the anticyclonic circulation does not extend across the country. The net change in the air flow over northern Europe, from winter to summer, is a tendency toward west or northwest winds rather than the winter southwest winds on the poleward side of the high-pressure axis (see Table 55).

The thermal low-pressure belt of Asia is prolonged across northern Africa. The North African winter anticyclone is then replaced by a low-pressure trough. In turn, the circulation over the Mediterranean is no longer dominated by a Mediterranean cyclone. This winter cyclonic circulation is replaced by the summer circulation around the extensive anticyclone. During the spring and summer months local conditions influence the surface winds at coastal stations as shown, for example, by the high frequency of sea breezes. Furthermore, with the general diminution of pressure gradients the summer winds are likely to be lighter and more variable than in winter. These factors should be borne in mind in analyzing the wind frequencies in Table 55. In summer, the influence of migratory cyclones and anticyclones upon the wind regime is less pronounced because of the definite tendency toward diminished cyclonic activity.

c. Upper Air. The mean January upper-air circulation at 4 km is given in Fig. 12. North of 45° north the mean wind directions veer toward the west with elevation; at the same time, there is a gradual increase in wind force. The cold surface anticyclone located over the mountainous region of southern Europe is quite shallow, and at 2 km there is a regime of weak westerly winds, which increase in force with elevation. For example, over the northeastern Mediterranean area the winds near the earth's surface have a decided easterly component [see Livorno (8), Venedig (10), and Athens (11), Table 55]. These easterly winds decrease in speed with elevation, and finally at 2 km the mean air flow is from the west. Occasionally, during the colder portion of the year, there are pronounced variations from the mean pattern; for example, on the north side of deep cyclones easterly winds may exist beyond 4 km. However, since the deepest cyclones are located north of Europe, the main variations from the mean pattern are usually changes in wind force, with relatively small changes in wind direction.

Consistent with the intense heating in summer, the surface thermal low pressure over North Africa is very shallow, and at 2 km this becomes a high-pressure region. The mean summer circulation is controlled by the pressure gradient between the African anticyclone and the polar cyclone. North of 50° north the change in wind velocity with altitude is essentially a gradual increase in force of the prevailing westerlies. From Fig. 13 it will be observed that the weak surface northeast winds over northern Norway give way to westerly winds aloft. South of 50° north to the coast of Africa, the circulation around the surface anticyclone gradually changes with elevation so that at 2 km there is a regime of weak westerly winds. They increase in speed with altitude as illustrated by the pressure gradients on the 4-km chart. Since cyclonic activity is less pronounced, the daily variations from the mean upper-air pattern are smaller in summer than in winter.

60. Ocean Currents. Since Europe is really a peninsula of Asia, narrow toward the west and projecting into the North Atlantic, the circulation of warm and cold water must have an important influence upon the climate of the adjacent land area. The essential features of the water circulation around Europe are given in Plate VI.

The Gulf Stream system dominates the circulation of the North Atlantic Ocean, with the result that warm water is transported northeastward from the southwestern Atlantic toward northwestern Europe. The North Atlantic Drift, the eastward extension of the Gulf Stream, is a broad warm current that divides into numerous branches before reaching the European coast line. The relatively high water temperatures observed near the coast of Scandinavia arise from the warm Norwegian Current, which extends from Scotland to Novaya Zemlya. To the south of this current, in the North Sea, a counterclockwise circulation dominates the flow. Here warm water flows southward along the east coast of the British Isles and is reinforced by the flow through the English Channel. An inflow of cooler surface water, from the Baltic, occurs through the Skagerrak, the straits between the Scandinavian peninsula and Denmark. Along the coast of France and Spain, a southward-moving branch of the North Atlantic Drift produces higher sea-surface temperatures than in the corresponding latitudes near the Atlantic coast of North America. South of 40° north the extension of this current results in relatively cool water over the eastern Atlantic, and it finally joins the cool Canaries Current off the northwest coast of Africa.

An exchange of water between the Atlantic and the Mediterranean Sea takes place through the Straits of Gibraltar. Within the Mediterranean Sea there are two main water circulations separated by the land masses of Italy and Sicily. Both currents rotate in a counterclockwise direction. In the east, surface water flows westward through the Bosporus and the Dardanelles. It is impossible to generalize about the currents of the Black, Adriatic, and Baltic seas because the water circulation is governed by the local wind conditions.

The influence of the various water circulations upon the sea-surface temperatures along the European coast line is demonstrated in the February and August temperature charts (Plates VII–VIII). The principal feature is the unusually warm water present off the northwest coast throughout the entire year. Since north of 45° north the prevailing air circulation is onshore, this warm water has a profound effect upon the climate of Europe. Throughout the major portion of the year maritime air masses dominate the weather of Europe, with the result that mild winter temperatures predominate, together with moderately high humidities. Similarly, because of the maritime influence, the mean summer temperatures are not particularly high and are associated with relatively high humidities.

61. Air Masses. Plates XVI and XVII show the source regions of the air masses that govern the weather and climate of Europe. The outstanding feature is the absence of a real air-mass source. Consequently, the major portion of Europe may be broadly classified as a zone of transition where the various air masses are subjected to transformations. The principal air-mass properties can be summarized briefly as follows:

a. Polar Maritime Air (*mP*). During the winter season the air that frequently invades Europe, north of 45° north, has come from the eastern Atlantic. As in the case of *mP* invasions along the west coast of North America, the properties of this air mass depend upon the water trajectory. Air that comes from the Greenland-Spitsbergen region reaches northwestern Europe as a seasonably cool unstable air mass because of its initial low temperature, as arctic air, and because of its comparatively short trajectory over the open water. On the other hand, *cP* air that leaves North America and travels southeastward over the western Atlantic and then northeastward on to Europe has had ample time to adapt its properties to the state of the underlying surface, *i.e.*, the ocean. Consequently, this air arrives as a relatively warm humid air mass that is in a neutral or slightly stable state of convective equilibrium. Hence, the trajectory of the air mass determines the stability of the air that reaches Europe. However, all *mP* invasions are characterized by their high temperatures, high relative humidities, and high cloudiness frequency.

While *mP* air crosses the land regions from west to east it is modified by the wintertime cooling from below, *i.e.*, the air mass becomes more stable. On occasion, in the rear of deep cyclones, the air may travel from north to south and therefore become unstable as a result of the surface heating. As shown by the mean air circulation, the latter trajectory is less frequent than a due easterly or northeasterly air motion. When *mP* air stagnates over the land, under the influence of an anticyclonic circulation, continual radiational cooling produces a decidedly stable moist air mass with widespread low cloudiness and fog.

During the summer season the *mP* invasions produce mild weather conditions over northwestern Europe, north of 45° north. However, with the diminished wind velocities, the air mass moves slowly across the land, and consequently the pronounced daytime heating causes a rapid modification of the source properties. While the air remains over land, the heating produces an increasingly unstable air mass. Cumuliform-type clouds are quite common. On those occasions when the air motion is offshore, the heated continental air is cooled from below and produces a stable air mass with widespread low cloudiness and fog over the neighboring ocean. This weather regime is common over the North Sea in early summer when the wind circulation may be favorable and when the sea-surface temperature is still low.

b. Tropical Maritime Air (*mT*). The winter source region of *mT* air is restricted to the tropical section of the Atlantic. Because of the long trajectories, the *mT* air masses have been subjected to considerable modification before reaching Europe. Consequently, they arrive over Europe with air temperatures appreciably lower than those observed in the tropics. Because of the land trajectory, modified *mT* air reaches northeastern Europe with temperatures about 40°F. For example, typical *mT* air at Berlin has a surface temperature of 43°F.[1] In every case, very high surface relative humidities are quite common. As a result of the continual cooling from below, *mT* air masses are characteristically stable in the lower layers. In general, since *mT* air originates in the eastern Atlantic, it is not convectively unstable like the *mT* air of eastern America.

Modified tropical maritime air invasions are observed during the summer season when the Atlantic subtropical anticyclone extends across Europe. However, as in winter, because of the circulation around the Atlantic cell, *mT* air masses are cooler and more stable over Europe than over eastern North America. Nevertheless, as a result of the continental heating, steep daytime lapse rates are common over the interior of Europe. Consequently, higher temperatures and greater instability are observed inland than along the west coast. With the pronounced stability in the eastern portion of the high-pressure cell, very stable *mTs* air is found south of 40° north along the coast of Europe.

c. Polar Continental Air (*cP*). A vast winter source region of continental polar air is located over the snow-covered regions of eastern Europe and Asia from about 45° north to the North Pole. Like North American *cP* air, these air masses are characterized by unusually low surface temperatures, low humidities, and stable lapse rates in their source region. These general properties are observed in *cP* air masses over eastern Europe,

[1] PETTERSSEN, S., "Weather Analysis and Forecasting," p. 184, McGraw-Hill Book Company, Inc., New York, 1940.

where the circulation is controlled by the westward extension of the Siberian high-pressure area. However, since the prevailing flow is from south to north and since the lower layers are stirred through turbulent mixing, the air is warmer than that observed in central Asia. When the *cP* air masses invade western Europe from Asia, they absorb heat and moisture from the warmer underlying surface; nevertheless, low temperatures still prevail. The absorption of moisture and turbulent mixing produce stratocumulus clouds.

Frequently, outbreaks of cold *cP* air occur in the rear of deep cyclones centered over northern Europe. The arctic air mass is rapidly modified by crossing the open water, and therefore the air arrives over Europe with steep lapse rates and can be classified as fresh *mP* air. For example, modified *cP* air from the Asiatic continent has a surface temperature of about 20°F[1] over the Rhineland, while modified *cP* air from the northwest has a temperature of about 35°F. On the other hand, at 4 km, both modified air masses have about the same temperature. This shows that the short trajectory of arctic air across the open ocean has resulted in a modification of the air-mass properties in only the lower layers of the atmosphere.

During the summer season the heating over the land results in the presence of warm air masses in middle latitudes; therefore, cool *cP* air masses are confined essentially to high latitudes. Over northern Europe, *cP* air is characterized by moderately low temperatures and low specific humidities. However, the summer *cP* air masses vary considerably with place and time. Thus, the air masses demonstrate a wide range of physical properties from cool, slightly stable air in the north to hot, dry, unstable tropical continental air in the south. Unlike the North American summer air masses, there is no distinct line of separation between *cP* and *cT* air masses over Europe and Asia.

d. Tropical Continental Air (*cT*). Tropical continental air masses influence only the Mediterranean region during the winter season and will be discussed in a later paragraph. In summer, the continental region south of 45° north is a source region for *cT* air masses. Because of the numerous seas in southern Europe, the humidity conditions within *cT* air vary considerably from one region to another. However, even though there is a

[1] *Ibid.*, p. 173.

wide range in specific humidity, all *cT* air masses are very warm and convectively unstable. Cumulus clouds and thunderstorms occur in the moist type of *cT* air.

e. Air Masses in the Mediterranean Region. In winter, the Mediterranean is a region that is invaded by air from adjacent sources. With the passage of cyclones across the Mediterranean from west to east, cool air masses from the north are transported southward toward Africa while warm African air is transported northward across the Mediterranean. Over the northwestern Mediterranean area, modified *mP* air reaches the sea as a cold, dry air mass, for example, the mistral. On crossing the warm water, the absorption of heat and moisture from below results in a cool and comparatively moist air mass over the southern Mediterranean Sea and the North African coast line. In a similar manner, cold modified *cP* air enters the eastern Mediterranean as a cold, dry air mass in the rear of a cyclone, for example, in connection with the bora. As in the case of the mistral, the latter air is heated from below and becomes more unstable on crossing the Mediterranean. In advance of a migratory cyclone, warm, dry air from Africa (the sirocco) invades the water region with a temperature higher than the sea-surface temperature. The air is therefore cooled and gains moisture from below. The modified air mass reaches Italy and the Aegean as a warm, moist air current. These three examples of pronounced air-mass modifications suffice to demonstrate that the Mediterranean is a transition zone. Naturally, on many occasions, the air-mass changes are less distinct than those already discussed.

In summer, when the temperature contrast between land and sea is comparatively small, the air-mass modifications are less pronounced than in winter. Furthermore, because of the rare occurrence of cyclones, there is no rapid transport of air from north to south, and vice versa. The summer circulation across the Mediterranean area is mainly from north to south. The warm air from central Europe crosses the northern Mediterranean as a dry air mass whose moisture content is increased after crossing the open water surface. Because the sea-surface temperatures are high, the air mass is cooled only slightly so that there is no tendency toward widespread low cloudiness and fog. Instead, the Mediterranean is a region of clear skies during the warm portion of the year.

62. Fronts, Cyclones, and Anticyclones. *a. Frontal Zones.* The main frontal zones in the vicinity of the European continent are indicated in Plates XVI and XVII for winter and summer, respectively. In view of the fact that cyclonic developments on the Atlantic polar, arctic, and Mediterranean fronts have a considerable influence on the weather of Europe, the essential features of each frontal system will be discussed in some detail.

During the colder half of the year the Atlantic polar front separates modified *cP* air from warm humid *mT* air. This front is most intense in the western Atlantic and usually dissolves as it moves eastward across the ocean. Consequently, no marked temperature changes occur over Europe, as the result of the passage of the Atlantic polar front, like those observed along the eastern section of the United States. Occasionally, with the splitting of the subtropical anticyclone, a second frontal zone may be located farther east. However, this front is generally weak. The Atlantic arctic front is located in the extensive mean-pressure trough from Iceland to Novaya Zemlya, where cold arctic air masses are brought into juxtaposition with warmer maritime air masses from the south. When the Icelandic Low is situated farther east than usual, frontogenesis may be particularly active since the strong flow of warm moist air from the southwest produces a zone of marked temperature contrast. On occasion, when cold European continental air spreads westward across the ocean, the Atlantic arctic front dissolves. Located in the mean-pressure trough across the Mediterranean Sea, there is a frontogenetical zone between the cold air of the European continent and the warmer air from the southern Mediterranean region. This mean frontal zone is called the "Mediterranean front." It is less persistent than either the Atlantic polar or the arctic front. Besides these three principal frontal zones, many regions of secondary fronts could be cited. The most important are those which form in western Europe when the wind field is favorable for intensifying the contrast between polar maritime and polar continental air masses.

During the summer, all frontal zones become somewhat indistinct because of the weaker temperature gradients and less favorable wind fields. The Mediterranean frontal zone entirely disappears, and it is only in high latitudes that conditions are favorable for frontogenesis. Consequently, the mean position of the Atlantic polar front is located farther north than during winter. Here it separates *mT* air masses from modified polar air masses. Throughout the year the mean position of the arctic front oscillates only slightly so that in summer it extends from Iceland to northern Scandinavia.

b. Cyclones and Anticyclones. The cyclones that develop along the main frontal zones and that have a profound influence upon the climate of the continent will now be discussed with regard to their seasonal frequency. The mean-pressure chart for January demonstrates that the region southwest of Iceland is a semipermanent low-pressure area. Cyclones that travel across North America or develop on the Atlantic front most frequently move into this region and account for the low mean pressure, below 1000 mb. These cyclones are generally deep occluded systems when they reach the longitude of Iceland. In most cases the lows move south of Iceland and follow a northeast trajectory toward Norway. During the summer season, in general, the values of barometric minima are higher than in the colder portion of the year. As a result, the mean July pressure in the Icelandic region is considerably higher than during January. Throughout the year, a great number of cyclones follow paths to the north of Iceland or across the mid-Atlantic as well as immediately south of Iceland. The frequencies of storm tracks across the North Atlantic and Europe have been discussed by numerous investigators. However, the researches have never been confined to the same periods, and consequently the percentage frequencies, given in the following, must be considered to be only approximate.

1. Cyclone Frequency in Eastern Atlantic. The latitudinal variation of cyclones that approach the west coast of Europe are given in Table 56 for winter (December–February), spring (March–May), summer (June–August), and autumn (September–November). The frequencies, after Schedler,[1] have been expressed as a percentage of all cyclones between 30° north and 70° north observed at longitude 15° west, with central pressures less than about 1013 mb. These values substantiate the previous statement that the majority of the cyclones are found about 60° north during the winter half of the year. However,

[1] Schedler, A., *Annalen der Hydrographie und maritimen Meteorologie*, January, 1924.

it is apparent that there are a large number of cyclones in lower latitudes which may subsequently move on to the coast of Europe from Spain to Norway. A short study of the direction of movement of cyclones across longitude 15° west indicates that north of 55° north lows are generally moving toward the northeast of east-northeast, whereas south of 55° north there is a general tendency toward a more easterly movement. An appreciable percentage of the cyclones between 40° north and 55° north appear to move from the west-northwest to the east-southeast following the paths indicated by 5*A* and 6*A* in Fig. 31.

TABLE 56.—APPROXIMATE CYCLONE FREQUENCY AT 15°W

Latitude	Winter	Spring	Summer	Autumn	Year
30°–35°N	1.7	2.0	0.4	2.0	6.1
35°–40°N	3.0	3.9	1.1	3.8	11.8
40°–45°N	3.3	4.2	2.3	3.1	12.9
45°–50°N	2.1	3.0	3.7	2.7	11.5
50°–55°N	2.3	3.7	4.5	2.4	12.9
55°–60°N	3.0	4.5	6.0	4.1	17.6
60°–65°N	3.2	5.1	5.0	5.4	18.7
65°–70°N	1.9	2.1	2.2	2.3	8.5
30°–70°N	20.5	28.5	25.2	25.8	100.0

2. CYCLONE TRACKS OVER EUROPE. Some of the common cyclone tracks across the continent have been indicated in Fig. 31. In this connection it should be remembered that a given cyclone may follow a very different trajectory from that suggested by the diagram; consequently, these storm tracks show only general trends. Because of the high frequency of cyclones immediately south of Iceland, the most frequently observed path is given by track 1, that is, 1, 1*A*, 1*B*, and 1*C*. Fewer cyclones follow paths suggested by 2, 4, and 5*B*, while tracks 3, 5*A*, 6, and 7 are still more infrequent. However, if only the winter half of the year is considered, the latter group of tracks may be as common as 2, 4, or 5*B*.

It is usually noted that cyclones have a greater tendency to follow the tracks shown in Fig. 31 in some months than in others. Cyclones are observed to follow track 1 in all months of the year, with the maximum occurrence in autumn and early winter. Likewise, they follow the general path suggested by track 2 in all seasons, but more lows are usually observed in winter than in any other season. Track 3 appears to show a definite winter and early-spring maximum of cyclone occurrence. On the other hand, track 4, according to van Bebber,[1] shows a summer maximum. It should be borne in mind, however, that even though lows may be more frequent during the late summer months the deepest cyclones occur in the colder half of the year.

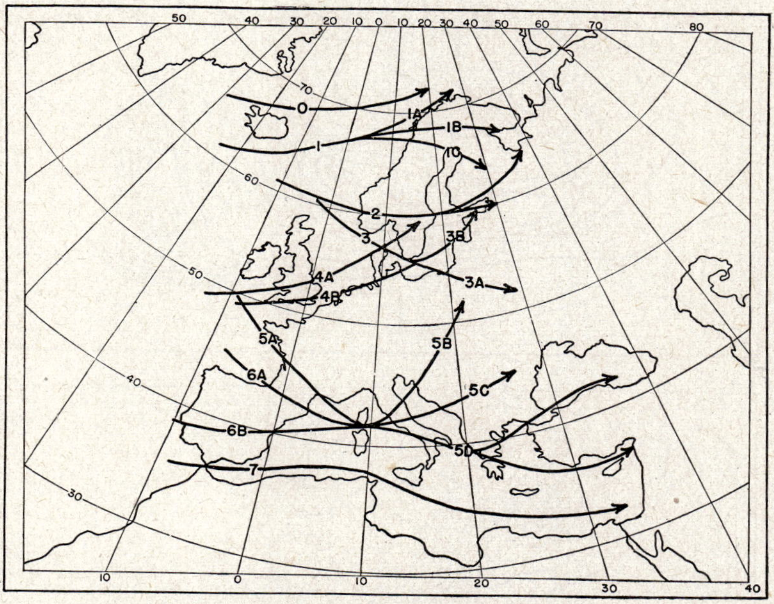

FIG. 31.—Cyclone tracks across Europe.

The cyclones that traverse the Gulf of Genoa and the Mediterranean Sea show a decided summer minimum; for example, tracks 5*A*, 6*A*, 6*B*, and 7 are most frequent in winter, 5*B* and 5*C* in spring and autumn, and 5*D* in all seasons except summer. In general, the Mediterranean lows are located farthest south during midwinter.

The cyclones that cross Europe have not necessarily traversed the Atlantic Ocean. In some cases, the lows may have developed over Europe or the adjacent seas. In others, secondary developments may take place while the old primary cyclone fills. Like-

[1] VAN BEBBER, W., *Archiv deut. Seewarte*, Vol. 5, 1882; Vol. 9, 1886.

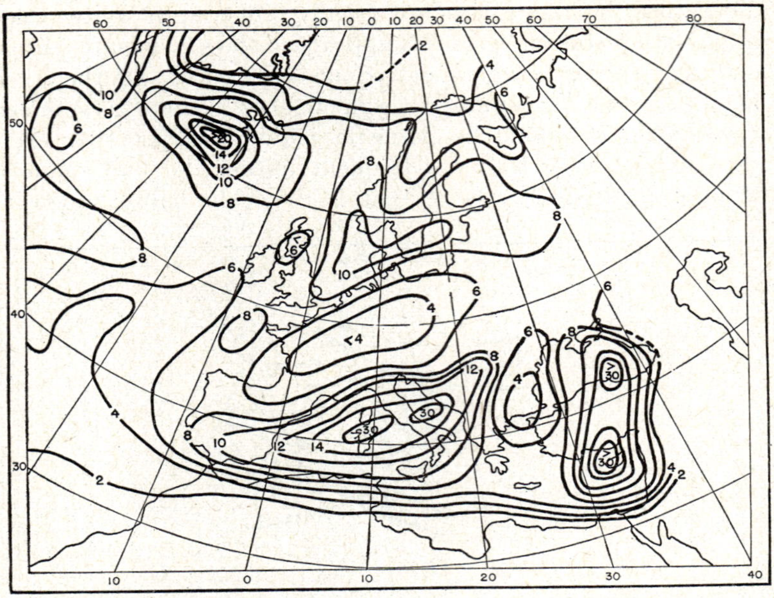

a

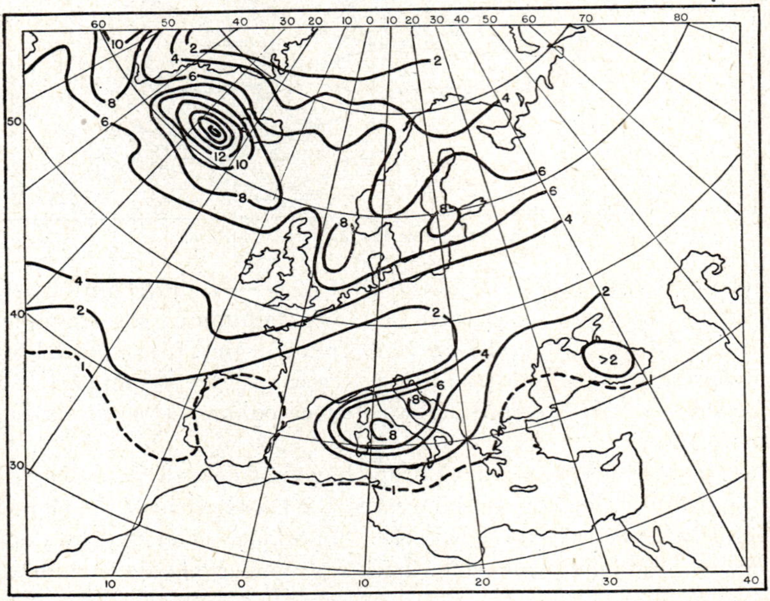

b

FIG. 32.—Yearly frequency of cyclones. *a*, Central pressures less than 1013 mb; *b*, central pressures less than 1000 mb.

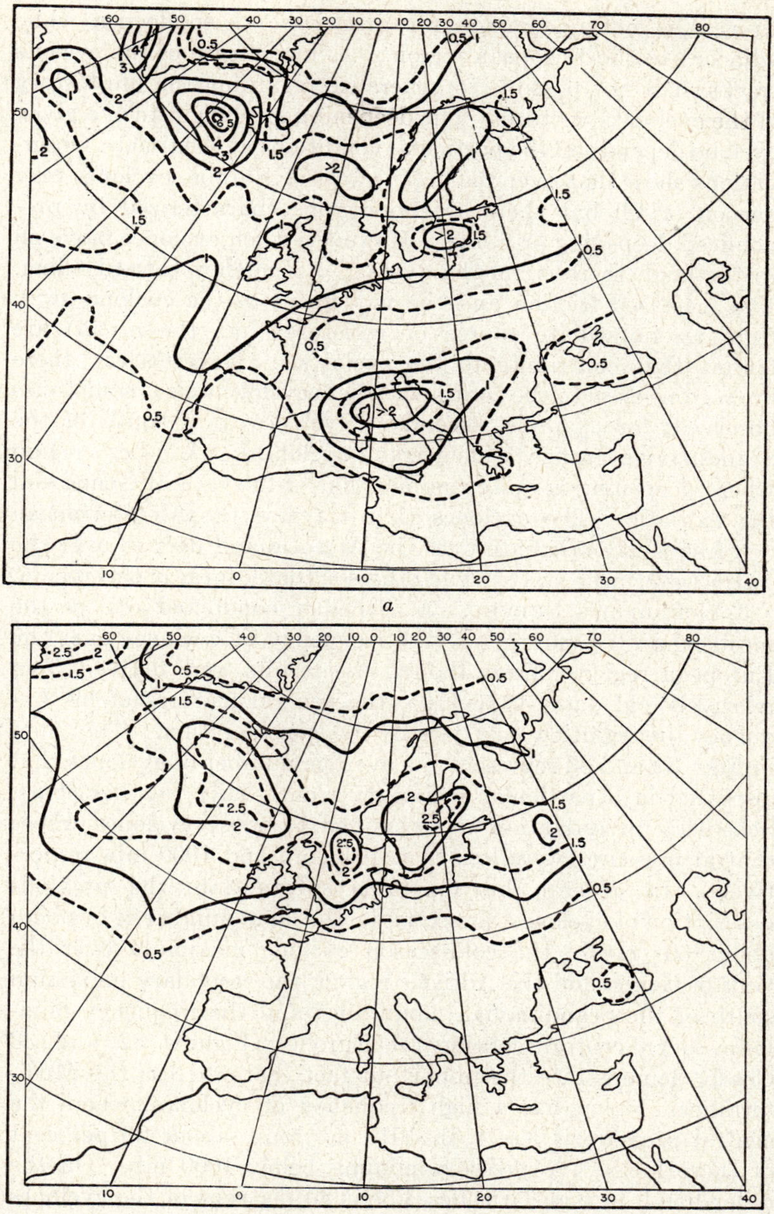

a

b

FIG. 33.—Frequency of cyclones with central pressures less than 1000 mb. *a*, Winter; *b*, summer.

wise, it is not uncommon for primary lows to deepen while adjacent to the European region. As in the case of the cyclone tracks, it is not possible to generalize regarding the life history of the cyclone structure. The deepening or filling rate of a given cyclone depends on the particular conditions at the time. Observations show that over the ocean, west of Europe, cyclones may deepen or fill but there are no regions characterized by pronounced deepening or filling. However, when cyclones move on to the land, there is, on the average, a tendency toward filling. The increased friction and the movement of the cyclone away from an immediate source of moisture (*i.e.*, the ocean) are factors that contribute to the cyclolysis. Nevertheless, there are regions close to the continent where cyclogenesis is sometimes observed; for example, secondary cyclones form south of the Scandinavian range (Skagerrak cyclones). Another region where deepening is quite common is over the Gulf of Genoa and the Adriatic. The cyclones that traverse the Mediterranean Sea frequently originate over the Atlantic and deepen over the warm water near Italy, while others actually form in this region.

3. Schedler's Survey. A. Schedler conducted one of the most exhaustive surveys of the frequency of cyclones over the European region. Even though the results were derived from relatively old data, 1881–1905, the general features of his frequency distribution are worthy of consideration. The lines in Figs. 32 and 33 represent the average number of cyclones that appeared in a particular region on a once-daily weather chart. Schedler's investigation was confined to those cyclones whose central pressures were less than 1013 mb and 1000 mb, respectively. In general, his diagrams substantiate the previous description of cyclones, for example, the large number of cyclones about 60° north, the cold-season cyclone maximum over the Gulf of Genoa and the Adriatic region, the secondary maximum south of the Scandinavian mountains, and the frequency minimum over central continental Europe. Figures 32 and 33 clearly demonstrate the important fact that whereas the Mediterranean region has a high frequency of cyclones, where the central pressure is less than 1013 mb, only about 25 per cent of these lows have pressure minima below 1000 mb. On the other hand, in high latitudes, about 80 per cent of the cyclones have central pressures below 1000 mb.

4. Anticyclones. Adjacent to western Europe warm high-pressure wedges are located ahead of the low-pressure centers, while cold wedges follow the cyclones. For reasons of continuity, the high- and low-pressure systems must have trajectories that are mutually consistent.

5. Weather Associated with Cyclones and Anticyclones. The weather that accompanies the systems of winter cyclones and anticyclones has an important influence on the climate. The deep autumn and winter cyclones that traverse the North Atlantic toward the Arctic Ocean do not produce heavy precipitation over the continent. However, because of the general southwesterly winds, the northwest section of Europe comes under the influence of comparatively warm, moist air. Low clouds and the drizzle type of precipitation are frequently associated with this moist air flow. Eastern Europe, which is far away from these cyclonic storms, may not come under the influence of the southwest flow. Instead, this region may be under the dominance of modified *cP* air from Russia. Those cyclones which traverse the British Isles and Scandinavia are accompanied by widespread cloudiness and precipitation. In advance of a cyclone low clouds may predominate within the relatively warm, moist maritime air. The passage of these low-pressure systems is not attended by vast temperature changes, for the cold air in the rear has crossed the warm North Atlantic. However, since this modified cold air is probably of arctic origin, the heating over the ocean produces a moderately unstable, moist air mass. Over northwestern Europe the northwest winds produce orographical precipitation within this *mPKu* air mass. The magnitude of the precipitation that accompanies these migratory cyclones depends on the moisture content of the warm air masses, the intensity of the frontal lifting, and the degree of convergence. Likewise, the instability of the air masses in the rear depends on the trajectory of the air; for example, fresh arctic outbreaks are characterized by greater instability than air that has had a long trajectory over the ocean.

Because the cyclones that cross continental Europe are often subject to cyclolysis, the usual weather conditions that accompany cyclonic activity are observed to a diminished extent over the continent. However, widespread frontal cloudiness can be expected in connection with the wintertime cyclones. The lows that traverse the Mediterranean have an important influence upon the cold-season weather. During midwinter, frontal precipitation is common over the southern Mediterranean, while in autumn and spring frontal weather is more common over the northern region. Because of the high temperature and relatively high moisture content of the air over the sea, moderate precipitation can be expected to accompany the deep cold-season cyclones. In advance of deep cyclones, warm, dry south winds cross the African coast—locally called the *sirocco.* These desert winds may traverse the sea as far as the coast of Europe. While traversing the open water, they gain moisture from below, with the result that they reach Europe as modified air masses. In contrast to this warm air flow in advance of the Mediterranean cyclones, the cold outbreaks in the rear may produce very low temperatures. These cold air invasions originate over the plateau region of central Europe, so that they arrive over the northern Mediterranean as abnormally cold air masses. As mentioned previously, west of the Gulf of Genoa the cold air flow is called the *mistral,* whereas at the head of the Adriatic the north wind is known as the *bora.* As the cyclones move eastward clear skies are produced by the downslope heating of the cold northerly winds. However, because of the gain of heat and moisture from below, these air masses are rapidly modified and arrive over the southern Mediterranean as cool, unstable air masses.

While precipitation usually accompanies cold-season cyclones, relatively fair weather is associated with the winter anticyclonic circulation. Over eastern Europe deep, cold anticyclones are regions of extremely low temperature and generally clear skies. As these anticyclones move westward, the increase in moisture may often result in widespread low cloudiness within the stirred *cP* air. These westward invasions are accompanied by low winter temperatures over western Europe. When Atlantic anticyclones extend across Europe, moderately fair weather may be expected. However, since the air is of maritime origin, turbulent mixing frequently causes low clouds. Near anticyclonic centers radiational cooling may result in ground fog. Over the plateau region of southern Europe, anticyclonic control usually predominates.

During the summer months cyclonic activity is less frequent and less pronounced. Summer cyclones that travel across northwestern and northern Europe produce precipitation and cloudiness similar to the winter migratory low-pressure systems. Because of the northward displacement and eastward extension of the Atlantic anticyclone, the principal cyclonic activity is confined to the northern portion of Europe. As in the case of North America, summertime precipitation over Europe is mainly the result of showers within the convectively unstable air masses rather than of the movement of deep cyclones. Weak fronts, however, often contribute to the precipitation by causing the release of the convective instability. The influence of cyclonic activity on the distribution of precipitation in the Mediterranean is illustrated by the rainfall regime over southern Italy. During December, cyclonic activity results in 4 to 5 in. of rain. During July, when cyclones are rare, the average rainfall is less than ½ in.

63. Climatic Regions. The principal climatological features of Europe are illustrated by Köppen's classification in the Frontispiece. The mean distribution of the meteorological elements is given in Plates I–V and IX–XVII. Europe, north of the Mediterranean region, can be divided into three main climatic zones, *viz.*, northwest Europe, with mild winters and moderate to heavy precipitation; central Europe, with cold winters and summer rainfall; and eastern Europe, with very cold, dry winters. The climatology of these areas will now be discussed in further detail.

a. Northwest Europe. This area comprises those sections of Europe which are exposed to the prevailing southwesterly winds of the mean Icelandic low-pressure circulation. As shown by Fig. 31, it is an area that comes under the direct influence of the migratory cyclones that traverse the eastern Atlantic. Because of the maritime influence, northwest Europe has a humid temperate climate (*Cf*).

The precipitation, which is the result of frontal and orographical lifting, can be directly attributed to the cyclonic activity. Since cyclones occur in all seasons of the year, it is to be expected that there is no dry season. Likewise, the presence of the deeper cyclones during the cold portion of the year would suggest a winter maximum of precipitation. The stations in Table 58, Tromsö (21) to Santander (33), definitely indicate an autumn and early-winter maximum. This can be attributed to the fact that during the autumn season the air masses, which come from the warm oceanic regions, have a considerably higher moisture content than during the spring, when the neighboring oceanic areas are relatively cold. Even though the number of days with precipitation is about the same in January as in October, the higher moisture content is favorable for heavier precipitation during October. As a result of the higher cyclone frequency the British Isles and western Norway show higher precipitation amounts than western France. The exposed coastal regions have more precipitation than the interior, since the former are directly exposed to the prevailing onshore winds and also since they come within the precipitation band of some cyclones whose centers do not traverse the land areas [for example, compare Greenwich (56) and Valencia (28) or Paris (64) and Brest (31)]. The excellent exposure and orographical effects contribute to the heavy precipitation areas north of Bergen, in the Highlands of Scotland, in the Welsh mountains, and in southwest Ireland.

With diminished cyclonic activity and a lower moisture content of the maritime air masses, spring, on the average, is the driest season. The summer minimum over southwestern France and northern Spain can be attributed to the stable air masses, which are under the influence of the eastward extension of the Atlantic anticyclone. The inland regions of northwest Europe show a secondary summer precipitation maximum, for example, Greenwich (56) and Lille (63). This is the result of the convective showers that always predominate over continental regions, particularly during the months of July and August. Stations like Paris (64) border on the climatic zone of central Europe.

Northwest Europe is one of the cloudiest regions of the earth, as clearly illustrated by the stations in Table 58. During the winter half year the high degree of cloudiness arises not only from frontal and orographical lifting but also from turbulent mixing within the moist air near the surface. The latter gives frequent stratocumulus and low stratus types of clouds. In the summer half year, convective clouds contribute to the high cloud frequency. As a result of the predominance of moist air masses, fog is not uncommon. Radiational cooling, during periods of light winds or calms, results in a high wintertime fog frequency over land; for example, Paris averages 8 days with fog during January. Frontal fog contributes to the autumn fog frequency over the western British Isles and Norway. The North Sea shows a spring and early-summer fog maximum. This arises from the cooling of warm continental air over cold water.

Since the air masses are generally convectively stable or indifferent, thunderstorms are relatively rare along the coastal regions. Summertime heating results in more frequent thunderstorms over the interior regions during June, July, and August; for example, Paris averages 5 to 6 days a month with thunder during these summer months.

The outstanding feature of the temperature distribution is the remarkably high mean temperature during the winter half year. This is the result of the prevailing onshore air flow from the warm oceanic regions. The coldest periods correspond to cold air outbreaks from Asia. They occur during a westward displacement of the cold Asiatic anticyclone. Stations nearer to this source region of cold air naturally have lower mean January temperatures than stations to the west [for example, compare Fanö (26), Denmark, with Valencia (28), Irish Free State]. Likewise, radiational cooling over the land results in lower January temperatures than in regions directly exposed to the onshore winds [for example, compare Paris (64) to Brest (31) or Greenwich (56) to Valencia (28)].

The same onshore circulation contributes to the cool summers along the coast line of northwest Europe. Continental heating results in higher temperatures over the interior; for example, Paris (64) and Greenwich (56) are warmer than Brest (31) and Valencia (28), respectively. Consistent with the maritime climate, autumn is warmer than spring, as discussed in Sec. 8.

b. Central Europe. Away from the coastal regions the maritime influence is less pronounced. On the other hand, the lack of a mountain barrier along the west coast means that the climate of Germany, Poland, Austria, Hungary, etc., is milder than that of eastern North America. The central European zone represents a transition between the very mild, wet winters of the northwest and the very cold, dry Russian winters. It is a region that comes under the influence of both types of weather regime. The following description will be confined to the land areas north of the Alps and of the mountains of Yugoslavia and Bulgaria.

Because of the filling of cyclones, there is a pronounced diminution of precipitation from the west toward the east as cyclones move eastward. Furthermore, the cyclone tracks in Fig. 31 demonstrate the decreased cyclonic activity over the central European region. Except for the mountainous region to the south, orographical precipitation is unimportant. Those areas which border on northwest Europe come under the influence of the autumn and early-winter precipitation regime, for example, Trier (62), Germany; Paris (64), northwest France; and La Tronche (71), southern France. These are regions of cyclonic precipitation. To the east, where frontal precipitation is less important and anticyclonic circulation predominates, winter is a season of precipitation minimum, for example, Warsaw (60) and Budapest (68). During the summer season, intense continental heating results in convective precipitation in the form of either heat showers or weak frontal convective showers.

Sweden and Finland show a marked decrease in precipitation in contrast to western Norway; this is a direct result of the sheltering effect of the Norwegian mountains. For example, Stensele (52) averages about 1 in. of precipitation during each winter month, while Trondheim (22) averages about 3 in. Here the convective type of precipitation produces a decided summer maximum. Farther south, along the eastern Baltic, migratory cyclones that pass south of the Scandinavian mountains produce a late summer and autumn precipitation maximum [see Libau (58)]. From the Baltic and North seas southward, central Europe is characterized by the summer precipitation regime.

Even though summer is the season of maximum precipitation, winter remains the cloudiest season. In central Germany, for example, the average January cloudiness is about eight tenths. This maximum can be attributed partly to an eastward extension of the frontal cloudiness of western Europe and partly to the low clouds that are frequently observed in the modified cold air masses. Even when the latter come from Asia, they frequently arrive as quite moist air masses in the lowest layers so that stratus and stratocumulus clouds are quite common. In summer the cumulus-type clouds result in a mean cloudiness of four tenths to six tenths. The cloudiness minimum occurs in the Danube lands as demonstrated by Budapest (68).

Wintertime radiation fogs show a high frequency over a large portion of central Europe; for example, Hamburg averages 11 days with fog during December. The cold half year is very favorable for fog formation because of the frequent stagnant anticyclonic circulation, the high surface moisture content of the air masses, and the favorable radiation conditions. Summer heating over the land causes a pronounced fog minimum. Like the North Sea, the Baltic shows a spring maximum of fog, whereas the neighboring land masses show an autumn maximum. The winter snow cover over Finland and Sweden is unfavorable for fog formation at temperatures about 15°F, as described by Petterssen.[1]

Thunderstorms are very rare during the winter months when stable air masses predominate. However, diurnal heating together with the general convective instability of summer continental air masses accounts for approximately 5 days with thunder during each of the summer months.

The mean January map shows that the longitudinal temperature variation is decidedly more significant than the zonal variation. The relatively high mean January temperatures in western Germany are the result of the modifying influence of the maritime air circulation that predominates over western Europe, whereas the low temperatures to the east arise from the predominating effect of the cold air masses that originate over Russia. Throughout central Europe the lowest winter temperatures occur during periods when cold anticyclones are displaced westward from Russia. Naturally these *cP* invasions are less frequent, less persistent, and less severe over western Europe. The higher temperatures observed in southwest Finland as compared with Sweden, at the same latitude, can be attributed to the maritime exposure of Finland. On the other hand, central Sweden is sheltered from the warm prevailing winds by the Norwegian highlands. Consequently, very low winter radiation temperatures are not uncommon; for example, Sveg in central Sweden has a mean yearly minimum temperature of −36°F.

[1] PETTERSSEN, *op. cit.*, p. 128.

As in eastern Germany, the low winter temperatures in Hungary and Rumania—Bucharest (66) has a mean January temperature of 24°F—are the result of the predominating influence of cold *cP* anticyclones.

During the summer, the continental heating results in higher mean temperatures over the interior than at stations adjacent to the cooler coasts; for example, Bucharest (66) has a mean July temperature of 73°F. The relative warmth of central Europe in spring and autumn depends on the continental or maritime exposure. Regions near the warm ocean show slightly higher temperatures in autumn than continental regions, whereas spring is slightly warmer in the east.

Since eastern Europe has a climatic regime very similar to that of Asia, the climate of eastern Poland and western Russia will be discussed in the following chapter.

c. Mediterranean Region. The lands south of the mountainous regions of southern Europe generally have a warm climate with dry summers. Nevertheless, the whole region is not entirely uniform; for example, northeastern Italy closely resembles central Europe.

As far as the annual rainfall variation is concerned, it may be broadly stated that the summers are dry and that wet periods occur during the winter half year. However, the rainfall varies quite definitely from place to place. The distribution can be discussed on the basis of cyclonic activity together with the topographical features of the Mediterranean basin. As mentioned in Sec. 62, the Sardinia–Gulf of Genoa region is an area where cyclones form or where cyclones from the Atlantic frequently deepen. During midwinter, when the belt of high pressure over Spain and the Alps is well developed, the cyclones follow a southern trajectory through the Mediterranean, and consequently cyclonic precipitation is more pronounced to the south. In spring and autumn the cyclones are displaced to the north and produce two rainy periods over the northern Mediterranean region. Since the sea is warmer in autumn than spring, the warm air masses have a higher moisture content in autumn, with the result that the precipitation is then heavier. The mountain ranges of Italy, Spain, and the eastern Adriatic region provide a barrier to moving cyclones so that orographical influences intensify the precipitation along the western slopes.

The presence of a larger number and of deeper cyclones east of 10° east causes heavier precipitation over the eastern Mediterranean than over the area around Spain. These factors, which influence the precipitation regime over the Mediterranean area, are clearly illustrated by Table 57.

TABLE 57.—MEAN PRECIPITATION DISTRIBUTION OVER THE MEDITERRANEAN

A. Northern stations

Station	Porto 41°N 8°W	Barcelona 41°N 2°E	Marseille 43°N 5°E	Livorno 44°N 10°E	Ancona 44°N 14°E	Mostar 43°N 18°E
Yearly rainfall....	47.6	20.7	22.3	35.8	25.4	55.7
Rainfall maximum	Nov., 6.5	Sept., 3.3	Oct., 4.0	Oct., 5.6	Oct., 3.9	Nov., 7.3
Secondary maximum..........	Jan., 5.7	Apr., 1.8	Apr., 2.7	Apr., 2.9	Jan., 2.6 Apr., 2.2	Mar., 5.2
Driest month.....	Aug., 0.8	July, 1.1	July, 0.7	July, 1.0	July, 1.1	Aug., 1.7

B. Southern stations

Station	San Fernando 36°N 6°W	Cartagena 38°N 1°W	Palermo 38°N 13°E	Syracuse 37°N 15°E	Argostoli 38°N 20°E	Larissa 40°N 22°E
Yearly rainfall....	28.1	13.4	29.5	25.1	33.9	19.8
Rainfall maximum	Dec., 4.5	Nov., 1.8	Dec., 4.5	Nov., 4.3	Dec., 7.0	Nov., 2.6
Secondary maximum..........	Mar., 3.5	Jan., 1.8	None	None	None	None
Driest month.....	July, 0.0	July, 0.1	July, 0.3	July, 0.2	July, 0.2	Aug., 0.9

With the anticyclonic control during the summer, the Mediterranean is generally a region with slight warm-season precipitation. The only area with summer precipitation is northeastern Italy. In winter this section does not come within the regime of cyclonic precipitation and resembles central Europe in that it has frequent convective showers during summer.

As compared with the other sections of Europe, the Mediterranean has a small degree of cloudiness. Even during the rainy seasons the mean cloudiness is less than six tenths of the sky. This may be attributed to two factors: (1) Cyclonic precipitation occurs less frequently than over central Europe. (2) The cool, dry northerly winds in the rear of eastward-moving cyclones are accompanied by clear skies over the northern Mediterranean area. Consistent with the anticyclonic control the mean summer cloudiness is very low, as shown by the stations in Table 58.

Radiational cooling in winter together with warm-front precipitation produces fog over the northern Mediterranean during the winter half year. It is a maximum at the head of the Gulf of Genoa, where both factors contribute to fog formation, and decreases southward, where there are diminished frontal precipitation and less favorable radiation conditions. For example, Genoa averages 4 to 5 days with fog in February as compared with 1 day at Rome. The southern Mediterranean is relatively free of fog, but very thick dust storms from northern Africa in advance of eastward moving cyclones may produce periods of low visibility. Summer fogs are quite rare over the northern Mediterranean.

Since summer is a dry season, severe thunderstorms are infrequent. However, light thunderstorms can occur occasionally; for example, in July, Marseille averages 2 days, Rome 1 day, Mostar 4 days with thunder. Frontal and orographical lifting cause infrequent thundershowers during the wet periods. As mentioned previously, northeastern Italy is characterized by frequent summertime convective showers, and consequently thundery days are common; observations indicate 5 to 9 days with thunder during July.

Because of the maritime influence the Mediterranean winters are generally mild and the summers cool as compared with the neighboring continental areas. The Iberian Peninsula is characterized by cool coastal regions, with the interior colder in winter and warmer in summer, as a result of radiational cooling and heating. The low yearly rainfall over the interior and the east produces semidesert conditions (*BS*) during the summer. The south coast of France, for example, Marseille (38, Table 58), has cool winters, and unusually cold periods accompany the mistral. Subfreezing temperatures are often associated with these strong winds. Summers are warm but slightly cooler than in central Europe. With the cold air drainage from the Alps, northern Italy has relatively cold winters [see Mantua (40)]. Continental heating gives rise to high summer temperatures, and so this area has a moderate continental climate like central Europe. Central and southern Italy, with their oceanic exposure, have a typical Mediterranean climate. At the head of the Adriatic the bora gives rise to abnormally cold winter periods with strong north winds and clear skies.

The Balkan Peninsula has a Mediterranean climate along the west and south coasts with a gradual increase in temperature to the south in all seasons of the year. Eastern Bulgaria, which borders on the Black Sea, has relatively cold winters as the result of cold air masses from the north. Hence, this region has a temperature regime like that of central Europe. Similarly, the central Balkan plateau with its cold winters resembles central Europe in climate. The presence of precipitation in all seasons of the year distinguishes these regions from Hungary and the central European countries. The winter precipitation is the result chiefly of migratory cyclones.

d. The Alps. Because of the elevation the Alps have a climatic regime somewhat different from the adjacent plains.

As a result of orographical lifting this is a region of relatively high precipitation amounts, the heaviest falls occurring along the ranges. To the north, summer is the rainiest season, as in central Europe, with frequent thundershowers; for example, Innsbruck averages 7 to 8 days with thunder during July. To the east and south, autumn precipitation is quite pronounced, as in northern Italy, and augments the summer rainfall. Throughout the entire region the predominance of anticyclonic circulation favors a winter precipitation minimum.

The subsidence within the cold anticyclones is responsible for the relatively small winter cloudiness along the ranges. The mean January cloudiness, about five tenths of the sky, is considerably smaller than over the plains to the north. Summer convective cloudiness accounts for a mean July value of about five tenths. In contrast to the summits, the stagnation of cold damp air in the valleys causes a high degree of winter cloudiness and radiation fog.

As would be expected from the elevation, the mean temperatures average less over the mountains than over the plains of central Europe. In sheltered regions, radiational cooling causes low ground temperatures in the winter season. Nevertheless, the narrow extent of the high ranges of the Alps results in only a small region with a typical snow-forest climate.

TABLE 58.—METEOROLOGICAL MEANS AT SELECTED STATIONS
(Temperature in degrees Fahrenheit, precipitation in inches, cloudiness in tenths)

Station	Lat.	Long.	Temperature				Precipitation				Cloudiness				Elevation, ft.
			Jan.	Apr.	July	Oct.	Jan.	Apr.	July	Oct.	Jan.	Apr.	July	Oct.	
West Coast															
21. Tromsö	70°N	19°E	26	32	52	36	4.3	2.3	2.2	4.6	7	6	7	7	150
22. Trondheim	63°N	10°E	27	38	57	40	3.4	1.8	2.2	3.4	8	7	8	8	130
23. Florö	62°N	5°E	35	41	57	46	9.0	4.4	5.2	9.2	8	6	6	7	30
24. Skudenes	59°N	5°E	36	42	57	47	4.4	2.5	3.2	5.6	8	6	7	8	10
25. Stornoway	58°N	6°W	40	43	54	47	5.1	3.1	3.0	5.1	8	7	8	8	50
26. Fanö	55°N	8°E	33	42	60	48	1.6	1.4	2.3	3.3	7	6	6	7	20
27. Helgoland	54°N	8°E	35	42	60	51	2.0	1.5	2.6	3.5	8	6	7	8	140
28. Valencia	52°N	10°W	45	48	59	51	5.7	3.8	3.9	5.7	8	7	8	7	50
29. Vlissingen	51°N	4°E	37	46	62	52	2.0	1.7	2.5	2.9	7	6	5	6	10
30. Scilly	50°N	6°W	46	48	60	54	3.0	2.0	2.2	3.6	8	6	7	7	130
31. Brest	48°N	4°W	45	50	62	55	3.2	2.2	1.8	3.5	8	6	6	7	110
32. Nantes	47°N	2°W	40	50	65	52	2.2	2.0	2.1	3.8	7	6	5	6	130
33. Santander	43°N	4°W	47	54	65	59	2.8	2.8	1.7	4.2	6	6	5	6	50
34. Porto	41°N	8°W	49	57	68	61	5.7	5.0	0.9	5.0	5	6	4	6	330
35. Gibraltar	36°N	5°W	55	61	73	66	4.9	2.7	0.0	3.2	4	4	2	4	50
Mediterranean															
36. Cartagena	38°N	1°W	51	59	75	66	1.8	0.8	0.1	1.5	4	3	1	4	40
37. Barcelona	41°N	2°E	48	57	75	63	1.4	1.8	1.1	2.8	5	6	4	5	140
38. Marseille	43°N	5°E	44	54	71	59	1.6	2.2	0.6	3.7	5	5	3	5	250
39. Genoa	44°N	9°E	46	57	75	62	4.1	4.4	2.0	6.8	5	6	4	6	180
40. Mantua	45°N	11°E	34	56	76	57	1.4	2.3	1.9	3.2	6	4	2	4	150
41. Trieste	46°N	14°E	39	54	74	58	2.2	2.9	3.7	4.7	6	6	4	6	90
42. Lesina	43°N	16°E	47	57	76	64	3.0	2.4	0.8	3.7	5	4	1	5	70
43. Valona	40°N	20°E	48	59	77	65	3.9	2.3	0.5	5.2	6	4	1	5	30

TABLE 58.—METEOROLOGICAL MEANS AT SELECTED STATIONS.—(*Continued*)

Station	Lat.	Long.	Temperature				Precipitation				Cloudiness				Elevation, ft.
			Jan.	Apr.	July	Oct.	Jan.	Apr.	July	Oct.	Jan.	Apr.	July	Oct.	
44. Kythera	36°N	23°E	53	61	81	70	4.1	0.9	0.5	2.8	6	4	1	4	530
45. Athens	38°N	24°E	48	58	80	66	2.1	0.9	0.3	1.4	6	5	1	4	350
46. Kavalla	41°N	24°E	42	57	78	62	2.7	1.9	0.7	0.8	5	5	2	4	40
47. Malta	36°N	14°E	53	59	77	69	3.1	0.9	0.0	2.8	6	4	1	5	0
48. Naples	41°N	14°E	47	57	76	62	3.5	3.1	0.5	4.8	5	5	2	4	490
49. Chieti	42°N	14°E	39	51	73	57	4.6	3.3	1.7	4.3	5	5	3	5	1120
50. Aquila	42°N	13°E	34	50	71	54	2.2	2.6	1.2	3.3	6	6	3	5	2410
51. Sassari	41°N	9°E	47	56	76	63	2.4	2.2	0.2	3.4	6	6	3	6	730
Inland															
52. Stensele	65°N	17°E	10	31	57	33	0.9	0.8	2.8	1.5	7	6	6	7	1080
53. Helsingfors	60°N	25°E	22	36	62	42	2.1	1.6	2.4	2.7	8	6	5	7	40
54. Linköping	58°N	16°E	28	40	63	44	1.1	1.3	2.8	1.8	8	6	6	7	210
55. Aberdeen	57°N	2°W	38	44	56	47	2.1	1.9	2.8	3.0	6	6	7	6	90
56. Greenwich	51°N	0°	39	47	63	50	1.7	1.5	2.2	2.5	7	6	6	6	150
57. Tartu	58°N	27°E	21	38	62	41	1.3	1.2	3.1	1.7	8	6	6	7	240
58. Libau	57°N	21°E	28	40	61	47	1.9	1.5	2.2	2.9	8	6	5	7	20
59. Vilna	55°N	25°E	23	42	65	44	1.3	1.5	3.1	1.5	8	6	6	7	450
60. Warsaw	52°N	21°E	26	44	65	46	1.3	1.6	3.0	1.2	8	6	6	7	400
61. Dresden	51°N	14°E	32	47	65	49	1.4	1.7	3.6	1.8	7	6	6	7	390
62. Trier	50°N	7°E	32	47	63	48	1.8	1.6	2.6	2.5	8	6	6	7	500
63. Lille	51°N	3°E	36	48	64	50	2.0	1.6	2.7	3.1	..	..	..	..	80
64. Paris	49°N	2°E	36	49	65	49	1.4	1.6	2.0	2.3	7	6	5	6	160
65. Sulina	45°N	30°E	31	49	71	55	0.9	1.1	1.6	1.3	6	4	2	5	10
66. Bucharest	44°N	26°E	24	51	73	53	1.3	2.1	2.2	1.7	6	6	4	5	280
67. Belgrade	45°N	20°E	31	53	71	54	1.3	2.4	2.4	2.2	7	6	4	5	450
68. Budapest	48°N	19°E	28	51	70	51	1.5	2.4	2.1	2.5	7	5	4	5	370
69. Innsbruck	47°N	11°E	26	48	64	48	1.5	2.3	5.0	2.3	5	6	6	5	1970
70. Milan	45°N	9°E	35	55	75	55	2.4	3.8	2.8	4.9	6	6	4	6	480
71. La Tronche	45°N	6°E	34	53	70	53	3.1	3.5	3.5	4.8	6	6	4	6	890
72. Madrid	40°N	4°W	40	53	76	56	1.4	1.9	0.4	1.7	5	5	2	5	2150
73. Monastir	41°N	21°E	30	52	72	56	1.9	2.6	1.5	3.0	6	6	3	6	2030

CHAPTER XII

ASIA (INCLUDING RUSSIA)

Although the vast continent of Asia is essentially located in temperate latitudes, the land extends from the equator to 80° north. Because of the similarity, the climate of eastern Europe will be discussed together with that of western Asia. Likewise, the Dutch East Indies will be included in the discussion of southeastern Asia. With these additions the land areas demonstrate a vast range of climatic types varying from tropical rain forest (*Af*) in the south to polar tundra (*ET*) in the north.

The principal features of the climate can be directly attributed to the vast land mass, located north of 30° north, which is sheltered from invasions of warm moist air from the south by the mountain barrier and of moist air from the west by the land barrier. During the cold season the continental cooling produces a semipermanent anticyclone over the interior of the continent. As a result of ideal radiation conditions the air masses are extremely cold and hence dry, especially at the surface. The outflow of the air masses across southern and eastern Asia is the one main determining factor of the winter weather—the "winter monsoon." Conversely, in summer, intensive heating produces a thermal low-pressure area about 30° north. The inflow of air from the south results in the flooding of the south and east with a warm and very moist air mass. This circulation is called the "summer monsoon." With these two wind regimes, winter is essentially a dry season and summer a very rainy period over the vast area that comprises southeastern Asia.

Because of the predominance of relatively dry air masses, central, northern, and western Asia are regions of low yearly precipitation amounts. The sources of moisture are the Arctic Ocean in summer and the North Atlantic in all seasons. With the continental exposure the summers are warm and the winters very cold. The net result is a continental forest type of climate.

ical Features. The heart of Asia consists of a vast ich comprises the following mountain ranges:

1. The Himalaya, which extends from west to east through the whole of northern India, includes the highest mountains in the world with elevations reaching 29,000 feet.

2. Between the Himalaya and the Kunlun to the north lies the very high plateau of Tibet, which is traversed by a number of smaller mountain chains.

3. A more or less continuous series of mountains, orientated from southwest to northeast, extends from the Arabian Gulf to Mongolia. The Hindu Kush and Tien Shan are the most prominent ranges in this region.

4. From the plateau of Tibet the country gradually slopes downward toward the Arctic Ocean. Various mountain chains are located in Mongolia and Siberia, for example, the Altai and Yablonovoi mountains.

5. To the southeast a minor ridge extends toward the Gulf of Siam.

The greater part of southwestern Asia exceeds 2000 ft in elevation and is interspersed with a series of mountain chains like the Caucasus, which extends from the Caspian to the Black Sea. The highest ridge stretches from the Black Sea to southeastern Iran and is flanked to the south by the Persian Gulf and the lowlands of the Euphrates and Tigris rivers.

The major portion of the Soviet Union consists of land whose elevation is less than 1000 ft, but it is crossed by one north to south mountain chain—the Ural mountains. Other prominent highlands of the Asiatic region are the series of small mountain chains along the east coast of the continent, the Japanese islands, the Dutch East Indies, southern India, and Arabia.

Additional physical features are the inland seas of southwestern Asia, *viz.*, the Black and Caspian seas, the Aral Sea, and Lake Baikal of Russia. The continent has numerous indentations such as the Persian Gulf, Arabian Sea, Bay of Bengal, Gulf of Siam, Gulf of Tonkin, Sea of Okhotsk, and the Kara Sea. However, in view of the vastness of the continent, these water bodies have a direct influence on the climate of only a small percentage of Asia. In contrast, the climate of Europe is strongly affected by the various gulfs and seas. Europe is not large as compared with the indentations, and consequently they tend to produce a maritime type of climate over a large percentage of the land west of Russia.

The controlling influences of the topography on the general climate of Asia can be summarized briefly as follows:

1. Although modified *mP* air masses may occasionally penetrate western Russia, the Ural mountains impede their eastward progress across Siberia. In a similar manner, warm air invasions from the south are prohibited from entering central Asia by the Himalaya and adjacent ranges. Likewise, the highlands of the northern part of eastern Asia separate the interior from the Pacific Ocean. As a consequence of this ring of mountain ranges, very cold and dry air masses can accumulate over the interior with the result that this is a region of extremely cold winters. For the same reason, it is also a region of very slight winter precipitation. In a similar fashion, abnormally high temperatures prevail during the summer season owing to the continental heating. Again, in summer, the moist air masses from the south cannot penetrate the interior. However, with the invasions of moist air from the arctic, summer precipitation is about 2 in. a month over the northern interior of the continent.

2. Besides sheltering central Asia from winter invasions of moist air masses, the high plateaus prevent direct outbreaks of cold *cP* air masses across India and southeastern Asia. Because of the downslope adiabatic heating, modified *cP* air masses arrive over the southern plains as mild air currents. However, as a result of their dryness India and southeastern Asia are regions of low winter rainfall. Conversely, in summer the southern slopes of the same mountain barriers are regions of excessive orographical precipitation.

3. The highlands of Turkey and adjacent areas, besides sheltering central Asia from invasions of Mediterranean air masses, contribute to the winter precipitation regime of the southwest by the orographical lifting of moist air masses from the west.

4. The precipitation over the Japanese islands and the Philippines is intensified by the orographical lifting of warm, moist air masses from the south in summer and of modified cool air masses from the northeast in winter.

5. The band of heavy summer precipitation along southwestern India, as compared with the relative dryness over southeastern India, can be attributed to the presence of the Ghats near the west coast.

6. The various inland seas have an important influence upon the weather over the adjoining country. They become sources of moisture for the normally dry, inland air masses.

Further minor influences of the topography on the climate of Asia will be discussed in connection with the detailed climatic subdivisions.

65. Mean Distribution of Pressure and Wind. The mean sea-level pressure maps for January and July illustrate the principal features of the general circulation. Because of the scarcity of reliable pressure observations, especially in the Tibet area, there may be some doubt as to the accuracy of the maps. Furthermore, as in the case of northwestern America, the pressure gradients over these mountainous regions do not give a true picture of the mean circulation. Here local orographical influences are very important. As a result of the extreme elevation, the wind regime above the deep friction layer cannot be demonstrated by the mean sea-level pressure charts. With these reservations, Plates IV and V will now be discussed in some detail.

a. Winter. The main characteristics of the January map may be summarized as follows:

1. As a result of the sheltering from maritime air masses, cold air accumulates over central Asia. In view of the vastness of the continent and the topographical features, the winter anticyclones become so well developed that the January chart shows a mean central pressure of about 1035 mb, south of Lake Baikal. The circulation around this anticyclone dominates the air flow.

2. Over northwestern Russia the circulation is controlled by the flow around the mean low-pressure trough to the north. This trough, which is an eastward extension of the Icelandic low-pressure system, exists over the warm water between Scandinavia and Spitsbergen. East of Novaya Zemlya the Arctic Ocean is frozen during midwinter, and through radiational cooling there is a tendency toward the development of high pressure. However, the pronounced continental cooling over the land to the south predominates; consequently, the pressure decreases northward.

As a result of the continental anticyclone and the lower pressure to the north, the prevailing winds north of 50° north are mainly southwest as illustrated by Kola (1), Hogland (10), Surgut (11), Vilyuisk (12), and Bulun (3) in Table 59. Occasional migratory cyclones and anticyclones cause pronounced changes in wind direction. In studying Table 59 it should be remembered that local factors always contribute to the wind variation.

3. The predominant air flow over the east coast is from a northerly direction, in agreement with the steep pressure gradient

TABLE 59.—PERCENTAGE FREQUENCY OF SURFACE-WIND DIRECTION

Station	Lat.	Long.	January								July							
			N	NE	E	SE	S	SW	W	NW	N	NE	E	SE	S	SW	W	NW
North Coast																		
1. Kola	69°N	33°E	5	2	2	8	20	*28*	*28*	7	*26*	18	17	8	13	8	5	5
2. Dudinskoe	69°N	86°E	5	14	16	*28*	22	6	3	6	22	*26*	10	7	8	6	7	14
3. Bulun	71°N	127°E	5	3	2	5	21	*41*	21	2	20	*29*	9	3	9	14	9	7
East Coast																		
4. Nikolaevsk	53°N	141°E	3	2	1	2	1	1	*61*	29	6	9	*36*	30	1	3	9	6
5. Rykovskoe	51°N	143°E	29	2	2	2	1	2	2	*60*	4	1	7	52	11	3	3	19
6. Vladivostok	43°N	132°E	42	6	0	1	0	1	1	*49*	5	1	8	*66*	7	5	3	5
7. Hanoï	21°N	106°E	18	18	13	*22*	10	7	2	10	11	8	12	*32*	17	7	5	8
8. Nhatrang	12°N	110°E	*37*	21	9	7	0	0	21	5	0	2	17	*65*	0	1	15	0
9. Yap	9°N	138°E	27	*66*	6	0	0	0	0	1	11	*21*	12	15	12	17	8	4
Inland																		
10. Hogland	60°N	27°E	8	8	7	15	13	*19*	18	12	6	9	10	14	8	*24*	21	8
11. Surgut	61°N	73°E	5	9	12	14	*24*	23	10	3	19	11	10	*15*	13	10	7	*15*
12. Vilyuisk	64°N	122°E	3	10	23	5	5	*29*	21	4	*18*	12	8	8	8	12	16	*18*
13. Sevastopol	45°N	34°E	8	*34*	12	8	11	12	6	9	5	9	19	5	9	15	17	*21*
14. Turkestan	43°N	68°E	9	*26*	28	7	6	6	14	9	17	*38*	6	1	3	7	13	15
15. Blagoveshchensk	50°N	128°E	14	5	0	2	11	5	4	*59*	9	13	6	13	*24*	11	6	18
16. Bombay	19°N	73°E	21	19	20	4	2	1	2	*31*	1	0	1	2	1	31	*54*	10
17. Calcutta	23°N	88°E	21	15	8	3	3	10	10	*30*	1	4	18	20	25	*26*	4	2
Islands																		
18. Aparri	18°N	122°E	5	*40*	30	9	8	4	2	2	8	14	6	8	*37*	16	5	6
19. Maety Miarang	8°S	128°E	4	0	0	0	3	11	24	*58*	0	0	30	*64*	6	0	0	0
20. Surigao	10°N	125°E	15	*46*	20	4	3	3	1	8	3	5	6	2	14	*41*	21	8

from the Asiatic anticyclone to the Aleutian low-pressure system. Unlike the condition in North America, the continental anticyclone has considerable permanence as a feature of eastern Asia, and therefore the general northwest winds are very persistent, as illustrated by Nikolaevsk (4), Rykovskoe (5), and Vladivostok (6). Over America, migratory pressure systems result in a more variable wind regime along the eastern portion of the continent. For example, whereas cyclones move northeastward across eastern America, the cyclones off Asia usually have an oceanic trajectory (see Fig. 35, page 288).

In accordance with the central high pressure about 50° north, the prevailing winds vary from northwest across northeastern Asia to northeast across southeastern Asia [see Yap (9, Table 59)]. The latter and the winds across India are commonly called the "northeast winter monsoon." In this respect it is important to note that the northwest winds of northeast Asia are an equally significant winter monsoon.

4. With the continental heating over northern Australia, a mean low-pressure area controls the circulation in neighboring regions. North of the equator the pressure gradient is favorable for persistent northeast winds over the Philippine Islands and the Dutch East Indies. As a consequence of the air motion across the equator the northeast winds become northwest in the Southern Hemisphere. This change of direction is demonstrated by the January frequencies at Aparri (18) and Surigao (20) in the Northern Hemisphere and Maety Miarang (19) in the Southern Hemisphere.

5. From the Persian Gulf to Indo-China the winter circulation is also controlled by the continental anticyclone. General northeast winds (winter monsoon) predominate at the surface except in the mountainous regions, where local orographical influences may produce winds entirely contrary to the mean sea-level pressure pattern. Day-to-day wind variations over the plains may arise from cyclones moving west across India.

6. Southwestern Russia is a region of somewhat variable winds. The mean high-pressure wedge, which is located about 48° north, causes a predominance of northeast winds over the Black and Caspian seas [see Sevastopol (13)]; but in the mountainous regions farther south the surface winds may be quite incompatible with the mean pressure gradient. Furthermore, this is a region that comes under the influence of moving cyclones, and therefore rapid fluctuations in wind direction may be expected during the winter season.

The warmth of the Persian Gulf, in contrast to the coldness of the land to the north and southwest, results in a narrow thermal low-pressure area, which may be considered to be an eastward extension of the Mediterranean Low. In view of this pressure distribution the prevailing surface air flow is toward the Persian Gulf. However, occasional deep winter cyclones cause pronounced variations.

b. Summer. Whereas the continental anticyclone dominates the winter circulation over Asia, the summer circulation is primarily controlled by a vast thermal low-pressure belt centered about 30° north. This mean low-pressure area is produced by the intense summer continental heating. The essential features of the air flow during the warm season will now be discussed briefly.

1. The circulation in northwestern Russia is dominated by a minor low-pressure belt, which extends across the continent about 65° north and which may be considered as a continental extension of the Icelandic low-pressure area. As demonstrated by Kola (1, Table 59), north winds predominate along the arctic coast. On the other hand, south of the pressure trough the prevailing air flow is from the southwest, as shown by Hogland (10).

2. Western and southwestern Russia come under the influence of the circulation around the eastward extension of the Atlantic high-pressure belt so that northwest and north winds predominate. In view of the weak pressure gradients, local conditions, for example, the sea breezes in the Black Sea area, become very important during the summer.

3. The persistent northerly winds along the arctic coast and the southeasterly winds along the North Pacific coast are a direct result of the continental low-pressure area and the oceanic highs. In summer, as in winter, southeastern Asia differs from the similar region of North America. Unlike the anticyclonic circulation in the latter area, the cyclonic flow around the continental low extends to the Pacific coast of Asia. Consequently, the monsoon circulation is decidedly more developed over Asia than over North America. Because of the continental low, the air flow across India and Burma is predominantly from the southwest, as illustrated by Calcutta (17). Farther west, with the mean low-pressure center to the northeast, west winds predominate at the surface, except in those regions where local factors cause erratic wind directions.

4. With the pressure gradient directed from north to south, the surface winds are essentially southeasterly in the Dutch East

Indies. However as the air crosses the equator, the prevailing direction becomes southwesterly in the Northern Hemisphere [for example, compare Maety Miarang (19) and Surigao (20), Table 59].

During the summer the doldrum low-pressure belt is located about 20° north, on the coast of eastern Asia, as compared with its winter location across northern Australia. This seasonal oscillation in the position of the doldrum belt is considerably greater than the similar variation in the eastern Pacific. Along the coast of western America the meridional oscillation is about 6° latitude. The reason for the slight shift of this low-pressure trough in the eastern Pacific lies in the relative persistence of the subtropical highs over the eastern portion of the North and South Pacific oceans. Since the highs oscillate over a small range, the equatorial low-pressure belt must also vary only slightly in position. On the other hand, there is a vast change from summer to winter in the western Pacific. This change is accompanied by a pronounced latitudinal shift of the doldrum belt.

Besides the pronounced changes in wind direction, the principal change in air flow from winter to summer is the decrease in the mean wind velocity over the continent, north of 30° north. On the other hand, across India, the southwest monsoon is usually stronger than the winter monsoon. For example, the mean velocity of the summer monsoon at Bombay is about 14 mph as compared with a mean December-February velocity of about 7 mph.

c. Upper Air. Even though the sea-level January pressure map shows a very pronounced high-pressure area, the upper-air chart for 4 km indicates no closed anticyclone aloft. This can be attributed to the fact that the anticyclone is very cold and that consequently the pressure decreases relatively rapidly with elevation. The main features of the January upper-air circulation are as follows:

1. The cold area in northern latitudes is a region of low pressure aloft, with the minimum pressure located over northeastern Siberia. As in the case of the American continent, the coldest air at upper levels is to be found immediately to the west of the upper-level cyclone. In general, the temperatures at 4 km are lower across eastern Asia than across the corresponding latitudes of western Asia, for the latter air masses have been modified,

even at upper levels, by their trajectory across the Atlantic Ocean. For example, outbreaks of cold air at Vladivostok (43° north) show at 4 km a temperature of −42°C as compared with −34°C at Moscow (56° north).

2. Between 20° north and 50° north the easterly surface winds become general westerlies at 2 km, in response to the pronounced south to north temperature gradient. In lower latitudes, where the temperature gradient is diminished, there is a considerably deeper layer of winds with an easterly component; for example, east winds predominate to 8 km at 10° north.

3. On the average, the westerly winds definitely increase in speed with elevation.

During midsummer the continental heating, which causes a thermal low-pressure area at the surface, is also responsible for relatively high upper-level pressures over central Asia. Therefore, the weak surface north and east winds beyond 50° north become west winds at 2 km, and the latter increase in velocity with elevation. Farther south the air over the land is warmer than the oceanic air so that, especially in the Indian region, the temperature decreases from north to south, contrary to the general condition. Whereas west winds predominate at the surface, the thermal effect causes a reversal of gradient aloft; therefore, at 6 km the air has an easterly component. This reversal is unusual, and consequently it is important to bear it in mind.

Figure 34, after Ramanathan and Ramakrishnan, demonstrates the changes in wind velocity with elevation over India. During the cold season, in response to the thermal gradient, the increase in westerly winds with elevation is quite noticeable. Likewise, the unusual reversal of summer winds is demonstrated by the mean chart for July.

As in the case of all continents, the day-to-day changes in wind velocity at upper levels are less frequent than near the earth's surface. With the smaller number of deep cyclones over continental Asia, these upper-air fluctuations are less important than over continental America, where deep surface cyclones cause appreciable variations from the normal distribution.

66. Ocean Currents. Since the continent is very large, the oceanic circulation directly affects only a very small portion of Asia. However, in view of the onshore summer circulation, the sea-surface temperature has an important influence upon the temperature and moisture content of the air masses that invade

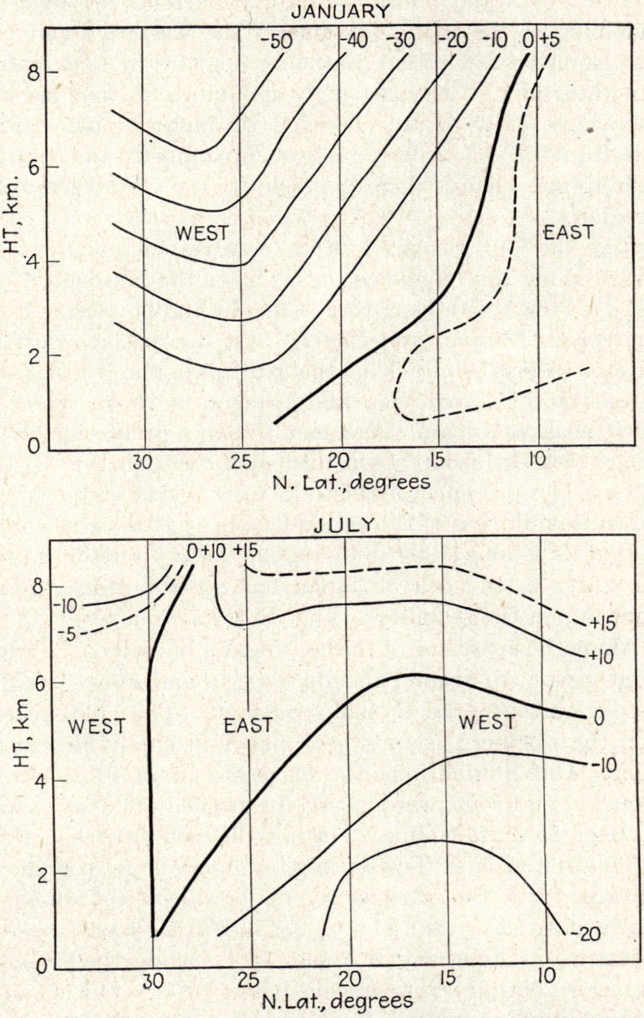

FIG. 34.—The variation of the normal west-east component of wind, in mph, along longitude 78° east. (*After Ramanathan, K. R. and Ramakrishnan, K. P., Mem. Indian Met. Dept., Vol.* 26, *part* 10, *Delhi*, 1939.)

Asia. Plate VI illustrates the principal elements of the oceanic circulation.

The main features of the surface currents and the distribution of water temperature may be summarized as follows:

1. The North and South Equatorial currents are deflected in the vicinity of the island systems of the western Pacific. The South Equatorial Current branches southward and transports warm water along the coast of New Guinea so that sea-surface temperatures of 28°C are observed throughout the year. The North Equatorial Current, in branching north and south near the Philippine Islands, is responsible for the very warm water in this region.

2. Like the Gulf Stream system, a warm current flows northeastward along the Pacific coast. This is the Kuroshio Current, which joins the North Equatorial Current and flows from Formosa to the coast of Japan and then almost due east toward North America. A small branch of the Kuroshio, the Tsushima Current, enters the Japan Sea and transports warm water along the west coast of Japan. As a result of this circulation the mean February sea-surface temperature about Japan, at 37° north, is 10°C. The presence of this relatively warm water is responsible for the mildness of the Japanese winter climate as compared with that of China, 1000 miles to the west. Furthermore, this warm water in the Sea of Japan has an important modifying influence upon the stability of the winter air masses.

3. Along the coast line of the continent a branch of the Aleutian Current circulates around the Bering Sea and moves southward as a cold current—the Oyashio Current. The cold water surrounds the northern islands of Japan and subdivides into two sections. One branch joins the Kuroshio about 40° north, while a second moves southward along the immediate coast, carrying cold water to Indo-China. During the winter season the cold air circulation around the continental anticyclone, together with cold water from the various rivers, produces abnormally cold coastal water. For example, at 23° north the mean sea-surface temperature in February is about 15°C, while the Yellow Sea has a mean February temperature below 10°C. During summer with the reverse monsoon circulation this cold water circulation is confined mainly to higher latitudes—north of 40° north.

Because of the proximity of the Kuroshio and Oyashio currents, steep temperature gradients are observed near northern Japan, about 40° north, in all seasons of the year. The cold coastal current is a very important contributing factor to the production of coastal fog. Unlike the condition off the east coast of North America, fog is observed quite frequently in low latitudes during spring when the coastal water is still abnormally cold, for example, at Hong Kong.

4. North of the equator the surface currents over the Indian Ocean vary considerably from summer to winter, in response to the pronounced change in the wind circulation. In general, with the winter monsoon, the sea-surface temperature increases southward. During the summer months a large area of the Bay of Bengal and the Arabian Sea has a water temperature in excess of 27°C. The flow of water through the Red Sea, southward in summer, probably contributes to the temperature minimum around the Gulf of Aden. However, this minimum primarily arises from upwelling.

5. As mentioned previously, the Mediterranean Sea has an important influence on the climate of the adjacent regions. During the cold season the sea surface has a relatively high temperature. This warm water provides a moisture source for the cold-season cyclones that traverse the region and that may cross southern Russia or northern India. The formation and maintenance of cyclones are also aided by the temperature contrast between land and sea.

6. Since they occupy relatively small areas, the Black and Caspian seas are subject to considerable temperature change from winter to summer. For example, with the general northerly circulation in winter the surface temperature of the Black Sea varies from 0°C in the north to 7°C in the south, while in summer the surface temperature exceeds 20°C. However, these changes are less than those over the adjacent land. The principal influences of the inland seas are to produce a moderation of air temperature in the immediate vicinity and to provide a source of moisture for air masses traversing the region.

7. In so far as Atlantic air masses invade Russia, the warm water off western Europe has an important influence upon the climate of Asia. These water masses were discussed in Sec. 60.

8. The Norwegian Current transports warm water into the Barents Sea so that there is open water to Novaya Zemlya for a relatively long period. Farther east the pack ice persists during the cold season but breaks up in summer. It is apparent,

therefore, that the winds from northwestern Russia should average warmer and moister than those from the cold northeast. This has an important bearing on the temperature and precipitation regimes of northern Asia.

67. Air Masses. The various air masses that are most frequently located over Asia are illustrated in Plates XVI and XVII. However, in so far as other air masses invade portions of the continent, the seasonal properties of all air masses will be discussed to some extent. The temperature and humidity values in this section are obtained from Petterssen.[1]

a. Polar Continental Air (*cP*). Since in midwinter the continent is snow-covered north of 40° north, and even to 30° north over central-eastern Asia, ground conditions are favorable for the production of vast cold continental air masses. Warm maritime air is prevented from entering central Asia, at the earth's surface, through the sheltering effect of the southern mountain chain that extends from the European Alps to the highlands of Indo-China. Likewise, the high ridges of eastern Asia shield the continent from invasions of warm or mild Pacific air masses. In the west, polar maritime air can enter Russia only as a considerably modified air mass, because of the trajectory across the European peninsula and the mountainous barrier of Scandinavia. The Ural mountains partly shield Siberia from surface invasions of even this distinctly modified air mass.

As a result of the snow cover and the sheltering influences, radiational cooling within the stagnant continental region produces a very cold and dry air mass. This *cP* mass comprises the air of the extensive and well-developed winter anticyclones. In view of the more pronounced sheltering and the more continental position, the deepest air masses are formed east of the Ural mountains. Here the typical air-mass characteristics are clear skies through anticyclonic subsidence and extremely low surface temperatures, with a pronounced increase in temperature to about 1500 m. Above this level the temperature decreases with elevation. Very low absolute humidity is a necessary characteristic in view of the low temperatures. West of 60° east the *cP* air masses are shallower and show the effect of *mP* invasions at upper levels. This air mass may be present as the

[1] PETTERSSEN, S., "Weather Analysis and Forecasting," McGraw-Hill Book Company, Inc., New York, 1940.

result of a westward displacement of a Siberian anticyclone or may form through the stagnation of *mP* air. The former case will be discussed later in connection with the modifications of *cP* air. When modified polar maritime air stagnates over eastern Europe, the effect of radiation is to produce a cold and relatively dry air mass. However, this air is warmer and moister than Siberian air, because of its more recent maritime origin. Furthermore, maritime polar air is generally observed at upper levels.

The extreme stability of all *cP* air masses in their source regions is a general characteristic. Even though clear skies are very common (for example, eastern-central Asia averages two tenths cloudiness or less in January), continued radiational cooling may produce ice-crystal fogs.

When *cP* air moves away from its source region, the air-mass properties change through the absorption of heat and moisture from the underlying surface, together with the turbulent transfer of heat and moisture along the vertical. The most abrupt modifications arise after the air leaves the snow cover. The continental polar air masses that invade eastern Europe and southwestern Asia may be subject to a variety of changes. For example, if the underlying surface is warm and moist, the absorption of heat results in an increase of the surface temperature and a disappearance of the ground inversion; the turbulent mixing and absorption of moisture may produce widespread low cloudiness—stratus and stratocumulus type clouds. On the other hand, if the air remains over a snow surface and the surface-wind velocities remain light, the principal modification is a gradual rise in surface temperature.

A stirred air mass is more frequently observed over northern Asia close to the coast than farther south. Here the principal modifications arise from the turbulent mixing, *viz.*, a rise in surface temperature and a disappearance of the temperature inversion at the ground. Over the northeast section of Asia there is a strong circulation around the Aleutian Low, especially at upper levels. Because of the deep cyclonic circulation this is a region of steep lapse rates and extremely cold air aloft. For example, the typical 3-km temperature in stirred *cP* air at Vladivostok is about $-35°C$ as compared with $-26°C$ in typical *cP* air at Moscow, which is 13° farther north. Because of the dryness of the underlying surface, modified *cP* air remains very dry along the northeast section of the country. The surface specific humidity is about 0.5 g/kg. Similar modifications are observed across China. Across southern China and Indo-China the *cP* air outflow produces abnormally low temperatures for this latitude.

The air transported by the so-called "northeast monsoon," which traverses southeastern Asia, India, and southwestern Asia, is a modified form of *cP* air. As a result of downslope adiabatic heating, especially across the Himalaya, the air arrives across southern Asia as a warm, dry air mass; for example, typical monsoon air at Poona has a surface temperature of about 73°F and a specific humidity of about 7 g/kg, with a relative humidity of less than 40 per cent.

When *cP* air crosses an inland open water surface, for example, the Caspian Sea, the heating and absorption of moisture from below produce a steep lapse rate and a high moisture content in the lower layers of the air mass. This modification causes low cloudiness and frequent snow showers on the leeward side of the body of water. The most abrupt modifications of *cP* air masses arise when the air crosses an extensive warm water surface such as the Sea of Japan, Yellow and East China seas, South China Sea, Pacific Ocean, Bay of Bengal, and Arabian Sea. The rapid absorption of heat and moisture results in unstable air masses over the Pacific islands and the Indian Ocean. The instability is further aided by the absence of subsidence.

During the summer the source region of *cP* air is confined to northern latitudes, north of 50° north, where the thermal heating is not too pronounced. Over the Arctic Ocean the underlying surface consists of melting snow and ice water. Consequently, the air is very stable in the lower layers. As a result of the moisture source, stratus and fog are frequently observed. Over the continent, summer *cP* air masses are characterized by moderately low temperatures and low humidities. The pronounced winter stability, which arises from anticyclonic subsidence, is no longer present during the warm season. Generally the air masses are convectively unstable or convectively indifferent. Because of the moisture contents, the cumulus clouds are characterized by relatively high bases. As *cP* air moves south, the increased continental heating produces an unstable air mass similar to tropical continental air. The presence of clouds depends upon the moisture content of the air mass. Across northwestern Asia, *cP* air is usually moister than across north-central Asia, because of the proximity of regions of warm water. The higher moisture content causes increased cloudiness and precipitation over the northwest as compared with the northeast.

b. Tropical Maritime Air (*mT*). In winter the source region of tropical maritime air is confined to the southern portions of the oceans. Over the continent, as a result of the pronounced anticyclonic circulation, *mT* invasions are rare. At upper levels tropical maritime air invades eastern Europe during periods of strong cyclonic activity. Since this air mass has come from the eastern side of the subtropical Atlantic anticyclone, it is characterized by only moderately high temperatures and is convectively stable. Surface air masses with properties similar to mild *mT* air occasionally invade southwest Asia, for example, across Turkey, ahead of eastward-moving cyclones. The relatively high temperature and moisture content arise from the air-mass trajectory across the eastern Mediterranean Sea. However, because of the short trajectory, these air masses show the maritime influence only in the lower layers. Similarly, with the occasional cold-season cyclones across the Persian Gulf and India, short periods of warm maritime air invasions are sometimes observed over the southern portion of the continent. These air masses also differ from typical *mT* air since they are usually modified or returning *cP* air. Their oceanic trajectory, however, results in an air mass with the surface characteristics of tropical maritime air.

As in the case of the Atlantic, the unstable type of *mT* air is observed in the southwestern Pacific. Besides being convectively unstable, this air mass has a very high moisture content as well as high surface temperatures. Unstable tropical maritime air prevails over the eastern Dutch East Indies and over the islands to the west, where it has originated from *cP* air. In the latter case, the instability will be very pronounced because of the rapid modification of the originally very cold air mass. Northward invasions of *mT* air accompany the winter cyclones that move northeastward from the southwest Pacific. However, these air masses are rarely observed over the land since the cyclones, unlike those of southeastern North America, usually follow a maritime trajectory.

In summer, modified tropical maritime air prevails over India and southeastern and eastern Asia. Throughout this region the air masses are an unstable type of tropical maritime air because of the cyclonically controlled circulation. The instability is intensified over the land areas through the continental heating. Besides being convectively unstable, the *mTKu* air is also very moist. The high moisture content is also present at upper levels through the absence of subsidence; for example, at 3 km typical monsoon air (modified *mT* air) has a relative humidity of almost 80 per cent at Agra in contrast to about 60 per cent in a similar section of the southeastern United States. The persistent flow of this unstable air mass across southern and eastern Asia results in heavy convective and orographical precipitation on the windward side of the Asiatic plateau.

If the air mass becomes located over central Asia, it has the properties of tropical continental air. The moisture content is decidedly diminished as a result of the orographical lifting, while the temperature is increased through the addition of the latent heat of condensation.

A more stable type of *mT* air is found over the eastern Dutch East Indies and New Guinea. In the Southern Hemisphere the subsidence in the southeast trades accounts for the increased stability. Nevertheless, the air is still convectively unstable. Farther north, toward the Philippine Islands, the air mass becomes more unstable because of the convergence in the doldrum low-pressure belt. Here the air mass is identical with that of the southwest monsoon over India. The importance of subsidence as a stabilizing factor in regions of anticyclonic control and of convergence as an unstabilizing factor in low-pressure regions is reflected in the seasonal distribution of precipitation over the islands off southeastern Asia.

c. Polar Maritime Air (*mP*). With the offshore circulation along the east coast and the ice cover north of Siberia, the polar maritime air that invades Asia, in winter, originates over the North Atlantic Ocean. Whereas *mP* air is warm and moist over western Europe, it arrives in eastern Europe and western Asia considerably modified as a result of the land trajectory. Because of the east or northeast flow it is gradually cooled from below and acquires the stable characteristics of *cP* air. However, modified *mP* air may be distinguished from true *cP* air through the higher moisture content of the maritime air. This air mass is frequently observed at upper levels over western Russia, while a shallow layer of *cP* air is present at the surface. The overrunning of the cold surface mass may result in widespread frontal-type cloudiness.

In summer, the coldest type of *mP* air originates over the melting ice area of the arctic. This air mass, originally very stable and cold at the surface, is rapidly modified as it moves southward across the warm continent. The trend toward instability is one of the most significant modifications of the air mass as it moves into central Asia. Over the continent the *mP* air is quite similar to *cP* air, especially in the northwest.

The summer onshore circulation across northeastern Asia results in the presence of *mP* air over this section of the country. With its trajectory across the cold water near the coast this air mass is accompanied by sea fogs in the Kamchatka Peninsula area. Although the air is stable over this cold water, when it moves inland continental heating produces an air mass similar to that of northwestern Asia.

d. Tropical Continental Air (*cT*). The presence of tropical continental air over Asia is confined almost entirely to the summer season. In summer the source region consists of central and southwest Asia. True *cT* air is found over the dry land surfaces where the summer heating is quite intense, for example, in central Asia. Its characteristic properties are high temperature and low humidity. Because of the latter, air-mass clouds have high condensation levels. Over southeastern Europe the properties vary considerably owing to the inland bodies of water, which provide a moisture source. Here the absorption of moisture increases the convective instability and accounts for the summer showers and thunderstorms. Likewise, as true *cT* air streams northward it absorbs moisture and arrives, for example, over eastern Europe, as a relatively moist, convectively unstable air mass.

In general, the dryness of *cT* air is very favorable for a large diurnal variation in temperature. Over those regions where there is an upper-level subtropical anticyclone the *cT* air is very dry and warm aloft as a result of subsidence.

68. Fronts, Cyclones, and Anticyclones. Although Asia covers a vast area, it is apparent from Plate XVI that, during the winter half year, only the borders of the continent come under the direct influence of the important frontal zones. However, cyclonic disturbances, which develop along these regions of air-mass contrast, influence the weather and climate of some sections of Asia.

a. Mean Frontal Zones. The Atlantic polar and arctic fronts have been discussed in Sec. 62. In connection with the arctic front it is important to note that the eastward extent of this frontogenetical region depends on the presence of a mild, modified maritime air mass to the south and a cold *cP* air mass to the north. The intensity and eastward extent are therefore controlled by the locations of the Icelandic low-pressure system and the cold Siberian or arctic anticyclone. When a deep cyclone lies east of its normal Icelandic location, maritime Atlantic air covers eastern Europe and western Asia. Consequently, a frontogenetical zone is likely to stretch across northern Asia. However, should the Icelandic Low be farther west and the Siberian or arctic anticyclone extend westward across northwestern Asia, the arctic front would probably be nonexistent east of northern Scandinavia.

The Mediterranean front influences the weather of southwestern Asia. In addition to the previous description (Sec. 62) it is important to note that the front becomes weak and generally indistinct about 50° east. This arises from the frontolytical wind and temperature fields associated with the anticyclonic circulation over southern Asia. For similar reasons, continental Asia is significantly devoid of winter fronts.

The main Pacific polar front, in the western Pacific, resembles the Atlantic polar front in so far as it arises from the frontogenetical wind and temperature fields between the cold continental and warm oceanic anticyclones. However, unlike the corresponding Atlantic front, the Pacific polar front is rarely located across the southeastern section of the continent. The persistence of the Asiatic anticyclone across China results in a mean position of the front farther east of the continent than in the case of the Atlantic polar front.

The intertropical front, in this general region, is everywhere south of the equator, in January, crossing southern Sumatra and the Coral Sea. It is essentially the mean position of the doldrum belt. The most important feature of the intertropical front is the convergence of unstable maritime air masses toward the low-pressure trough. Generally the intertropical front is quite diffuse so that the region of convergence cannot be restricted to a narrow zone; for example, in the vicinity of southern New Guinea the zone may be considered to extend across approximately 15° latitude from the equator to northern Australia. For this reason the term "front" is a misnomer.

As mentioned in previous chapters, there is a pronounced northward shift of the main polar fronts with the approach of summer. In contrast, the arctic front is located farther south in summer than in winter. It separates the cold air located over the melting arctic ice from the warmer continental air. In summer, because of the less favorable wind field, the front is weaker but may, nevertheless, be very important in view of the general instability of summer continental air masses. Whereas, in winter, the eastward extension of the front is hindered by the cold anticyclone, in summer the front usually extends across northeastern Asia. Farther east it gradually merges into the Pacific polar front. The latter may be prolonged southward to northeastern China. It separates the moist *mT* air from the drier and slightly warmer air located over the continent. Again, even though the frontogenetical wind and temperature fields are weak, the front is significant because of the unstable air masses. Weak frontal lifting is often sufficient to release pronounced convective instability. Over the Mediterranean Sea during the summer season, anticyclonic circulation results in a frontolytical wind field.

In response to the continental heating the diffuse intertropical front lies well north of its January position. During July the convergent zone is located in the vicinity of the mean low-pressure trough, *i.e.*, across northern India, southern China, and the northern Philippine Islands. As a result of the meridional extent of the low-pressure trough the doldrum belt is even more diffuse in summer than in winter.

b. Extratropical Cyclones. Even though the major fronts are important climatological factors, the extratropical cyclones that develop in the general frontal regions have perhaps a greater effect on the weather regime. In Sec. 62 the cyclones of the eastern Atlantic were discussed in some detail. When these lows move eastward, they affect the weather of Russia. A review of the movements and modifications of cyclones that cross western Asia has been given by Rykatschew,[1] and the statistics that follow have been based upon this study of 16 years of cyclones. Since these data were compiled before 1900, it should be borne in mind that more recent observations may

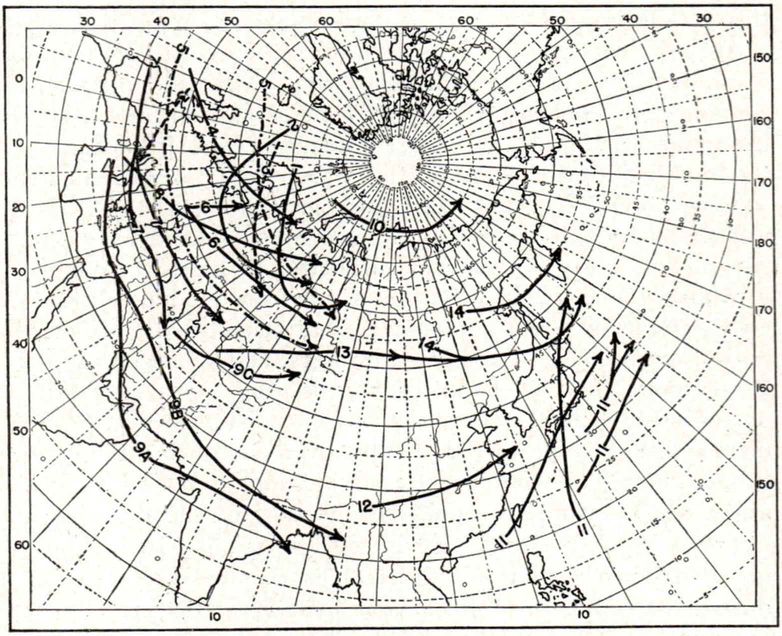

FIG. 35.—Extratropical cyclone tracks across Asia.

change the distribution somewhat. However, the major features are probably reliable.

The most noticeable change in the cyclone distribution is the decrease of cyclonic activity toward the east. This arises from the absence of air-mass contrasts over the continent as well as the absence of an immediate source region of moisture. Table 60 demonstrates the longitudinal and latitudinal distribution of cyclones across Europe, based upon 100 cyclones between 40° north and 70° north at longitude 10° east.

[1] RYKATSCHEW, M. I., *Met. Z.*, 14, Literaturbericht p. 9, 1897.

TABLE 60.—APPROXIMATE CYCLONE FREQUENCY

Lat., °N	10°E			30°E			50°E		
	Oct.–Mar.	Apr.–Sept.	Year	Oct.–Mar.	Apr.–Sept.	Year	Oct.–Mar.	Apr.–Sept.	Year
40–45	8	4	12	3	2	5	1	1	2
45–50	5	3	8	8	4	12	0–1	1	1
50–55	9	6	15	7	6	13	5	4	9
55–60	15	12	27	12	6	18	10	6	16
60–65	12	9	21	14	13	27	6	4	10
65–70	12	5	17	14	7	21	3	2	5
40–70	61	39	100	58	38	96	25	18	43

This table shows the secondary cold-season cyclonic maximum over the Gulf of Genoa as well as the cyclonic minimum over the mountainous region from 45° to 50° north. In general, it will be observed that the land region is a region of few cyclones as compared with the neighboring oceanic areas.

Except for those systems which develop over the Mediterranean, either the cyclones which invade eastern Europe are of Atlantic origin or they develop in regions adjacent to the ocean. The principal extratropical cyclone tracks across eastern Europe and Asia are illustrated, in an idealized manner, in Fig. 35. The distribution of cyclones that traverse western Asia is given in Table 61. On the average, the greatest number of cyclones occur in November and December, with the least number in June. Type 6 cyclones, of continental origin, are the main exceptions to this rule. Deepest cyclones are observed

TABLE 61.—APPROXIMATE DISTRIBUTION OF CYCLONES THAT INFLUENCE WESTERN ASIA
(After Rykatschew)

Season	Type								All types
	1	2	3	4	5	6	7	8	
Dec.–Feb.	6	3	5	1	8	1	2	3	29
Mar.–May	3	3	3	2	4	3	2	6	26
June–Aug.	2	1	1	2	4	3	0	4	17
Sept.–Nov.	4	3	4	2	7	2	1	5	28
Year	15	10	13	7	23	9	5	18	100

in January and shallowest in early summer. With the stronger winter circulation the low-pressure centers move faster in winter than summer.

When the cyclones move inland, they gradually fill because of the weaker temperature contrasts over the continent and because of the lower moisture content of the continental air masses. The presence of a semipermanent winter anticyclone prevents the movement of many cyclones beyond the Ural mountains. The migratory cyclones usually avoid the mountainous regions.

The displacements of high-pressure wedges and anticyclones are consistent with the cyclone movements. For example, cyclones of types 1 and 2 can move across northwestern Europe only when the continental anticyclone is displaced away from central Russia.

When deep winter cyclones move eastward, comparatively warm maritime air is displaced eastward across the continent and produces a short period of mild winter weather; for example, the deep cyclones of type 7 produce a flow of warm Mediterranean air across Turkey. Cold air masses move southward in the rear of the cyclones. It is evident, therefore, that the migratory pressure systems produce pronounced short-period temperature fluctuations.

Since the majority of the winter cyclones are in their dying stages, there is only relatively slight precipitation connected with these pressure systems. The rare occurrence of deep cyclones across Russia as compared with western Europe also accounts for the smaller amounts of winter precipitation over Russia than over western Europe. The cyclones, however, cause widespread frontal cloudiness as well as air-mass stratocumulus clouds within those regions of stirred, moist air masses. The large continental anticyclones are generally associated with clear skies because of the pronounced subsidence. However, when the cold air streams over a moist surface the accumulation of moisture from below results in air-mass cloudiness along the extremities of the anticyclones, for example, cold air moving westward across northern Russia. The importance of the cyclones and anticyclones is clearly illustrated by the winter distribution of cloudiness. The average cloudiness diminishes eastward across northern and central Asia.

The shallower summer cyclones and frontal displacements have an important influence upon the warm-season weather. In view of the instability of the summer air masses, these migratory systems are associated with convective showers. This is particularly noticeable across northern Russia and Siberia where the convective activity in the vicinity of the arctic front contributes to the west-east zone of maximum precipitation across the continent. In contrast, the absence of cyclonic and frontal displacements in the southwest is associated with very slight rainfall over the lands east of the Mediterranean Sea.

During the cold portion of the year some of the more intense Mediterranean cyclones move across southwest Asia to India, for example, types 9*A* and 9*B*, Fig. 35. The path across Syria, northern Iraq, Persia, and Afghanistan is one of the more important cyclone paths. These cyclones occur, on occasion, from October to May and may arrive over northwestern India as primary lows or secondary depressions associated with the primary Mediterranean cyclones. During midwinter, when the continental anticyclone is well developed, the cyclones follow a southern trajectory, while during the fall and spring their trajectories, similar to 9*C*, may be to the north of the western part of the Himalaya. The open-water areas of the Caspian Sea, Persian Gulf, and Arabian Sea provide the moisture sources for the warmer air masses connected with the depressions. Because of the predominant influence of the Asiatic anticyclone, these cyclones are rare, but they occur in sufficient number to provide the chief source of winter rainfall for southwest Asia and northern India. Their climatic significance will be discussed in further detail in Sec. 69.

Along the northern border of the Siberian mainland, infrequent summer cyclones (type 10) affect the weather of the Arctic Region. These cyclonic storms develop on the arctic front and have a warm-season maximum, especially to the east, since the winter circulation is dominated by a high-pressure area. The lows contribute to the late-summer precipitation maximum.

The families of cyclones that develop on the Pacific polar front, in winter, do not influence the climate of eastern Asia to the same extent as the similar families of cyclones that develop along the Atlantic polar front. This is the result of the general eastward location of the Pacific polar front. With the semipermanent anticyclone over southeastern Asia the development of deep cyclones over land is rare. In general, the majority form over the ocean, deepen, and move northeastward toward the Aleutian Islands region. Because of the marked air-mass contrasts as well as the high moisture content of the tropical maritime air, conditions are favorable for the development of deep cyclones accompanied by widespread precipitation. Some of the cyclones that move northeastward toward Japan are of tropical origin. They develop into extratropical disturbances after moving into higher latitudes.

This system of cyclones (type 11) is always accompanied by heavy precipitation. Since their trajectory is near the Asiatic coast and Japan, they contribute to the winter rainfall of these areas. During midwinter their paths are farthest from the continent. During spring and autumn, when the continental anticyclone is less pronounced, the lows are nearer to the coast. Consequently, frontal precipitation is most intense along the coast line in the months of April and October. The movement of cold anticyclones from the northwest results in very cold dry weather in the rear of the northeastward-moving cyclones. However, when this air crosses the ocean, it arrives over the Japanese islands as a moister and more unstable air mass. The instability and orographical precipitation over western Japan are just as significant as the frontal precipitation.

The summer cyclones of low and middle latitudes are principally of tropical origin and will be discussed later. In higher latitudes extratropical cyclones move across the Pacific coast toward Alaska during the summer months. They develop in the vicinity of the polar front and follow paths suggested by type 14 or, farther north, by type 10. The instability precipitation associated with these cyclones contributes to the summer and early-autumn precipitation maximum of the entire northeast—especially over the Amur region, the Sea of Okhotsk, and the Kamchatka Peninsula.

Previously it has been suggested that no winter cyclones traverse central Siberia and China because of the semipermanent anticyclone. However, during periods of weak anticyclogenesis some lows may traverse the entire continent to the Pacific coast. For example, in late winter and spring, occasional deep lows follow a path indicated by type 13. The cold air that sweeps southward in the rear of these systems produces severe blizzard conditions called the *buran*. Also, to the south, infrequent winter cyclones produce slight precipitation in China. In general, they follow paths, like type 12, along the Yangtze valley. This path is more frequent, however, in summer than winter. Type 12 cyclones finally move northeastward across the Pacific Ocean.

c. Tropical Cyclones. Cyclones of tropical origin influence the weather of southern and southeastern Asia during all seasons. This whole region is traversed by more tropical cyclones than any other section of the world. It is difficult to compare cyclone frequency as between one region and another because of the varying definitions of tropical storms and because of the uncertainty as to the correct charting of intense cyclonic systems. Besides these deep cyclones, there are a large number of weaker tropical lows, which traverse the same general areas. The main characteristics as well as the monthly distribution of the Asiatic tropical cyclones will now be discussed briefly.

1. INDIAN REGION. Tropical cyclones generally develop in regions of light and variable winds where the air is quite moist and unstable. Since the summer and winter monsoons are associated with steady winds from the equator northward, it is apparent that tropical cyclones are unlikely to form over the Indian Ocean either in midsummer or in midwinter. However, during the advance and retreat of the summer monsoon the wind circulation and air-mass properties are favorable for tropical cyclogenesis north of the equator. The relative monthly frequency of cyclones across the Arabian Sea and the Bay of Bengal is given by Table 62. The double frequency maxima correspond to the times of a weak monsoon circulation. This

TABLE 62.—APPROXIMATE PERCENTAGE DISTRIBUTION OF TROPICAL CYCLONES

Location	Months					
	Jan.–Feb.	Mar.–Apr.	May–June	July–Aug.	Sept.–Oct.	Nov.–Dec.
Arabian Sea.....	1	11	50	1	11	26
Bay of Bengal...	1	2	17	29	34	17

seasonal distribution differs from the frequency over other areas of the world where tropical cyclones are observed mainly in late summer. Over the Arabian Sea most tropical cyclones are observed during the spring. The absence of an autumn maximum is perhaps explained by the relative stability and low moisture content of the air that flows across the cool ocean from the hot, dry regions of northeastern Africa and Arabia.

Typical cyclone tracks are indicated by types 1 and 2 in Fig. 36. Again unlike the condition in other regions, the trajectories across the Arabian Sea are not parabolic. This fact arises

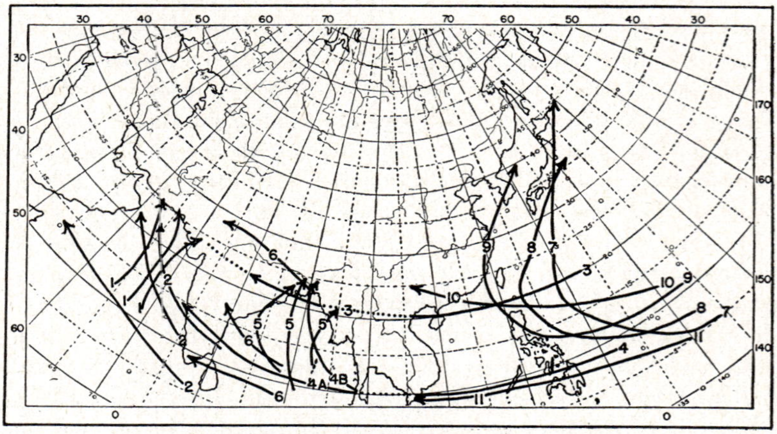

FIG. 36.—Tropical cyclone tracks of Asia.

from the absence of a subtropical anticyclone over the North Indian Ocean. During the cooler months, for example, April and November, they develop slightly south of 10° north. During the warmer months, their origin is farther north. The trajectories are quite erratic, but two general types may be distinguished, *viz.*, type 1 toward the north and type 2, which follows an east-southeast to west-northwest path. Cyclones of the latter class may have developed over the Bay of Bengal or, in the extreme, over the Philippine region. In general, these cyclones fill on crossing land but may redevelop on reaching the Arabian Sea.

The violent storms are accompanied by heavy precipitation, high seas, and destructive winds. Minor lows develop at the head of the Arabian Sea in July and August, but they usually remain shallow. Convective phenomena generally accompany these tropical disturbances.

The Bay of Bengal has at least twice as many tropical cyclones as the Arabian Sea area. Three main source regions and storm types may be distinguished, *viz.:*

i. China Sea or Pacific tropical cyclones (typhoons), which move due westward across Indo-China and Burma, then redevelop over the Bay of Bengal, and continue to move westward over India (types 3 and 4). These storms have been described by Iyer.[1] Of every 100 China typhoons, approximately 33 travel due west and strike the coast of southern China or the Indo-Chinese peninsula; about 12 of the latter type finally progress westward over India. Less than 20 per cent of types 3 and 4 finally develop into deep cyclones over the Bay of Bengal, whereas 50 per cent move on to India as very flat lows.

Tropical disturbances of this category are observed only during periods of well-established east winds aloft, *i.e.*, from July to October. In response to the latitudinal shift of the easterlies, type 3 paths are common in July and August, while type 4 paths occur from September to October. In the latter case, two main paths are observed across the Bay of Bengal. Type 4*A* lows move westward and may become type 2 lows of the Arabian Sea, and type 4*B* lows recurve northward or even northeastward on to Burma.

TABLE 63.—BAY OF BENGAL CYCLONES
(After Iyer)
(25 years' observations)

Item	Jan.–Feb.	Mar.–Apr.	May–June	July–Aug.	Sept.–Oct.	Nov.–Dec.	Total
Total number of cyclones......	3	7	56	92	109	54	321
Number from Far East.....	..	..	1	10	31	7	49

Heavy rainfall generally accompanies these cyclones whether they are shallow or deep pressure systems. Table 63 shows the number of deep or moderately deep cyclones in the Bay of Bengal that have a far-eastern origin. The cyclone paths across

[1] IYER, V. DORAISWARNY, *Mem. Indian Met. Dept.*, **26**, part 6.

India may vary within wide limits; for example, some fill shortly after moving inland, while others may be traced to northwest India.

ii. A second class develops over the southern section of the Bay of Bengal, about 10° north, and follows a somewhat erratic path toward the north (type 5). In general, the paths are parabolic and skirt the east coast of India. Type 5 cyclones have a maximum frequency in September and October. The heavy rainfall associated with these tropical disturbances contributes to the large precipitation amounts in the Ganges delta.

iii. Type 6 cyclones move principally in a westward direction and may appear anywhere from southern India to the Ganges valley. Possibly some of these disturbances belong to types 3 and 4. The July and August lows north of 20° north are rarely deep but nevertheless are accompanied by precipitation.

2. China Sea Region. This section is traversed by more tropical disturbances than any other region of the world. This high cyclone frequency is probably the result of a vast, yet distinct, area of converging air currents east of the Philippine Islands. The tropical air masses that invade this general low-pressure region come from (*a*) the Indian Ocean (the southwest monsoon), (*b*) the North Pacific Ocean (the northeast trades), and (*c*) the South Pacific Ocean (the deflected southeast trades). The pronounced instability of all air masses is undoubtedly favorable for the development of violent cyclonic storms.

Table 64.—Monthly Frequency of Tropical Cyclones in the Philippines

	Jan.	Feb.	Mar.	Apr.	May	June	July	Aug.	Sept.	Oct.	Nov.	Dec.	Year
Number of cyclones	0.5	<0.5	<0.5	0.5	1.0	2.0	3.5	3.5	4.0	3.0	2.5	1.0	22

The monthly distribution of the far-eastern tropical cyclones, or typhoons, is given in Table 64, which is based on observations in the vicinity of the Philippine Archipelago (see Algué[1]). Obviously, cyclones are rare during late winter when the northeast monsoon controls the atmospheric circulation. If the

[1] Algué, José: "The Cyclones of the Far East," Philippine Weather Bureau, Manila, 1904.

typhoons are subdivided according to the trajectory of the center, two main classes may be distinguished, as follows:

i. Typhoons with a parabolic path. The parabolic path arises from the decided tendency of a tropical disturbance to move around the subtropical anticyclone. Type 7 contains the rare winter storms (December–March), which recurve between 15° north and 20° north and move in a north-northeast direction across Japan as extratropical disturbances. In general, they move farthest east during February.

Type 8 cyclones occur during the months of April, May, October, and November. With the northward displacement of the subtropical anticyclone they recurve about 20° north and move northeastward across Japan. With the westward extension of the upper-level anticyclone and the development of the continental thermal low, type 8 typhoons are observed farther west than those of type 7.

The cyclones of the warmest months (June–September) recurve nearer 25° north and move along the Pacific coast of the mainland—type 9.

ii. Typhoons that move due west. As mentioned previously, a large number of the far-eastern tropical cyclones move directly toward the mainland. During July and August, when the upper-air easterlies appear at relatively high latitudes and when the doldrum belt is farthest north, the storms move on to the China coast about 23° north (type 10). As the winter season approaches, their trajectories are farther to the south so that by January and February the storms appear on the coast about 10° north (type 11).

In view of the high frequency of all tropical cyclones regardless of their intensity, these disturbances are very important; for example, the secondary August precipitation maximum of central China can be attributed to the heavy rainfall that is usually associated with tropical cyclones. Also, destructive winds and high seas may cause widespread damage, especially during August and September.

The meridional distribution of cyclones in the Philippine region shows that approximately 60 per cent of those crossing the archipelago pass north of Manila. Besides those cyclones, described above, others form and recurve farther to the east and consequently have no direct influence on the weather of the Philippines or Japan. In addition to the typhoons, shallow summer depressions move east across China, for example, down the Yangtze river, and then northeast in the direction of Japan (type 12, Fig. 35). All shallow cyclones of this nature contribute to the summer precipitation regime of China.

3. Cyclones of the Dutch East Indies Area. Since the Dutch East Indies and New Guinea are essentially confined to an area from 5° north to 10° south, this is not a region of violent tropical disturbances. However, some of the cyclones of the Australian area originate along the southern periphery of this region. For example, tropical cyclogenesis occurs over the Timor Sea from January to March. Since the main portion of the Dutch East Indies and New Guinea is unaffected by tropical cyclones, the discussion of these Southern Hemisphere systems will be undertaken in Sec. 80.

In conclusion, it should be borne in mind that tropical disturbances vary considerably in intensity and that individual cyclones may follow very irregular paths.

69. Climatic Regions. The principal climatic features of Asia are demonstrated by Plates I–V and IX–XVII, which indicate the mean distribution of the various meteorological elements. The various climatic types are summarized in Köppen's classification scheme (Frontispiece). The following subdivision has been based partly on the climatic regime and partly on the geographical and political boundaries.

a. Eastern Europe, Russia, and Siberia. In the discussion of central Europe it was shown that there is a gradual change in climate from west to east. Consequently, the climate of the eastern Baltic states and Finland will be treated together with that of Russia. Because of the absence of high mountain barriers there is a slow transition from the humid temperate climate (*Cf*) of western Europe to the cold continental climate (*Db*, *Dc*) of central and northern Siberia. North of 50° north, the boundary between the *C* and *D* climates is generally about 20° east.

Since the major portion of Russia is sheltered from invasions of warm, moist tropical air masses, the yearly precipitation is relatively low, generally less than 20 in. Furthermore, except over small regions, it is reasonably uniform across the continent. This distribution may be attributed to the lack of extensive north-south mountain barriers. As might be expected, the continental type of summer rainfall regime predominates, except in the extreme southwest, where there is a Mediterranean precipitation regime.

The region with a winter precipitation maximum is quite restricted. It includes the Black and Caspian Sea areas and extends as far as Lake Balkhash. This precipitation results from the cold-season cyclones illustrated by types 7, 9*C*, and 13 in Fig. 35. Many of the storms that originate over the Mediterranean do not traverse the land east of the Caspian; therefore, the winter rainfall or snowfall is generally considerably less than that of the Mediterranean countries. Sochi (29, Table 65) clearly demonstrates the autumn and winter rainfall regime over the Black Sea. Farther east, Krasnovodsk (34), Kyzyl Arvat (41), and Khorog (42) show a winter and spring maximum. However, there is a noticeable decrease in the total precipitation, as might be expected from the dryness of the air masses and from the scarcity of migratory cyclones.

The stations in the Turkestan region, for example, Khorog (42), Dzhizak (40), Krasnovodsk (34), Kopal (39), and Irgiz (38), definitely indicate a precipitation maximum in spring. This is the result partly of the cyclones of types 9*C* and 13 and partly of air-mass instability. The instability is probably caused by surface heating when the upper-air temperatures are still low. In summer, because of the dryness of the air masses and the absence of cyclones, the rainfall is very slight. The net result is a desert region from the eastern Caspian to Lake Balkhash.

The remainder of Russia is characterized by a late-summer precipitation maximum, mainly in the form of convective showers that result either from diurnal heating, especially toward the south, or from frontal activity, especially along the arctic front. Across the continent the maximum summer rainfall occurs about 55° north, for example, Bialystok (25), Zlatoust (31), Yeniseisk (36), Irkutsk (49), and Blagoveshchensk (50). Farther north, although the air masses are moist, the heating is insufficient to produce appreciable convective precipitation. Farther south, because of the dry air masses, the convective precipitation is necessarily slight. The band of maximum precipitation therefore coincides with the region of sufficient heating and sufficiently moist air masses to produce convective showers. Frontal lifting definitely aids in the production of showers and thunderstorms.

Over northwestern Russia the eastward-moving cyclones of fall and winter (types 1 to 5) result in slight cold-season precipitation, mainly in the form of snow. In view of the decreased cyclonic activity to the east this winter precipitation is more pronounced over the Baltic States than over central Russia. Nevertheless, even in the west, the summer convective type of precipitation is still the more important. In general, heavier precipitation amounts are observed in autumn than in spring. This is mainly the result of the increased cyclonic and frontal activity during autumn.

Along the Pacific coast the northeastward-moving cyclones produce heavier precipitation on the coast than over the interior. Since the cyclones of late summer have a trajectory near the continent, June to October is the season of maximum rainfall [see Novo-Mariinsk (53), Okhotsk (55), Tigil (56), Medny Island (57), Nikolaevsk (58), and Vladivostok (60) in Table 65].

The mean cloudiness varies mainly from northwest to southeast. In the northwest the stratuslike clouds during winter produce a mean cloudiness of seven or eight tenths of the sky. In the regions of anticyclonic subsidence and very dry air masses the winter cloudiness is only two or three tenths of the sky, for example, Blagoveshchensk (50) and Verkhoyansk (52). With the dry air masses even the Pacific maritime provinces, for example, the Vladivostok (60) region, have a very low mean cloudiness. West of 100° east, almost without exception, the average cloudiness is less in summer than in winter, despite the fact that summer is the season of maximum precipitation. During the warm season, convective clouds do not persist like low winter stratus or stratocumulus, and therefore the mean cloudiness is less in summer than in winter. East of 100° east, where the winter cloudiness is abnormally low, summer cumuli result in an average cloudiness of about six tenths of the sky. Considering the entire year, the region of lowest cloudiness is in the steppes and deserts of southwest Russia—for example, Irgiz (38), Khorog (42), etc., in Table 65—where the average July cloudiness may be only one tenth of the sky. One of the cloudiest periods of the entire continent is observed along the arctic coast in summer and early autumn, through the invasions of moist, foggy maritime air [see Dudinskoe (35) and Bulun (45)].

As mentioned previously, summer is the season of maximum convective activity. However, the thunderstorms are confined to those regions of intense heating where the air masses are sufficiently moist to produce towering cumulus clouds. For example, in the vicinity of the Caucasus, stations average about 10 thunderstorms during the month of June. Near the north coast of the continent the heating is insufficient; for example, Kem (21) averages less than 2 thunderstorm days a month in June and July. In Turkestan and adjacent dry-summer regions, the summer air masses are too dry to produce many thunderstorms; for example, Irgiz (38) averages about 3 a month in July. On the other hand, in the mid-continent region of appreciable summer precipitation, it is quite common to have 6 or more days with thunderstorms during July.

Two seasons of fog maxima may be detected. Over the continent, especially in the moister air to the west, cold-season radiation fogs are frequent. Except at very low temperatures, the snow cover tends to dissipate the midwinter fogs so that autumn is generally the period of maximum water-droplet fogs. In winter, over the continent, ice-crystal fogs occur at very low temperatures, mainly in the anticyclonic calms of Siberia. Along the coast lines and over the neighboring water surfaces, spring and summer are periods of maximum fogginess. Two regions notable for fogs are as follows:

1. Along the arctic coast the cold water produces a very pronounced summer fog maximum. Fifteen to 20 days of fog are common during July. As a result of the continental heating, this fog does not persist very far inland.

2. An equally foggy region is found over the cold-water areas around northeast Siberia, for example, near the Sea of Okhotsk and the Kamchatka Peninsula. This fog arises mainly from the cooling of maritime air from the south and is a maximum in July.

During the summer, thermal heating precludes long periods of fog over the continent.

Russia is characterized by the coldest winters of any section of the globe, except perhaps the Antarctic Continent. During the winter the prevailing southwesterly winds produce relatively mild temperatures in the west as compared with central Siberia. As a result of the maritime influence, the mean isotherms of

January have a general northwest to southeast orientation from 20° east to 130° east. The region of lowest temperature is found in the vicinity of Verkhoyansk (52), where clear skies and calms favor continued radiational cooling. The anticyclonic circulation over Siberia is everywhere accompanied by very low surface temperatures. In regions of moderate or fresh winds the surface temperatures are higher through the destruction of the surface inversion. However, even so, the cold winds produce very bitter weather. The increase in surface temperatures toward the Pacific coast is partly the result of the turbulent mixing associated with the fresh winds and partly the result of invasions of warmer air masses, even though they occur quite rarely. When occasional winter cyclones traverse the continent, the strong winds cause the most extreme of weather conditions, for example, the cold winds in the rear of type 9*C* or 13 cyclones. The latter winds are called the "buran" or "purga." The eastward bending of the isotherms along the north coast essentially arises from the following causes, of which the third is the most important:

1. The Atlantic Drift toward Novaya Zemlya accounts for warm water along northwestern Europe. Because of this warm current the Barents Sea is frozen for a shorter period than the Gulf of Archangel.

2. Conduction of heat through the ice produces milder temperatures over the frozen Arctic Sea than over the land.

3. Surface inversions are not common because of the lack of calms over the ice.

Continental heating produces very high summer surface temperatures over the southern sections of Russia. Since this heating is the most important factor determining the summer temperature, the July mean isotherms are orientated approximately from west to east. The warmest section coincides with the desert area in the southwest, as illustrated by Kyzyl Arvat (41). Over southeastern Russia the varying topography causes an erratic isothermal pattern.

The various inland water areas have a local effect upon the temperature regime. They produce warmer cold-season weather and cooler warm-season weather. For example, the mean July temperature at Irkutsk (49) near Lake Baikal is low as compared with neighboring stations. The Black, Caspian, and Aral seas

TABLE 65.—METEOROLOGICAL MEANS AT SELECTED STATIONS
(Temperature in degrees Fahrenheit, precipitation in inches, cloudiness in tenths)

Station	Lat., °N	Long., °E	Temperature				Precipitation				Cloudiness				Elevation, ft
			Jan.	Apr.	July	Oct.	Jan.	Apr.	July	Oct.	Jan.	Apr.	July	Oct.	
Longitude 20°–40°															
21. Kem	65	35	12	30	58	34	0.8	0.9	2.6	1.9	7	7	6	8	30
22. Helsingfors	60	25	22	36	62	42	2.1	1.6	2.4	2.7	8	6	5	7	40
23. Tartu	58	27	21	38	62	41	1.3	1.2	3.1	1.7	8	6	6	7	240
24. Moscow	56	38	13	38	64	39	1.4	1.7	3.1	2.3	8	6	5	7	550
25. Bialystok	53	23	24	44	65	45	1.1	1.7	3.4	1.3	7	6	6	6	450
26. Voronezh	52	39	14	42	69	42	1.3	1.4	2.2	1.5	6	5	4	5	400
27. Tarnopol	50	26	21	44	65	46	1.0	1.7	3.5	1.9	7	5	5	6	1050
28. Sulina	45	30	31	49	71	55	0.9	1.1	1.6	1.3	6	4	2	5	10
29. Sochi	44	40	42	52	72	61	6.6	3.6	3.4	4.6	6	6	3	4	260
Longitude 40°–60°															
30. Pustozersk	68	52	−1	19	54	28	0.7	0.7	1.7	1.3	7	7	7	8	40
31. Zlatoust	55	60	4	33	61	32	0.6	0.9	4.0	1.5	7	6	6	8	1500
32. Fort Aleksandrovsk	45	50	25	49	79	53	0.3	0.2	1.0	0.6	7	6	4	5	80
33. Vladikavkaz	43	45	24	46	69	49	0.9	3.1	4.3	2.0	7	7	6	6	2230
34. Krasnovodsk	40	53	36	56	83	63	0.6	0.8	0.1	0.2	6	5	3	3	−50
Longitude 60°–100°															
35. Dudinskoe	69	86	−22	3	55	15	0.2	0.3	1.2	0.7	6	7	7	8	70
36. Yeniseisk	58	92	−8	29	66	30	0.8	0.7	2.4	1.5	5	3	5	8	260
37. Minusinsk	54	92	−3	36	68	34	0.4	0.5	2.0	0.8	..	..	..	..	840
38. Irgiz	49	61	4	43	77	43	0.4	0.6	0.2	0.7	6	4	4	4	420

TABLE 65.—METEOROLOGICAL MEANS AT SELECTED STATIONS.—(*Continued*)

Station	Lat., °N	Long., °E	Temperature				Precipitation				Cloudiness				Elevation, ft
			Jan.	Apr.	July	Oct.	Jan.	Apr.	July	Oct.	Jan.	Apr.	July	Oct.	
39. Kopal	45	79	19	45	68	44	0.5	1.8	1.1	1.1	5	6	4	4	4060
40. Dzhizak	40	68	30	59	83	56	2.5	2.7	0.2	1.1	6	5	1	3	1270
41. Kyzyl Arvat	39	56	31	60	88	60	1.1	0.9	0.2	0.6	6	5	2	3	330
42. Khorog	37	72	18	48	72	51	1.1	1.1	0.1	0.4	5	4	2	3	6880
Longitude 100°–130°															
43. Khatanga	72	102	−31	−2	53	9	0.2	0.5	1.6	0.8	4	6	7	8	160
44. Sagastyr	72	124	−34	−7	41	6	0.1	0.0	0.3	0.1	4	5	8	7	20
45. Bulun	71	127	−42	4	54	10	0.3	0.6	1.3	1.0	6	6	8	9	110
46. Vilyuisk	64	122	−39	18	65	17	0.4	0.3	1.3	0.7	7	6	7	8	390
47. Markhanskoe	62	130	−47	18	65	17	0.3	0.6	1.7	0.6	4	6	7	8	320
48. Tunka	52	103	−14	34	63	30	0.1	0.4	2.4	0.3	..	..	..	..	2380
49. Irkutsk	52	104	−6	33	63	31	0.4	0.6	3.0	0.7	5	6	7	6	1530
50. Blagoveshchensk	50	128	−12	36	70	35	0.1	1.0	5.0	0.7	3	5	6	5	440
East of 130°															
51. Nizhne-Kolymsk	69	161	−36	4	53	12	0.5	0.2	1.5	0.5	5	5	6	6	20
52. Verkhoyansk	68	133	−58	8	59	6	0.2	0.2	1.1	0.3	3	5	6	6	400
53. Novo-Mariinsk	65	177	−10	5	51	23	0.3	0.2	1.2	0.5	5	6	7	7	20
54. Ust Maiskaya	60	134	−44	17	66	19	0.1	0.3	1.2	0.3	2	3	4	6	330
55. Okhotsk	59	143	−13	21	53	28	0.1	0.4	2.3	0.9	4	6	8	5	20
56. Tigil	58	158	−3	27	50	31	0.6	0.7	3.1	1.6	..	..	..	..	70
57. Medny Island	55	167	25	32	47	40	1.8	1.6	2.0	2.7	..	..	..	..	50
58. Nikolaevsk	53	141	−12	27	62	36	0.6	1.1	2.0	1.9	4	6	7	6	70
59. Korsakovsk	47	143	12	34	58	45	1.0	1.5	2.1	2.2	6	5	7	6	120
60. Vladivostok	43	132	7	40	65	49	0.2	1.3	2.9	1.6	3	6	8	5	100

and the Sea of Azov as well as the inland Lakes Balkhash and Baikal produce local climatic regimes.

Except for southern and southwestern Russia, the remainder of Russia is essentially a region of cold climate. From Köppen's classification it is seen that types *D* and *E* predominate. Over southwestern Russia there is a large area of dry climates—the desert climate of Turkestan flanked on all sides by the steppe type of climate.

b. Southwest Asia. Southwest Asia comprises those land areas, south of Russia, from Turkey and Arabia to Afghanistan and Baluchistan. Unlike Russia it is a region of varying topography, and moreover it flanks the warm waters of the Mediterranean and Arabian seas. Consequently, rapid weather changes may be expected, depending on whether the air streams south from Russia or north and east from the warm humid seas. The winter weather is controlled by the extensive high-pressure system of central Asia, together with the migratory cyclones which originate over the Mediterranean Sea (types 7, 9*A*, and 9*B*, Fig. 35). These low-pressure systems result in cold-season precipitation, with a maximum near the coast lines and smaller cyclonic precipitation over the high interior. Rapid temperature changes also accompany the passage of the cyclones; for example, the warm dry sirocco of Palestine and Syria arises from the southerly winds in advance of a cyclone, while the cold northerly winds in the rear of a deep cyclone (especially during the spring) produce extremely cold weather. In summer the greater part of southwest Asia is under the influence of dry air masses so that clear skies and intense sunshine control the weather. The coastal regions are cooler and moister than the interior. In the absence of moist air masses and migratory cyclones, the summer precipitation is generally very light except immediately south of the Caucasus, where convective rainfall is quite common. The dryness and upper-level stability of the summer air masses result from the subsidence associated with the high-level anticyclone.

Asia Minor is a region of varying climate. Heavy autumn and winter precipitation occurs along the north coast in connection with the Black Sea cyclones, while summer precipitation results from convective showers. As shown by Batum (71), Tiflis (72), and Baku (73) in Table 66, the winter cyclonic precipitation is confined mainly to the coast; likewise, the summer convective rainfall decreases rapidly toward the east. Over the interior, for example, at Mazara (62) and Urfa (63), the precipitation is confined to the cold season. A similar rainfall regime (*Cs*) exists over southern Turkey (Mediterranean type of climate). With the oceanic influence, the coastal areas have mild winters and relatively cool summers as compared with the continental temperature extremes of the interior; for example, compare Istanbul (61) and Mazara (62). Fog occurs during winter and spring over northern Asia Minor as the result of either frontal precipitation or radiational cooling. However, the fog frequency is relatively low. Radiational ground fog is observed in the valleys of the interior during the cold half year. Along the south coast of Turkey, fog is not common. Except in regions of appreciable summer rainfall, thunderstorms are infrequent. For example, Tiflis (72) has 10 days with thunder in June as compared with only 1 day at Baku (73).

Syria and Palestine have a Mediterranean type of climate with winter cyclonic rainfall and summer dryness. The latter is the result of the persistent anticyclonic control, especially at upper levels. The winter rainfall is higher over the western plateaus than along the coast since orographical lifting contributes to the frontal precipitation; for example, Jerusalem (66) (2460 ft) averages 6.5 in. in January as compared with 5.0 in. at Jaffa (70 ft). Toward the interior the climate changes rapidly from the Mediterranean type to the desert or steppe type with cold, dry winters and very warm, dry summers.

The major portion of Arabia has a desert climate. Summer convective precipitation occurs in the mountainous region of Yemen; consequently, this is an area of mild climate with warm summer rain and winter dryness. As demonstrated by Muscat (68) in Table 66, the cyclones that traverse the Persian Gulf cause only slight winter rainfall along the north coast. Except over the mountains of southern Arabia, very high temperatures prevail during summer.

Iraq has a steppe type of climate. The relatively slight precipitation, which occurs in winter and spring, is produced by the cyclones that move eastward from Cyprus through the Persian Gulf. The spring rainfall is often associated with thunderstorms; for example, Baghdad (69) has about 2 or 3 days

TABLE 66.—METEOROLOGICAL MEANS AT SELECTED STATIONS
(Temperature in degrees Fahrenheit, precipitation in inches, cloudiness in tenths)

Station	Lat., °N	Long., °E	Temperature				Precipitation				Cloudiness				Elevation, ft
			Jan.	Apr.	July	Oct.	Jan.	Apr.	July	Oct.	Jan.	Apr.	July	Oct.	
Turkey															
61. Istanbul	41	29	42	53	74	63	3.4	1.7	1.0	2.5	6	6	3	5	250
62. Mazara	39	39	21	53	77	57	1.4	2.4	0.1	1.4	6	5	1	3	3540
63. Urfa	37	39	39	62	89	69	2.7	1.2	0.0	0.4	4	4	1	2	1850
Syria															
64. Beyruth	34	35	56	64	80	75	7.5	2.1	0.0	1.9	5	4	2	4	110
Cyprus															
65. Nicosia	35	33	49	62	83	71	2.9	1.0	0.1	0.7	4	3	1	2	500
Palestine															
66. Jerusalem	32	35	45	59	73	66	6.5	1.6	0.0	0.4	5	4	1	2	2460
Arabia															
67. Aden	13	45	76	83	88	84	0.3	0.2	0.0	0.1	..	..	..	..	90
68. Muscat	24	59	71	83	91	85	1.1	0.4	0.0	0.1	..	..	..	..	20
Iraq															
69. Baghdad	33	44	49	71	94	78	1.3	0.9	0.0	0.1	3	2	0	2	200
70. Basra	30	48	52	74	93	78	1.4	0.5	0.0	0.0	..	..	..	..	20
South of Caucasus															
71. Batum	42	42	43	52	73	62	9.3	4.9	6.4	9.4	5	6	5	4	10
72. Tiflis	42	45	32	53	76	57	0.6	2.3	1.8	1.6	6	6	4	5	1340
73. Baku	40	50	38	51	77	62	0.7	0.7	0.2	0.9	6	5	4	5	−40
Iran															
74. Teheran	36	51	34	61	85	66	1.8	1.4	0.0	0.4	7	6	2	4	3810
75. Meshed	36	60	34	57	78	57	0.3	1.9	0.0	0.6	6	5	1	3	3050
76. Kerman	30	57	45	61	81	64	0.5	0.7	0.0	0.0	2	..	1	..	6100
77. Jask	26	58	67	79	90	82	0.6	0.0	0.0	0.0	..	..	..	..	20
Afghanistan															
78. Kabul	35	69	31	59	77	59	1.0	2.2	0.2	0.1	5	4	0	1	6230
Baluchistan															
79. Quetta	30	67	40	61	78	56	1.9	1.1	0.5	0.1	..	..	..	..	5510
80. Kalat	29	67	36	57	73	52	1.8	0.5	0.5	0.0	..	..	..	..	7150

with thunderstorms in April and May as compared with none in July. Because of the predominant influence of hot, dry air masses, summer is a season of intense heating and slight cloudiness. During the winter the eastward movement of Mediterranean cyclones causes abrupt temperature changes with warm air in advance and very cold air in the rear of the moving low-pressure systems. The cold air mass originates over the cold plateau of Iran.

Because of their elevation, Iran, Afghanistan, and Baluchistan have milder summers than Iraq. Precipitation occurs mainly in connection with the winter cyclones of types 7 and 9*B*. However, the total fall is relatively low since the moving low-pressure systems are not very frequent and since the air masses do not have a high moisture content. Spring thunderstorms also contribute to the scant yearly rainfall. Except for the Arabian Sea coast of Baluchistan, clear skies predominate in summer. Along the eastern coast line the southwest monsoon produces appreciable summer rainfall.

c. Central Asia. Central Asia comprises the high plateau of Tibet and the vast basin that extends from the Pamirs in the west to the Khingan mountains in the east.

Almost the entire country of Tibet exceeds 12,000 ft in elevation, with the highest mountains in the south—the Himalaya. This high mountain chain shields the country to the north from invasions of moist air masses. In view of the high elevation the temperature is low in all seasons of the year as shown in Table 67, in which the doubtful meteorological means have been included in order to illustrate the broad climatic features. In general, two rainfall regimes may be detected over Tibet. In the west, wintertime precipitation results from the cyclones that traverse Afghanistan and northwest India. Summer rainfall also occurs in this region; but, as with the winter precipitation, the total fall is slight. The dryness of western Tibet is a result of the sheltering effect of the Himalaya. Over eastern Tibet the mountains are lower; consequently, moderate precipitation occurs during midsummer. The total rainfall—see Gyangtse (82)—is considerably smaller than along the southern part of the Himalaya but is sufficient to form a source of water for the Yangtze, Makong, and Salween rivers and to maintain vegetation. Actually, the river valleys of southeastern Tibet are quite fertile. Because of the sheltering from the west and the cold anticyclone to the immediate north, eastern Tibet has very dry winters.

The elevation precludes very high summer temperatures and is favorable for long periods of low winter temperatures. The seasonal changes are illustrated by Leh (83) and Gyangtse (82) in Table 67. The highest portions of Tibet have therefore a polar tundra climate (*ETH*), while the remainder of the area has a snow-forest type of climate.

Situated in the heart of Asia, Mongolia, Sinkiang, and western Manchukuo are surrounded on all sides by high mountain ranges. The climate is characterized by slight precipitation, large seasonal variation in temperature, and very dry air. The large desert region includes the Gobi, the Tarim and Dzungari basins, and smaller desert areas. The rainfall in the west averages less than 4 in. and occurs principally in the form of summer convective showers within the modified *cPK* air masses from the north. The Kashgar (85) and Pamirski Post (86) means in Table 67 demonstrate the effect of the sheltering of the high mountain ranges. The Gobi also receives its slight rainfall in the form of summer thundershowers. The original moisture source is probably the southeast monsoon. On rare occasions extra-tropical cyclones yield slight amounts of precipitation. The temperature regime is characterized by a very large annual and diurnal range. Excessively high daytime temperatures occur in the low-lying regions during the summer, for example, in the vicinity of Lukchun (88). The violent spring dust storms contribute to the severity of the climate.

A desert-steppe type of climate surrounds these desert regions. The rainfall is slight, about 6 in. a year, and occurs almost entirely in the form of summer convective showers, while the temperature regime resembles that of the desert. The true steppe climate is farther removed from the extreme desert areas and comprises north and northwestern Mongolia, inner Manchukuo, the slopes of the Khingan mountains, and the mountain slopes of Sinkiang. Here the yearly rainfall averages about 8 to 12 in., nearly 80 per cent occurring between July and September. The total rainfall fluctuates considerably from year to year. As shown by the stations in Table 67 the summers are warm, while the continental cooling results in extremely cold winters.

TABLE 67.—METEOROLOGICAL MEANS AT SELECTED STATIONS
(Temperature in degrees Fahrenheit, precipitation in inches, cloudiness in tenths)

Station	Lat., °N	Long., °E	Temperature				Precipitation				Cloudiness				Elevation, ft
			Jan.	Apr.	July	Oct.	Jan.	Apr.	July	Oct.	Jan.	Apr.	July	Oct.	
Tibet															
81. Lhasa	30	91	27	..	62	..	...	...	...	...	..	..	..	..	12,890
82. Gyangtse	29	90	27	44	59	45	0.0	0.1	2.2	0.1	1	2	8	2	13,090
83. Leh	34	78	17	43	63	43	0.4	0.2	0.5	0.2	..	..	..	..	11,520
Kashmir															
84. Srinagar	34	75	31	56	74	55	2.7	3.8	2.2	1.2	8	6	4	2	5,200
Eastern Turkestan															
85. Kashgar	39	76	22	60	80	56	0.2	0.8	0.2	0.0	5	5	3	2	4,250
The Pamirs															
86. Pamirski Post	38	74	1	33	56	32	0.2	0.2	0.5	0.1	5	5	4	5	11,980
Russia															
87. Karakol	42	78	23	45	62	43	0.7	1.7	2.4	1.4	5	6	5	4	5,820
Sinkiang															
88. Lukchun	43	90	13	66	90	56	...	...	...	...	3	5	4	3	−50
89. Yarkand	38	77	21	64	82	56	...	...	...	...	..	..	..	..	4,120
90. Urumchi	44	87	−3	46	75	44	0.2	1.6	0.9	0.7	..	..	..	..	3,000
Mongolia															
91. Urga	48	107	−15	34	64	29	0.0	...	2.4	...	2	..	4	..	3,800
Ningsia															
92. Liangchow	38	103	20	53	73	49	...	...	...	...	..	..	..	..	1,640
Northwest Manchukuo															
93. Hailar	49	120	−19	33	70	32	0.2	0.4	3.1	0.4	..	..	..	..	2,030
94. Manchouli	49	117	−15	33	70	32	0.1	0.2	2.9	0.3	..	..	..	..	2,150
Suiyuan															
95. Saratsi	41	110	5	47	73	44	0.1	0.3	4.0	0.6	..	..	..	..	3,360

Data doubtful.

Since the climate is not favorable for habitation a large section of the heart of Asia is devoid of meteorological data. The values included in Table 67 are short-term means and therefore are subject to question. However this scant observational evidence serves to illustrate the principal climatic features, *viz.*, very cold dry winters and warm summers with slight or moderate rainfall.

d. India, Ceylon, and Burma. In its broadest sense the climate of India is controlled by the monsoon circulation. January and February are principally cool and dry as a result of the offshore circulation. The continental anticyclonic conditions produce slight precipitation and cloudless skies. However, in the northwest, the cyclones of type 9 in Fig. 35 produce appreciable precipitation for this region, as shown by Srinagar (100) in Table 68. Furthermore, cold-season precipitation occurs in the extreme south because of the instability showers connected with equatorial low-pressure areas.

From March to mid-June the weather is very hot and dry. This type of weather arises from the strong thermal heating with the approach of summer while there is no strong onshore circulation. During this period an equatorial low-pressure area is located to the south of the continent, so that there is no deep current of air from the Indian Ocean toward northern India, as in July and August. The inland regions of northern and central India are generally unaffected by cyclones and are under the influence of warm, dry air masses. Clear skies predominate, as shown by the mean cloudiness at Agra (99). Dust haze is very common. Along the coastal areas and over Ceylon the sea breezes favor lower temperatures than those observed in the interior. Some rainfall occurs in the dry northwest principally in the form of convective showers and thunderstorms. These are connected with the weak cyclones of type 9. The rainfall of Bengal, Assam, and Burma is also convective and arises from the heating of the moist air from the Bay of Bengal. Thunderstorms are common and are occasionally very violent, like tornadoes.

From mid-June to mid-September there is a strong onshore circulation from the Indian Ocean toward the thermal low-pressure region of northwest India—the southwest monsoon. The presence of this warm, moist, and convectively unstable air mass results in a period of general rains and cooler weather than that of May and June. This is the so-called "monsoon" type of annual temperature variation (see Sec. 8). Orographical lifting favors very heavy precipitation along the Western Ghats and over Assam, western Burma, and the southern slopes of the Himalaya. Particularly across northern India, the rainfall is associated with the westward movement of shallow cyclones (type 6, Fig. 36). These cyclones move along the low-pressure trough toward Rajputana. Tropical disturbances from the Bay of Bengal, discussed in Sec. 68, help to intensify the precipitation. In general, north of 20° north, the rainfall decreases from Assam to Baluchistan. In the northwest the air is drier and more stable since it has originated as an Arabian or African *cT* air mass. The stability is caused by the surface cooling over the Arabian Sea, while the relatively low moisture content is a result of the short oceanic trajectory.

The period from mid-September to December is characterized by clear skies, mild temperatures, and light winds. The cyclones from the Bay of Bengal, some of hurricane intensity, contribute to an October and November precipitation maximum in Madras. Other land areas bordering on the Bay of Bengal also receive precipitation in connection with the late-summer tropical disturbances.

This brief discussion indicates that the climate of India differs from that of other tropical countries. There is a very pronounced variation in the rainfall regime of different sections of the country. These variations will now be discussed in more detail by considering the prominent features of Fig. 37. This diagram shows the average rainfall for 33 subdivisions of India and Burma and indicates the season of maximum rainfall (over 75 per cent of the yearly total) for each section of the country. The solid lines have been drawn around the provincial boundaries, for no attempt has been made to distinguish between variations within a subdivision. More detailed data would probably modify the diagram somewhat; for example, Lahore in the northern Punjab has appreciable late-winter rainfall like the Northwest Frontier. The Northwest Provinces, Punjab, Rajputana, and Sind comprise the most arid region of India. The summer monsoon is the principal rain-bearing wind; as already mentioned, it produces only slight precipitation. In the west and north the cyclones from the west produce late-winter precipitation, for example, the rainfall regimes at Srinagar (100), Lahore (101), and Jacobabad (108). The hottest season is from April to June, when clear skies predominate. After the summer rainy period, *i.e.*, from October to December, clear skies favor pleasant mild weather. The rainfall and temperature distributions indicate that the western section has a steppe climate while the eastern section has a desert climate.

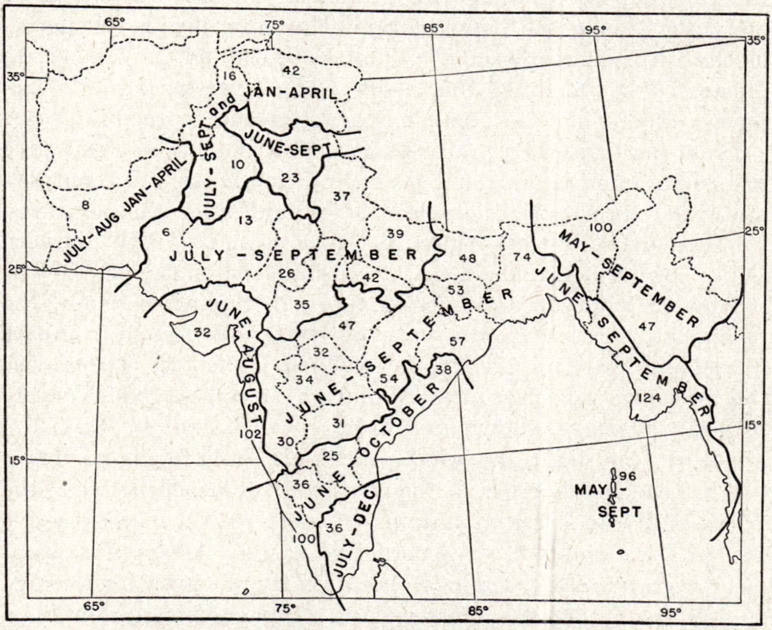

FIG. 37.—Rainfall variation over India. Numbers represent the average yearly amount, in inches, in each subdivision. Months represent the period during which there occurs at least 75 per cent of the yearly total.

South of the Tropic of Cancer, the major part of India has a tropical savanna or tropical rain-forest climate. The rainfall of the middle and late summer months is associated with the southwest monsoon. The heaviest falls occur along the western Ghats because of orographical lifting, while Bay of Bengal cyclones produce another region of heavy precipitation in Bengal and Orissa. With the sheltering of the western mountains, the western central provinces and the Deccan have a yearly rainfall between 20 and 40 in. Relatively high temperatures prevail during all seasons of the year, with the maximum in the early summer. The maximum cloudiness is observed in late summer and the minimum in midwinter, when clear skies prevail.

Because of different rainfall regimes two areas in southern India are worthy of note. Immediately east of the Ghats, about 15° north, the sheltering of the mountains produces a region of relatively slight rainfall. Even during late summer the rainfall does not exceed 5 in. a month. Over southeast India and east Ceylon the heaviest rainfall occurs during the period of the retreating monsoon, *viz.*, in October and November, as illustrated by Madras (105, Table 68). Even western Ceylon, for example, Colombo (109), has heavy autumn precipitation.

From the United Provinces to Assam the shallow cyclones give rise to summer rainfall and persistent cloudiness. In contrast, midwinter is a period of clear skies and mild temperatures. Consequently, this is a region of *C* type climate. The southern slopes of the Himalaya have a similar climate. Since the southwest wind does not cross the high mountain chain, the precipitation is less than over some sections of Assam [compare Darjeeling (98) and Cherrapunji (97) in Table 68]. Moreover, the precipitation over the Himalaya decreases considerably toward the west, as shown by Srinagar (100). Leh (83, Table 67) illustrates the slight precipitation at high levels to the north.

With the convergence of the southwest monsoon between the Khasi Hills and the mountains of western Burma, excessive precipitation amounts are recorded in eastern Assam; for example, Cherrapunji (97, Table 68) has a yearly precipitation average of 428 in. Even though this area of phenomenal rainfall is very restricted, the remainder of northeastern India is characterized by heavy late-summer precipitation. The moist winds and cyclones from the Bay of Bengal are the cause of this heavy rainfall. As in the remainder of northern India, clear skies and mild temperatures predominate during the winter. Besides the monsoon rainfall, convective showers and severe thunderstorms produce appreciable springtime precipitation.

Like that of India, the climate of Burma is controlled mainly by the monsoon circulation. Very cloudy skies and heavy precipitation prevail from June to September, in contrast to clear skies and slight precipitation during winter. The geographical distribution of precipitation is principally the result of the

TABLE 68.—METEOROLOGICAL MEANS AT SELECTED STATIONS
(Temperature in degrees Fahrenheit, precipitation in inches, cloudiness in tenths)

Station	Lat., °N	Long., °E	Temperature				Precipitation				Cloudiness				Elevation, ft
			Jan.	Apr.	July	Oct.	Jan.	Apr.	July	Oct.	Jan.	Apr.	July	Oct.	
India															
96. Calcutta	23	88	65	85	83	80	0.4	1.8	12.9	3.9	2	3	9	5	20
97. Cherrapunji	25	92	..	..	..	..	0.7	33.1	99.5	13.8	..	..	..	..	4310
98. Darjeeling	27	88	40	56	62	55	0.6	3.8	32.3	4.5	..	..	..	..	7380
99. Agra	27	78	58	86	86	76	0.4	0.2	8.8	0.7	2	2	7	1	550
100. Srinagar	34	75	31	56	74	55	2.7	3.8	2.2	1.2	8	6	4	2	5200
101. Lahore	32	74	52	80	88	75	0.9	0.5	5.1	0.3	4	4	4	1	700
102. Karachi	25	67	67	80	85	81	0.6	0.2	3.1	0.0	3	2	8	1	10
103. Bombay	19	73	74	82	81	81	0.1	0.1	25.2	1.8	2	1	9	2	40
104. Bangalore	13	77	67	79	71	71	0.3	1.3	4.2	5.9	4	5	9	8	3020
105. Madras	13	80	75	84	85	81	1.0	0.7	3.9	10.9	6	5	9	7	20
106. Hyderabad	17	78	72	89	80	79	0.2	0.8	6.1	3.2	..	..	..	..	1720
107. Trivandrum	9	77	77	82	77	78	0.7	4.5	7.4	10.6	5	6	8	7	200
108. Jacobabad	28	68	57	86	95	79	0.3	0.2	1.0	0.0	..	..	..	..	190
Ceylon															
109. Colombo	7	80	79	82	80	79	3.5	7.8	6.5	13.1	..	..	..	..	20
Burma															
110. Myitkyina	25	97	62	78	81	80	0.4	2.0	19.2	6.8	4	5	8	5	480
111. Mandalay	22	96	68	89	85	81	0.1	1.2	2.7	4.3	2	3	8	5	250
112. Akyab	20	93	70	84	81	82	0.1	2.1	52.8	10.9	1	4	9	5	30
113. Yamethin	20	96	71	88	82	82	0.1	1.5	3.9	5.6	2	2	8	5	640
114. Rangoon	17	96	77	87	81	82	0.2	1.6	17.8	6.9	4	5	10	6	20

topography. Coastal regions and the western slopes of the mountains have heavy falls as shown by Myitkyina (110), Akyab (112), and Rangoon (114), while sheltered sections have relatively light amounts of precipitation, as, for example, Mandalay (111) and Yamethin (113). As in the case of India, the highest temperatures are observed during early summer before the cloudy monsoon season. Occasional cyclones from the Bay of Bengal affect the northwest coast and are associated with heavy precipitation. On the other hand, China Sea typhoons become quite indistinct by the time they reach Burma. They may give rise, however, to widespread cloudiness.

e. China, Japan, and Southeast Asia. Since the inland areas of China have been mentioned in connection with central Asia, the following discussion will be confined mainly to the maritime provinces.

As in the case of India, the climate of China is controlled by the monsoon circulation. With the cold air flow from central Asia, Chinese winters are abnormally cold and dry. The south and southeast winds in summer result in a period of warm, humid weather.

North China, from 35° north, has essentially a snow-forest type of climate. About 90 per cent of the precipitation occurs from May to September, when the modified unstable mT air masses predominate. Heat convection, weak frontal lifting, and convergence within occasional cyclones are the principal causes of summer precipitation. As shown by the stations in Manchukuo, Korea, and North China (115 to 121, Table 69), the winter precipitation is considerably smaller than that of the corresponding region of North America. In the latter, cyclones cross the eastern section of the continent and produce appreciable precipitation. However, winter cyclones rarely cross the eastern Asiatic continent; instead, they have a trajectory across the western Pacific (type 11, Fig. 35). Even Korea has slight cold-season precipitation as shown by Gensan (118) in Table 69. The small increase over the inland stations is probably the result of its location nearer to the Pacific cyclones. Rare winter cyclones from the Yangtze valley also pass near the Korean peninsula. The seasonal distribution of cloudiness is very similar to the precipitation distribution, with a summer maximum and a winter minimum. Fog commonly occurs along the coastal regions, as a result of the advective cooling of warmer maritime air masses over the cold coastal water. Similar to the fog of eastern North America, this sea fog is most frequent during spring and summer. Midwinter fogs are rare.

North China is very cold in winter, particularly during periods of strong cold outbreaks. The fresh winds carry large masses of dust from the interior across the maritime provinces, the so-called "Peking dust." The summer heating produces hot weather from June to August (*Dwa* climate).

Central China has a temperate climate with cold winters and hot summers. In the Szechwan province the winters are warmer than over the coastal regions. This unusual temperature distribution is caused by the mountains, which shelter the Red Basin of Szechwan from the direct outbreaks of cold continental air. For example, the mean January temperature at Chungking (124) is 8°F higher than at Shanghai (122).

Even though over 60 per cent of the precipitation of central China occurs in the months of May to September, there is no month without some precipitation. The cold-season rain and snow are connected with the occasional cyclones that follow the Yangtze valley (type 12, Fig. 35). Since moist air masses rarely appear over the inland regions, this winter precipitation is a maximum along the coast. The summer rainfall is mainly convective and arises either from diurnal heating or from cyclonic activity; thunderstorms and heavy convective showers are not uncommon. Besides the spring and summer coastal fog, central China has cold-season fog through radiational cooling of the moist air located along the Yangtze valley. Except along this valley, the winter skies are generally clear. Cloudiness is at a maximum during the warmer months.

South China is subtropical except for the abnormally cold winters, which are caused by outbreaks of cold *cP* air. Uniformly high temperatures are observed during late spring, summer, and early autumn. Because of warm, moist, unstable *mT* air masses, rainfall is very heavy during the summer monsoon season. The frequent tropical cyclones of the China Sea are generally accompanied by very heavy rainfall. The winter rainfall is light and, as shown by Canton (126) and Yünnanfu (127), the precipitation amounts decrease inland toward the dry interior. Except for the Yünnan province, the winter cloudiness is greater than over the plains to the north. However, the maximum cloudiness occurs during the period of heavy rainfall, *i.e.*, during the summer. Since there is abnormally cold water along the coast during the spring, low stratus clouds and fog are very common. However, they are definitely restricted to the coast line and are not observed inland; for example, Hong Kong has a decidedly higher degree of fog and stratus clouds than Canton, a short distance inland.

Indo-China, Thailand, and the Malay States have a tropical rainy climate. However, with the cold air flow from the north, the winters of Indo-China and Thailand are cool. Hot, sultry weather persists during the summer. The physical features help to produce an irregular distribution of precipitation. The east or northeast winds on the Annam coast are moist through having crossed the ocean. Therefore, heavy orographical rainfall is observed along the east coast during the early winter months. To the west, Thailand is somewhat sheltered from this cold-season rainfall; for example, compare the October and January rainfall at Bangkok (131) with that of Nhatrang (129) on the east coast. During the warm months heavy precipitation is quite general. The tropical cyclones that reach the mainland are always accompanied by very heavy precipitation.

Because the Malay States are near the equator and almost completely surrounded by ocean, heavy rainfall occurs in all months of the year. Nevertheless, a short dry season may be detected over the northwest section of the Malay Peninsula, as shown by the Penang (132) rainfall regime. Singapore (133), which is almost on the equator, has no dry season. Heavy convective showers and thunderstorms are very common throughout the entire peninsula.

The weather of the Japanese islands is quite different from that of northern China. The difference arises principally from the oceanic exposure and the frequent passage of cyclones across or near the archipelago. Consequently, Japan has no pronounced dry season, and the winters are warmer than in northern China. Because of the mountainous topography, there are decided differences in climate between low-lying and high-level areas as well as between the east and west coasts.

In winter the temperature varies from east to west as well as from north to south. The former variation arises partly from the presence of warmer water along the east coast (Kuroshio Current) and partly from the adiabatic heating of the air that moves downslope toward the east coast. The cold continental air is heated from below and gains moisture while traversing the Sea of Japan, and on reaching the west coast the water vapor is condensed through orographical lifting. The addition of the latent heat contributes to the production of milder air masses across eastern Japan. Besides showing the general temperature distribution the stations in Table 69 demonstrate the variation from east to west. Despite the advective heating, it is evident that the northerly winds produce a few months of cold winter weather. During the summer the temperatures are uniformly high from Formosa to 40° north. The northern islands, however, remain cool during summer because of the presence of cold water, which has been brought south by the Oyashio Current.

In winter, precipitation is connected with the passage of cyclones across Japan or near the east coast. Along the west coast the cold air outbreaks in the rear of these cyclones are accompanied by orographical precipitation. The importance of this air-mass precipitation is demonstrated by the east- and west-coast stations in Table 69; for example, compare Kanazawa (141) with Tokyo (140). The summer *mT* air masses and occasional cyclones produce heavy rainfall, especially over Formosa and southeastern Japan. Orographical lifting of the prevailing southeasterly winds favors heavier summer precipitation along the south and east slopes of the mountains than on the west and north slopes. Naturally, the mountain ridges are regions of heavy precipitation.

In comparison with China, the mean cloudiness is high. Table 69 shows that, in general, the distribution of cloudiness follows the precipitation distribution; *i.e.*, the west coasts are decidedly cloudier than the east coasts in winter. Low stratus clouds and sometimes fog accompany the frontal precipitation. The spring and summer sea fogs are more prevalent to the north, where there is a steep gradient in the sea-surface temperatures. Actually, in any season, the advection of warm air across the cold water of the Oyashio Current produces an advection type of fog, similar to the fog observed so frequently off Newfoundland.

f. East Indies. The island groups between Australia and the Asiatic mainland lie entirely within the tropics; consequently, a

TABLE 69.—METEOROLOGICAL MEANS AT SELECTED STATIONS
(Temperature in degrees Fahrenheit, precipitation in inches, cloudiness in tenths)

Station	Lat., °N	Long., °E	Temperature				Precipitation				Cloudiness				Elevation, ft
			Jan.	Apr.	July	Oct.	Jan.	Apr.	July	Oct.	Jan.	Apr.	July	Oct.	
Manchukuo															
115. Tsitsihar	47	124	−5	41	74	40	0.1	0.4	4.0	0.4	3	6	6	4	490
116. Mukden	42	123	9	46	76	48	0.2	1.1	5.5	1.5	..	..	..	..	200
117. Antung	40	125	15	48	75	52	0.4	1.5	10.6	2.4	..	..	..	..	300
Korea															
118. Gensan	39	127	26	51	73	57	1.6	2.8	7.9	1.6	..	..	..	..	10
North China															
119. Peiping	40	116	23	57	79	54	0.1	0.6	10.0	0.7	2	..	5	..	130
120. Kaifeng	35	115	30	58	84	58	0.4	1.0	12.1	1.3					
121. Tsingtao	36	120	31	51	74	61	0.4	1.5	6.1	1.3	..	..	..	..	300
Central China															
122. Shanghai	31	121	38	56	81	63	2.0	3.7	5.9	2.9	6	7	6	6	30
123. Hankow	31	114	38	62	85	65	1.9	5.9	7.3	2.9					
124. Chungking	30	107	46	66	84	66	0.7	3.9	5.5	4.3	..	..	..	..	850
South China															
125. Foochow	26	120	51	64	84	71	1.8	5.0	6.6	1.9	..	..	..	..	300
126. Canton	23	114	56	71	83	71	1.8	6.1	10.4	2.3	..	..	..	..	300
127. Yünnanfu	25	103	48	68	70	63	0.5	0.7	9.4	3.6	..	..	..	..	6500
Indo-China															
128. Hanoï	21	106	62	74	84	78	1.0	3.6	11.1	4.3	8	8	7	6	50
129. Nhatrang	12	109	74	80	82	79	2.6	0.8	1.9	13.4	7	5	6	7	10
130. Saïgon	11	107	78	84	80	80	0.7	1.7	12.0	11.0	5	5	8	7	40

TABLE 69.—METEOROLOGICAL MEANS AT SELECTED STATIONS.—(*Continued*)

Station	Lat., °N	Long., °E	Temperature				Precipitation				Cloudiness				Elevation, ft
			Jan.	Apr.	July	Oct.	Jan.	Apr.	July	Oct.	Jan.	Apr.	July	Oct.	
Thailand															
131. Bangkok	14	100	76	85	83	81	0.3	1.7	6.1	8.1	3	4	7	6	10
Malay Peninsula															
132. Penang	6	100	79	81	80	79	3.7	7.0	8.2	16.3	5	6	6	6	10
133. Singapore	1	104	78	80	80	80	9.7	7.8	6.7	7.9	7	6	6	6	20
Japan															
134. Kamikawa	44	142	12	38	66	45	2.6	2.0	4.7	4.4	8	7	8	8	370
135. Nemuro	43	146	24	37	58	51	0.9	2.8	3.3	3.6	5	6	8	6	80
136. Sapporo	43	141	21	41	66	49	2.8	2.0	3.4	4.2	6	6	7	6	50
137. Aomori	41	141	27	45	69	54	5.2	2.4	5.7	4.3	8	7	7	6	10
138. Miyako	40	142	31	47	68	55	2.6	4.3	4.5	6.8	5	6	8	6	100
139. Akita	40	140	30	47	71	54	4.5	4.5	8.1	6.6	9	8	8	7	20
140. Tokyo	36	140	37	54	75	61	2.0	5.3	5.6	7.2	4	7	7	6	70
141. Kanazawa	37	137	36	52	75	60	10.8	6.7	8.2	7.5	9	8	6	7	100
142. Osaka	35	136	39	56	79	62	1.9	6.2	6.9	5.0	5	6	7	5	20
143. Hiroshima	34	132	39	55	78	62	1.7	7.4	8.5	4.2	5	6	7	5	20
144. Kagoshima	32	131	45	60	78	66	3.5	9.1	11.2	5.1	5	6	8	5	390
145. Oshima	28	129	59	67	81	73	9.0	9.8	10.3	12.9	8	8	7	6	10
146. Naha	26	128	62	70	82	75	5.9	5.5	7.4	6.7	7	7	8	6	30
Formosa															
147. Taihoku	25	122	61	69	82	73	3.6	5.8	9.6	4.2	7	8	8	6	30
148. Koshun	22	121	69	76	82	78	0.9	2.2	12.6	7.1	6	5	7	6	80

tropical rain-forest climate predominates. Their exposure to warm, moist air currents ensures uniformly high temperatures with a very slight seasonal variation. The stations in Table 70 show that elevation causes the only significant change in temperature from place to place. The mountainous regions are characterized by pleasantly cool temperatures throughout the year, as compared with the sultry weather of the low-lying stations.

In its broadest sense the monthly rainfall variation follows the meridional displacement of the doldrum low-pressure belt. North of 5° north there is a definite trend toward a precipitation maximum from June to October, with a precipitation minimum from December to April. On the other hand, south of 5° south, June to October is a period of diminished rainfall, while the maximum occurs from December to April. In both hemispheres the maximum occurs during the period of maximum instability within the *mT* air masses, *i.e.*, within the vicinity of the doldrum low-pressure area. Even though dry and wet seasons may be distinguished, there is no month without appreciable rainfall. About the equator the rainfall is distributed approximately uniformly throughout the year.

Many variations from this general precipitation regime may be observed. These arise principally from local causes, for example, the orientation of the mountain ranges and coast lines. In the Northern Hemisphere, during the period of the northeast monsoon, the land areas on the east coasts have heavier precipitation than those on the west coasts. This is demonstrated by comparing the January rainfall at Legaspi (150) (east coast) with Manila (151) (west coast), or Surigao (152) (east coast) with Iloilo (153) (west coast), or Tarakan (158) (east coast) with Jesselton (157) (west coast). Similarly, in the Southern Hemisphere, during the period of the southeast winds, the east coasts have slightly heavier precipitation than the west coasts. This is a less pronounced variation since the principal mountain chains of the Southern Hemisphere islands, for example, New Guinea, are orientated in a general southeast to northwest direction. Moreover, the air from the Australian continent is initially dry; for example, Koepang (164) has a July average rainfall of only 0.2 in. The stations in Table 70 further demonstrate the importance of the influence of topography on the rainfall distribution.

TABLE 70.—METEOROLOGICAL MEANS AT SELECTED STATIONS
(Temperature in degrees Fahrenheit, precipitation in inches, cloudiness in tenths)

Station	Lat.	Long.	Temperature				Precipitation				Cloudiness				Elevation, ft
			Jan.	Apr.	July	Oct.	Jan.	Apr.	July	Oct.	Jan.	Apr.	July	Oct.	
Philippine Islands															
149. Aparri	18°N	122°E	73	80	82	79	6.8	1.9	7.6	11.7	7	3	6	6	10
150. Legaspi	13°N	124°E	78	82	81	81	15.2	5.4	10.1	13.0	6	4	6	6	10
151. Manila	15°N	121°E	77	83	81	80	1.0	1.3	16.2	7.3	6	4	8	7	50
152. Surigao	10°N	125°E	78	80	81	80	20.2	9.7	6.3	10.2	7	6	7	7	20
153. Iloilo	11°N	123°E	78	82	80	80	2.4	1.6	16.3	10.3	6	5	8	7	20
Sumatra															
154. Takengon	5°N	97°E	67	68	67	67	6.1	7.6	3.0	8.3	..	..	..	..	3950
155. Padang	1°S	100°E	79	80	79	79	13.4	14.3	10.9	19.4	..	..	..	..	20
156. Tandjungpandan	3°S	108°E	78	78	78	78	10.9	10.7	6.9	10.4	..	..	..	..	10
Borneo															
157. Jesselton	6°N	116°E	78	78	78	78	5.5	5.2	7.9	12.9	6	6	6	6	30
158. Tarakan	3°N	118°E	78	78	78	78	10.4	13.5	9.0	12.2	..	..	..	..	10
159. Balikpapan	1°S	117°E	78	78	78	79	7.7	7.6	7.8	5.7	..	..	..	..	20
160. Pontianak	0°	109°E	78	79	80	79	10.6	11.1	6.5	14.7	..	..	..	..	10
Celebes															
161. Tomohan	1°N	125°E	70	70	70	71	12.3	8.7	6.1	7.8	8	7	7	6	2620
162. Macassar	5°S	119°E	78	80	77	79	26.6	6.0	1.4	1.7	..	..	..	..	10
Miscellaneous															
163. Amboina	4°S	128°E	80	79	76	78	5.0	11.3	22.9	6.0	..	..	..	..	10
164. Koepang	10°S	124°E	79	79	77	81	15.2	2.6	0.2	0.7	..	..	..	..	10
165. Batavia	6°S	107°E	78	79	78	79	10.6	5.2	2.8	6.7	8	6	5	6	30
166. Christmas Island	10°S	106°E	80	80	78	80	8.8	7.9	4.9	3.2	5	4	4	4	20
New Guinea															
167. Manokwari	1°S	134°E	78	79	79	80	10.7	11.1	6.0	4.2	..	..	..	..	60
168. Hatzfeldthafen	4°S	145°E	80	80	78	81	15.8	14.8	6.9	6.1	7	7	6	6	10
169. Port Moresby	10°S	147°E	82	82	78	80	7.5	3.3	1.0	0.8	..	..	..	..	130

Manokwari (167) on the north coast of New Guinea, although near the equator, has a decided precipitation maximum during the season of onshore northerly winds, *i.e.*, December to April. On the mountain slopes the yearly rainfall may be excessive, as shown by Padang (155) at the base of the high mountains along the south and west coasts of Sumatra.

The general heavy precipitation of the East Indies is caused by the high moisture content and instability of the *mT* air masses. The convective instability results in a large number of thunderstorms—probably the highest thunderstorm frequency of any region of the world. Very heavy rain showers are quite common, and it is only in regions of orographical lifting or tropical cyclones that long periods of rain are observed. The convective showers are produced either by convergence toward the equatorial low-pressure belt (doldrums) or by diurnal heating over land. Nocturnal cooling over the oceanic areas produces early morning showers. Squalls of short duration are very common during the wet season, when the instability is more pronounced. Waterspouts over the sea are further examples of the instability. The Philippine Islands sometimes have heavy rainfall in connection with the tropical hurricanes discussed in Sec. 68.

Even though the East Indies are a region of heavy precipitation, the mean cloudiness is not particularly high. The amount of cloudiness is consistent with the convective nature of the rainfall since it favors alternate periods of clear and cloudy skies. The maximum cloudiness occurs during the wet season. Within the vicinity of a pronounced low-pressure area, overcast skies persist for many days. In general, cumuliform cloud types predominate, having a diurnal maximum over the land in the afternoon and over the ocean in the early morning, consistent with the time of maximum instability.

CHAPTER XIII

AFRICA

Since it extends approximately from 35° north to 35° south, the continent of Africa lies in the low temperate and tropical latitudes. To the north, the climate is affected to some extent by Europe and Asia, whereas South Africa protrudes into the southern oceans. Despite these contrasting influences, there is an approximately symmetrical distribution of climatic types about the equator. A vast area near 0° has a tropical rain forest or savanna type of climate. The *C* type of climate of Angola and Rhodesia has a small counterpart in Abyssinia; the vast arid Sahara area is duplicated to a minor degree in southwestern Africa; finally, the Mediterranean climate of northwestern Africa is duplicated around Cape Town in South Africa.

In so far as the continent is located in low latitudes and almost entirely surrounded by water, there is no immediate source region of cold continental air. Consequently, no portion of Africa experiences a snow-forest or tundra climate. In contrast to South America, there are no invasions of cold air from Antarctica; the 2000 miles of ocean between the Cape of Good Hope and the southern continent ensure that the cold polar air masses are considerably modified when they reach 35° south.

Even though it is located in similar latitudes, the climate of Africa differs appreciably from that of South America. Whereas the tropical rain-forest climate of the latter extends from the interior to the east coast, the same climatic belt stretches from the interior to the west coast of Africa. In general, the principal differences arise from the absence of a high mountain chain along the west coast of Africa similar to the Andes of South America. Gradual transitions rather than abrupt changes of climate are typical over Africa.

Because the continent is very wide between 20° north and 30° north, the desert type of climate occupies an extremely large area—the largest area of desert climate in the world. As a result of the tapering of the continent to the south, the desert region from 20° south to 30° south is small (compare the vast desert of Australia).

70. Physical Features. Unlike the previously discussed continents, Africa is devoid of extensive high mountain chains. However, even though there are few mountains above 10,000 ft, only a small section of the country has land below 1000 ft in elevation. Instead, especially toward the south, vast plateaus are located above 3000 ft. The most prominent topographical features may be summarized as follows:

1. In the northwest, the Atlas mountains and plateau of Algeria separate the coast from the relatively low-lying desert to the south. Occasional peaks in Morocco extend beyond 10,000 ft.

2. The vast Sahara Desert varies considerably in elevation, having isolated mountains whose peaks tower above 8000 ft, for example, the Ahaggar and Tibesti mountains of the interior.

3. The principal mountain regions of Africa are irregularly located between Northern Rhodesia and the Red Sea; in the vicinity of Lake Victoria and in Ethiopia many of these isolated peaks extend above 12,000 ft.

4. Smaller mountainous areas appear in the southeast (Drakensberg mountains), in southwest Africa (Auaz mountains), in the Cameroons (Cameroon mountains), in Madagascar (Ankaratra mountains), and elsewhere.

Africa is very compact and has no large indentations like those in Europe. Minor variations in the coast line are found in the vicinity of the Gulfs of Guinea, Gabès, Sidra, and Aden. Except for Lakes Rudolf, Victoria, and Nyasa the continent is singularly devoid of inland water bodies.

This summary illustrates that there is no elongated mountain chain that can form a climatic barrier, similar to the Andes of South America or the Rockies of North America. However, the vast plateaus exert a profound effect on the climate of the eastern and southern portions of Africa. Despite the low latitude, the relatively high elevation ensures a comparatively dry and mild climate, which is favorable for European habitation. Consequently, from 10° north to 20° south, Africa differs from South America because of the smaller area of true tropical rain-forest type of climate.

The principal effects of the topography on the weather of the continent are:

1. The winter rains, which accompany the migratory Mediterranean cyclones, are prevented from invading southern Morocco and Algeria by the Atlas Mountains. Besides restricting the frontal precipitation to the coastal area these mountains cause the total precipitation to be augmented by orographical lifting. The net effect of the Atlas chain on the distribution of precipitation is clearly demonstrated by Plate IX.

2. The moderate rainfall of eastern Africa from the Cape of Good Hope to Ethiopia may be partly attributed to orographical lifting over the high plateaus.

3. Since the mountains are orientated normal to the prevailing wind circulation, heavy orographical precipitation occurs along the east coast of Madagascar.

4. Because the air masses are extremely dry, the mountain chains of the Sahara interior have slight influence on the precipitation regime.

Besides influencing the rainfall distribution, the elevation of the land results in lower mean temperatures than over the low-lying sections of the country; for example, the winter temperatures average about 10°F colder over the eastern plateaus than along the western coast lines of South Africa.

71. Mean Distribution of Pressure and Wind. As a result of the paucity of meteorological observations the mean air circulation over Africa is not precisely determined. However, the mean sea-level pressure maps for January and July demonstrate the large-scale elements of the general circulation.

a. January. The main features of the January sea-level chart may be summarized as follows:

1. The country north of the equator is subjected to winter cooling in contrast to the thermal heating south of the equator. This radiational cooling results in the formation of a high-pressure center near the Atlas Mountains and Algerian plateau, which may be considered to be an extension of the Atlantic subtropical anticyclone. The high-pressure ridge stretching across the northern part of the Sahara is separated from the Asiatic thermal anticyclone by a low-pressure belt over the warm Mediterranean Sea. In view of this mean pressure distribution, westerly winds predominate across the coastal areas of northern

Africa, while light winds and calms prevail from about 20° north to 30° north—the mean position of the continental anticyclonic ridge.

2. Pronounced heating produces a mean low-pressure area south of the equator. Actually, the trough extends north of the equator to about 10° north, since the land areas are still warmer than the South Atlantic or Indian oceans. This vast cyclone separates the well-developed oceanic subtropical anticyclones to the east and west. The relatively steep mean-pressure gradient along the west coast, south of the equator, produces persistent southerly winds. Even though the isobars show an offshore gradient wind, frictional inflow and sea breezes ensure, at low levels, a frequent flux of maritime air inland toward the low-pressure center as shown by Swakopmund (6, Table 71).

3. On crossing the equator the southeast trades are deflected and reach the Guinea coast as southwest winds, which are partly directed toward the inland low-pressure trough. North of this low-pressure trough the prevailing surface air flow across the southern Sahara, the Sudan, and northern Ethiopia is from the east or northeast. These northeast trades are illustrated in Table 71 by Kayes (11) and Mogador (4).

4. Along the east coast the anticyclone over the Indian Ocean and the continental cyclone produce a mean surface flow from a northeasterly direction, *i.e.*, an onshore component. Along this coast, frictional and sea-breeze effects definitely favor onshore winds. Consequently, maritime air is drawn inland toward the central portion of the continent.

b. July. With the northward movement of the sun, the thermal influences result in a low-pressure area over the Sahara in July, together with a high-pressure area over the Union of South Africa. Consequently, from 30° south to the equator the southeast trades predominate, as shown by the east-coast stations in Table 71; these trades are deflected to southwest winds from the equator to 15° north. It will be observed that the prevailing sea-level air flow favors onshore winds along the east coast south of the equator and along the Guinea coast in all seasons of the year. On the other hand, along the west coast, south of 10° south, the onshore flow is restricted to shallow sea breezes, which affect only the immediate coast line.

During the Southern Hemisphere winter the Cape of Good Hope comes into the westerly flow on the south side of the subtropical high-pressure belt. However, owing to the proximity of the mean high-pressure ridge, the westerly winds are weak as compared with those of the "roaring forties" farther south.

With the extension of the North Atlantic anticyclone across southern Europe and the Mediterranean, the prevailing air flow across northern Africa is predominantly from the northeast. It is important to realize that this air originates over Europe as a relatively dry air mass. In the west, from 15° to 30° north it will further be observed that the prevailing air currents are offshore in all seasons of the year.

Because migratory pressure systems are rarely observed in tropical latitudes, the day-to-day wind variations from 20° north to 20° south are relatively unimportant. Local factors, such as sea breezes or mountain and valley winds, are the principal causes of surface-wind fluctuations. Those regions which are influenced by extratropical cyclones would be expected to show the greatest wind variability, for example, Morocco, the Mediterranean coast, and the Union of South Africa. The wind frequencies at a few isolated stations have been included in Table 71.

TABLE 71.—PERCENTAGE FREQUENCY OF SURFACE-WIND DIRECTION

Station	Lat.	Long.	January								July							
			N	NE	E	SE	S	SW	W	NW	N	NE	E	SE	S	SW	W	NW
North Coast																		
1. Tripoli	33°N	13°E	10	11	5	12	10	*20*	16	16	14	*29*	16	12	4	6	6	13
2. Bengasi	32°N	20°E	12	2	4	7	*38*	8	22	7	*71*	3	4	2	5	1	4	10
3. Alexandria	31°N	30°E	11	7	7	12	13	*23*	12	15	28	6	2	1	1	1	11	*50*
West Coast																		
4. Mogador	31°N	10°W	18	*33*	16	10	8	8	3	4	42	*49*	1	0	1	3	1	3
5. Conakry	9°N	14°W	4	1	2	6	20	22	*27*	18	2	0	5	11	*29*	21	19	13
6. Swakopmund	22°S	16°E	12	2	0	1	14	*30*	23	18	10	13	9	5	20	*26*	10	7
East Coast																		
7. Zanzibar	6°S	39°E	30	*38*	10	5	4	3	2	8	0	0	9	*53*	36	2	0	0
8. Beira	20°S	35°E	13	*39*	15	13	13	4	0	3	1	15	17	*36*	25	6	0	0
9. Lourenço Marques	26°S	33°E	6	17	*24*	15	19	12	2	5	13	14	11	8	12	*19*	8	15
Interior																		
10. Fort-Lamy	12°N	15°E	*52*	37	4	2	2	0	0	3	6	7	1	16	*32*	31	6	1
11. Kayes	14°N	12°W	5	*36*	30	5	1	1	3	19	6	7	4	6	2	6	14	*55*
12. Salisbury	18°S	31°E	13	*35*	21	16	8	1	3	3	1	14	20	*43*	18	2	1	1

c. Upper Air. Since, in the Northern Hemisphere, the mean January temperature increases toward the equator, the pressure gradient between 0° and 25° north changes with elevation. The persistent easterlies in the lower layers decrease in speed and finally become westerly at 3 km. At 3 km the mean January air flow is therefore from a general westerly direction across northern Africa. The mean velocities are relatively light at this level. However, above 3 km they gradually increase in speed. The absence of extensive aerological data prohibits a precise discussion of the upper-air circulation.

South of the equator the January upper-air circulation is probably weak to 30° south, in view of the diffuse surface cyclone and the indeterminate temperature gradient. In all likelihood, the surface thermal low gives way to an upper-level anticyclone about 20° south to 25° south, as shown by Salisbury in Table 72. Consequently, light winds are probable from 0° to 25° south, while farther south the easterly winds decrease gradually with elevation, eventually becoming westerly at higher levels. These west winds increase with elevation.

TABLE 72.—UPPER WINDS OVER SOUTH AFRICA
(Wind force indicated by Beaufort numbers)

Station	Lat., °S	Long., °E	2000 m		4000 m		6000 m	
			Dec.–Feb.	June–Aug.	Dec.–Feb.	June–Aug.	Dec.–Feb.	June–Aug.
Cape Town....	34	18	W 3	NW 4	W 5	W 3	WSW 6	WSW 5
Durban........	30	31	WNW 2	WSW 2	WSW 4	WSW 3	WSW 4	WSW 3
Bloemfontein...	29	26	NW 3	NW 3	WSW 3	WSW 3	SW 2	SW 3
Port Nolloth...	29	17	N 1	N 1	WNW 3	WNW 4	SW 2	SW 4
Pretoria.......	26	28	N 2	NW 1	S 2	SW 2	SSW 3	SW 4
Walvis Bay....	23	14	NE 1	NNW 2	N 2	NW 2	E 1	W 3
Salisbury......	18	31	NE 2	ESE 3	ESE 2	SSE 2	ESE 2	SW 4

The mean July circulation is approximately the converse of the January pattern. North of the equator the thermal low-pressure area becomes a high-pressure center at 2 km and is centered about 20° north. Consequently, the surface easterlies north of 25° north decrease with elevation and become westerly; these west winds increase in speed with elevation. South of 20° north the surface west winds are replaced by weak east winds above 2 km. In view of the July temperature gradient, the surface anticyclone about 30° south is displaced northward with increasing elevation. Consequently, the west winds about the Cape of Good Hope increase in speed aloft, while the surface easterlies north of 25° south gradually decrease in speed and become westerly aloft, as shown by Salisbury (Table 72).

Because migratory cyclones are infrequent, the deviation from the mean upper-air pattern is of minor importance. The greatest fluctuations probably occur along the Mediterranean coast in January and near the Cape of Good Hope in July. In these regions extratropical cyclones produce at all levels short-period changes in wind speed and direction.

72. Ocean Currents. As in the case of all continents, the climate of the African coast line is directly affected by the prevailing oceanic circulation. In so far as the properties of the maritime air masses that invade the land area are affected by the underlying surface, the ocean currents indirectly influence the inland regions.

The most important elements of the oceanic circulation follow:

1. The Canaries Current is a cool current that flows southward along the west coast of northern Africa. The relatively low sea-surface temperatures from Gibraltar to Dakar are the result partly of the southward-moving water and partly of the upwelling of cold water. The latter is caused by the persistent northerly winds, as in the case of the California region. This prevailing wind and oceanic circulation contributes to the frequent coastal fogs, low stratus clouds, and mild temperatures, especially in summer. Furthermore, as in North America, the interior is very arid.

2. The counterpart of the Canaries system in the Southern Hemisphere is the Benguela Current, which flows northward along the west coast of South Africa. The upwelling again favors low coastal temperatures. For example, the sea-surface temperature off southern South-West Africa is less than 16°C throughout the year. As in the case of California, northern Africa, and Chile, this oceanic condition favors low cloud and fog along the immediate coast. This is in sharp contrast to the cloudless desert of the interior (Kalahari Desert).

3. These two cold currents swing westward about 15° north and 10° south, respectively. Near the equator the Equatorial Countercurrent, or Guinea Current, flows eastward toward the

mainland in such a manner that the land areas, bordering the Gulf of Guinea, are washed by very warm water. This circulation, together with the onshore air motion, accounts for the very warm and moist air masses from Dakar to Libreville. The indirect influence of the oceanic circulation is clearly demonstrated by the mean-precipitation chart. Those areas adjacent to cold currents directed toward the equator are characteristically dry, whereas the equatorial west coast experiences heavy rainfall.

4. Since the principal ocean currents are driven by the prevailing winds, the east coast differs markedly from the west coast. The circulation around the subtropical anticyclone causes warm water to flow poleward along the east coast. South of the equator, the southward-flowing current is known as the Mozambique Current off southeastern Africa and the Agulhas Current about 35° south. In the vicinity of 40° south, it encounters the cool West Wind Drift of the Southern Hemisphere. Because the water is warm, the sea-surface temperature off southeastern Africa is approximately 8° to 10°C warmer than near the southwest coast of Africa. The presence of the warm current favors warm, moist air masses along the southeast coast and, therefore, moderate precipitation.

5. The ocean currents in the North Indian Ocean vary from summer to winter, in response to the Asiatic monsoon circulation. In summer, the southwest monsoon produces a warm northward-flowing current along the coast of eastern Africa; this later moves eastward about 10° north. Consequently, the immediate coast line from Somaliland to Arabia is relatively cool. Cool water also prevails off the east coast in winter. The northeast wind causes a current to flow toward the equator. Therefore, in all seasons of the year, relatively cool coastal water is found north of 10° north. The presence of this cool water is probably a contributing factor to the aridity of Somaliland and Arabia.

6. The currents of the eastern and western Mediterranean Sea have a tendency to circle in a counterclockwise direction.

The indirect climatic significance of the sea-surface conditions will be discussed in later paragraphs.

73. Air Masses. The nonavailability of aerological observations prohibits a precise description of the various air masses that affect Africa. Nevertheless from the broad-scale features of the atmospheric circulation, a qualitative discussion is possible.

a. Tropical Maritime Air (mT). Since the major portion of the continent is located within the tropics, *mT* air masses have an important effect on the climate. The general source regions of this warm, moist air mass coincide with the ocean areas north and south of the equator. However, the properties of the air mass vary appreciably with region, wind circulation, and season. The stable type of *mT* air is found on the eastern side of the subtropical anticyclone, for example, along the coast of northwestern Africa from Gibraltar to Cape Verde and along the coast of southwestern Africa from the Cape of Good Hope to 15° south. The subsidence within the anticyclonic circulation favors a moist surface air mass with a warm, dry air mass above a pronounced temperature and moisture inversion. The unstable type of *mT* air occurs either on the western extremity of the subtropical anticyclone, for example, along the coast of southeastern Africa, or within the convergent wind regime adjacent to the intertropical front. Besides being warm and moist, this air mass is typically convectively unstable and shows little or no subsidence characteristics.

1. January. In the Northern Hemisphere winter the *mTs* air is confined to the western coast lines of North and South Africa except from 10° north to 10° south. The wind circulation, parallel to the coast, favors the restriction of this air to the immediate coast line. The presence of *mTs* air principally accounts for the light amounts of precipitation, from December to March, between 10° north and 30° north. Likewise, from 10° south to 30° south, there is slight precipitation from June to September.

Across the east coast, from the Cape of Good Hope to 5° south the unstable *mT* air invades the continent in response to the wind circulation toward the thermal low. Continental heating increases the instability. This *mTKu* air mass accounts for the moderate to heavy precipitation over southeastern and south-central Africa. Similarly, in the Gulf of Guinea the *mTu* air results in relatively heavy precipitation from 5° north to 5° south. Here the instability arises from the convergent flow toward the equatorial low-pressure belt. North of the equator, the east coast is invaded by a modified type of *cP* air from Asia. Its surface characteristics resemble *mT* air in so far as it has been modified by the Arabian Sea. However, since the maritime trajectory is relatively short, the air remains dry aloft. Hence, this air mass is not a true *mT* mass but rather a modified type of *cP* air. The low moisture content accounts for the slight precipitation along the east coast, north of the equator.

2. JULY. During the Northern Hemisphere summer, with the shift of the doldrum belt northward *mTu* air is most frequently observed from Dakar to Cape Lopez. The prevailing winds cause a flow of this air mass eastward. The increased instability, which arises from the continental heating, is favorable for heavy convective precipitation, as observed over a wide continental band from 15° north to 5° south. The tropical maritime air that invades eastern Africa, south of the equator, is more stable in July than in January. This greater stability arises from the relatively straight wind flow in the former month as compared with the convergent flow in January. Furthermore, the absence of continental heating favors a relatively stable air mass over the land. Even though the air is less stable than *mTs* air along the coast of southwestern Africa, it is sufficiently stable to produce only slight precipitation from 0° to 30° south. North of the equator the prevailing circulation along the east coast is offshore. Hence *mT* air masses are rarely observed over northeastern Africa. As in January, *mTs* air masses predominate along northwestern Africa and along the west coast south of the equator during July. Again the presence of this stable air favors slight precipitation.

b. Tropical Continental Air (*cT*). There are two main source regions of this warm, dry air mass, *viz.*, the vast arid region from 30° north to 15° north and a smaller arid region in the vicinity of the Kalahari Desert. During the Northern Hemisphere winter, *cT* air dominates the weather of almost the entire continent north of the equator. With the exception of the Mediterranean coastal area and the Gulf of Guinea, this dry, comparatively cool air mass causes a vast area of clear skies and no precipitation. Although, during January, the air mass is quite unstable in southwestern Africa, the low humidities favor only slight precipitation.

In the Northern Hemisphere summer the Sahara is still dominated by *cT* air. In contrast to winter the region south of 15° north frequently comes under the influence of modified *mTKu* air masses. Even though the strong continental heating from

15° north to the Mediterranean makes the *cT* air masses unstable, the low humidities ensure meager cloudiness and precipitation. Along the Mediterranean coast the continent is invaded by air that has crossed the open sea. Over southern Europe the summer air masses are similar to *cT* air except for the slightly higher moisture contents. Even though this air crosses the open water, it still arrives over Africa as a relatively dry air mass. The subsidence over the Mediterranean prevents a vertical transport of moisture. Consequently, despite the higher surface humidity, the upper-level stability favors clear skies within this modified *cT* air mass, similar to those within the true *cTKu* air mass of the central Sahara. During the Northern Hemisphere summer, the dry *cT* air over southwestern Africa is subject to continental cooling. Hence, it is cooler and more stable than that observed over North Africa.

c. Polar Maritime Air (*mP*). Because of the low latitude, invasions of polar maritime air are somewhat infrequent. During the Northern Hemisphere winter the principal *mP* invasions occur over extreme northwestern Africa, in the rear of the eastward-moving cyclones that traverse the Mediterranean Sea. Since the *mP* air has moved southward in a cyclonic circulation, it arrives as a cool, moist, convectively indifferent air mass. The steep lapse rates, together with a high moisture content, favor appreciable orographical precipitation in the mountainous regions of Morocco and Algeria.

Polar maritime air masses invade the extreme southern portions of South Africa, especially during the Southern Hemisphere winter. Again these invasions consist of relatively warm modified *mP* air that has moved northward from high latitudes.

d. Polar Continental Air (*cP*). It has been noted that *cP* air masses or modified *cP* air masses have been observed over South America. In contrast to this continent, there are no real cold air invasions over South Africa. The vast oceanic area between the Cape of Good Hope and Antarctica ensures that all cold air invasions must be polar maritime air. Nevertheless, during the Southern Hemisphere winter the continental cooling of modified maritime air tends toward the creation of a mild type of *cP* air over the highlands of South Africa. This air mass represents the closest approximation to a South African *cP* air mass.

In the rear of eastward-moving cyclones, outbreaks of cold *cP* air cross the Mediterranean toward the African coast. The addition of heat and moisture from the underlying surface results in a very modified air mass over Africa. This modified *cP* air is characterized by mild surface temperatures, a high moisture content in the lower layers, and a relatively steep lapse rate. Instability and orographical precipitation from this air mass partly account for the winter rainfall of the Mediterranean section of Africa.

Over the highlands of the Sahara the winter *cT* air resembles *cP* air. The elevation ensures relatively low surface temperatures.

74. Fronts, Cyclones, and Anticyclones. *a. Frontal Zones.* In view of the low latitude, the African continent is relatively free of polar-front disturbances. However, since some sections are subject to frontal weather, the principal frontal zones will be discussed briefly. In January, the three main frontal regions are as follows:

1. The Mediterranean front extends southwestward into the Atlantic. With the southward displacement of the Atlantic subtropical anticyclone, this front, in the west, separates modified maritime polar air from maritime tropical air. Farther east, over the Mediterranean, the warm air mass is either modified maritime or continental tropical air.

2. Through the equatorial low-pressure belt the intertropical front is quite diffuse since, as usual, there is a wide area of converging winds.

3. One of the mean positions of polar fronts in the Southern Hemisphere appears to extend southeastward from the Cape of Good Hope. As in the case of the Northern Hemisphere, this front oscillates considerably from time to time. Nevertheless, because its mean position is adjacent to the continent, cyclogenesis about 30° south can affect the weather of the extreme southern sections of Africa.

During July the most prominent feature is the dissolution of the Mediterranean front. However, occasionally a front is formed across northwestern Africa between the maritime air, moving in a southerly direction around the Atlantic anticyclone, and dry continental air from the desert interior. The broad intertropical front is displaced northward in summer. The converging wind systems result in the juxtaposition of tropical continental air from the north and modified tropical maritime air from the south. Again, as in the Southern Hemisphere summer, the mean position of the polar front is immediately off the coast of South Africa. It separates the tropical maritime air of the Indian Ocean from the polar maritime air of higher latitudes. One important difference between this front and the North Atlantic polar front lies in the trajectory of the cold air mass. Whereas fresh or modified *cP* air is located to the rear of the North Atlantic polar front, the cold air south of the African continent is definitely *mP* air. Consequently, very cold temperatures never occur, north of 30° south to 40° south, through the passage of Southern Hemisphere polar fronts. Another difference between the two hemispheres is the indefiniteness of the mean frontal zones in the Southern Hemisphere. This arises from the absence of land barriers and distinct, or isolated, source regions of cold air masses in temperate latitudes. In view of the variability of the frontal location, any region, even in low temperate latitudes, may be subject to frequent cold-front passages.

b. Cyclones and Anticyclones. The cyclones that develop in the main frontal zones affect only a relatively small area. During the Northern Hemisphere winter, cyclonic activity along the Mediterranean front is the principal cause of the cold-season precipitation from the Canary Islands to Cairo. The low-pressure systems develop either over the Atlantic Ocean or over the Mediterranean Sea. In the former case, they usually move in a southwest to northeast direction and, on entering the Mediterranean, follow an easterly path. Tropical maritime air generally constitutes the warm sector when the cyclones are west of Gibraltar, whereas tropical continental air is found in advance of the center east of 0° longitude. In agreement with the respective moisture contents, cloudiness is common within the *mT* air masses whereas the *cT* air currents are relatively cloudless. Sandstorms accompany the latter. The local name for this warm, dry air is "sirocco." Besides the frontal weather, considerable cloudiness and precipitation are observed within the *mP* air in the rear of these cyclones. This rainfall arises partly from the orographical lifting along the northwest coast line. The mean path of the disturbances coincides approximately with the water areas. However, some cyclonic centers cross southern Morocco and southern Algeria. Because of the presence of drier air masses these lows are accompanied by diminished cloudiness; the precipitation is due to air-mass instability along the coast rather than to frontal lifting. Besides being a region where cyclones from the Atlantic frequently deepen, the Mediterranean is a favored area for cyclone formation. In either case, cold-front cloudiness and precipitation occur along the African coast in the rear of the eastward-moving disturbances. Again, in the rear of the cold front, clouds and rain are caused by air-mass instability and orographical lifting. With the dissolution of the Mediterranean front from May to October, cyclonic activity is rather unimportant during the summer months. However, occasional weak disturbances develop on the quasi-stationary summer front that is sometimes located across Morocco and Algeria. Here the main cloudiness and precipitation are the result of the unstable cool, moist air that moves southward in the rear of the cold front. Thundershowers are quite common. In general, these summer cyclones are considerably less intense than their winter counterparts.

Any disturbances that develop about the intertropical front differ from the extratropical cyclones in their horizontal extent. In general, the tropical systems are very narrow. Shallow pressure waves migrate along the front and cause a local intensification of the general cloudiness and precipitation of the doldrum belt. The most violent storms are the thundersqualls, or tornadoes, which usually travel from east to west. They are accompanied by excessive rainfall and electrical phenomena. Apparently they are due to the convergence of *mTKu* and *cTKu* air masses; consequently, they occur most frequently at the beginning and the end of the rainy season. During the latter season only *mTKu* air masses are present. These violent local storms are observed throughout the doldrum regions where tropical maritime air masses predominate, for example, in the Guinea lands, Nigeria, and the Congo basin.

Tropical hurricanes are rarely experienced along the African coast. The only region of important cyclonic activity is the southwestern Indian Ocean. The tropical cyclones are most frequent during the months from January to April. They generally develop east of the Seychelles about 10° south and follow a parabolic path. The majority recurve about the latitude of Madagascar, more frequently to the east of the island than to the west. Consequently, the accompanying violent weather affects the African mainland only occasionally.

In autumn (March and April) some of the tropical cyclones move southward along the east coast as extratropical lows and produce heavy coastal precipitation. As mentioned previously, the cyclones that develop on the polar front move southeastward and consequently do not cross the continent. However, frontal precipitation does occur in British South Africa with the passage of cold fronts in the rear of deep cyclones that are centered about the "roaring forties." The winter rainfall of the southwest coastal belt, for example, at Cape Town (53, Table 75), results almost entirely from cold-front passages or instability showers in the rear of the cold fronts.

The above discussion suggests that, considering the entire continent, only a very small section is affected by true frontal precipitation. Hence, the principal cause of precipitation is instability within the *mTKu* or humidified *cTKu* air masses. The instability may arise either from local heating or from convergence toward the continental low-pressure belts. In the latter case, heating aids the production of rainfall.

Consistent with the relatively low frequency of migratory cyclones, the weather is only slightly affected by the movement of temperate-latitude anticyclones. Along the northwest coast and in the extreme south, high-pressure wedges follow the passage of cold fronts. These wedges cause an influx of unstable polar maritime air. With the exception of these two regions, the continental stations do not experience the rapid day-to-day weather changes observed in temperate latitudes during the passage of highs and lows.

75. Climatic Regions. Even though a dense network of reliable meteorological stations is lacking, there are sufficient data to allow a general description of the weather and climate.

a. Northernmost Africa. Included in this area are the coastal lands from Morocco to Cairo together with the inland mountains and plateaus from Morocco to Tunisia. The rainfall results largely from cyclonic activity along the Mediterranean front. Because this front is nonexistent in summer, the precipitation regime is definitely one of a winter maximum. Furthermore, the heaviest rainfall occurs where the front is situated most frequently and where at the same time the air masses are moist, *viz.*, on the coast from Tangier to Bizerte. The coast of southern Morocco, on the other hand, has slight rainfall as illustrated by Cape Yubi (14) and Mogador (15) in Table 75. The inland regions of Morocco and Algeria have slight summer rainfall as well as winter precipitation. This summer precipitation [see Marrakech (16) and Geryville (19)] occurs mainly in the form of convective showers. The continental heating favors increased convection over the interior. East of Bizerte (21) the influence of the Sahara is more pronounced, since there are no mountain ranges to shield the coast from invasions of hot, dry Sahara air. As a result of the lower humidities, in both the warm air to the south and the cold air from the north, the Mediterranean cyclonic activity produces less precipitation than is observed west of Bizerte. Furthermore, the precipitation extends only a short distance inland since there are no mountain barriers to provide orographical lifting.

The mean cloudiness follows the precipitation regime, *viz.*, a winter maximum and a decided summer minimum. The main exception is the coastal region south of Cape Mogador where low stratus produces a summer maximum as illustrated by Cape Yubi (14). This cloudiness closely resembles California stratus.

The temperature distribution shows that the coastal areas have milder winters and cooler summers than the interior. The Atlantic coast of Morocco has decidedly cool summers as a result of the cold Canaries Current. The mean values in Table 75 illustrate the influence of the Sahara on the temperature distribution. The stations east of Bizerte show high mean temperatures for the month of July. However, the coastal strip is cool as compared with the interior. This is because of the prevailing onshore wind and the cooling influence of the sea. For example, Cairo, only a short distance inland, has a mean July temperature 5°F higher than Alexandria (24), which is on the coast.

Very warm dry periods accompany the fresh southerly winds that occur in advance of a cyclone—the so-called "sirocco." This hot, dry wind from the Sahara is generally accompanied by clouds of dust.

b. Sahara. The Sahara includes the vast arid land mass from the Mediterranean coast to about 15° north. During the Northern Hemisphere winter the prevailing northeast winds to 12° north prohibit the presence of any moist air masses over the entire Sahara. Except in the north the Sahara is almost entirely free of precipitation. As illustrated by Ghardaïa (25, Table 75), a southward extension of the Mediterranean rainfall regime produces occasional precipitation north of 30° north. At the same time the northeast trades cause the winter dryness to extend southward over the Sudan. During the summer the converse is true; the northern portion of the Sahara is devoid of precipitation, whereas the southern portion borders on the summer rainfall regime connected with the northward displacement of the low-pressure belt. The appreciable summer precipitation at Timbuktu (36) suggests that occasional rain showers may be expected to 20° north. The vast area of the Sahara between 28° north and 18° north therefore has a total mean annual precipitation that is less than 2 in. In the east, adjacent to the Red Sea, the moist surface air is responsible for light precipitation, as illustrated by Suakin (39) in Table 75. Consistent with the slight precipitation the average cloudiness is extremely low over the entire Sahara. The only exception is the Atlantic coast, where low summer stratus over the cool coastal water contributes to a mean cloudiness of about four tenths.

The very low relative humidity of *cT* air favors excessively high daytime temperatures together with a large diurnal variation. The mean January temperatures are about 55°F, with high daytime values and very cool nights. With the pronounced heating the Sahara has hotter summers than any other part of the world; mean maximum temperatures of 120°F have been recorded during the summer months. The dust storms connected with the *harmattan* (fresh northeast trade) provide one of the most disagreeable weather conditions of the Sahara.

The high mountains of the central part of the Sahara experience almost a steppe type of climate. Violent thundershowers produce slight precipitation; a yearly total of about 10 in. occurs over the Ahaggar range. Besides the increased precipitation, the elevation ensures cooler temperatures than those observed over the neighboring desert.

c. Sudan and Guinea Coast. The rainfall regime of western Africa from 15° north to the equator depends on the meridional displacement of the doldrum low-pressure area. During midwinter it is farthest south but remains north of the equator west of 15° east, because of the continental heating immediately north of the Gulf of Guinea. The Guinea coast line may be considered to be the southern limit of the low-pressure belt. As summer approaches, this belt moves northward until July and August, when it is about 20° north. Of the two converging air masses, *cT* from the north and *mT* from the south, only the *mT* air mass contains the high moisture content necessary for the production of precipitation. Since the convergence is strongest near the pressure trough, the heaviest precipitation is to be expected immediately south of the intertropical front. At the same time, heavy air-mass precipitation occurs within the *mTKu* air mass.

Table 73 clearly verifies the following general conclusions: The northernmost stations have a single precipitation maximum about July or August, while the more southern stations show double maxima corresponding to the twice-yearly passage of the low-pressure belt (equatorial type of annual rainfall variation). In general, heavy precipitation occurs during the summer months when the *mTKu* air mass predominates. Furthermore, as a result of orographical lifting, the coastal stations experience heavier rainfall than the inland stations, with the exception of the Bight of Benin. In the latter region, the orientation of the coast line parallel to the prevailing southwest winds, together with the upwelling of cool water off the coast, favors slight precipitation. The data for Lomé clearly verify this partial rain shadow. Besides reducing the rainfall, the cooling of *mT* air over the cold water results in summer fog and low stratus clouds.

Throughout the entire region the summer rainfall is in the form of heavy showers or thundershowers. On occasion, the heavy downpours result from the westward motion of tornadoes, as described in Sec. 74.

The dry season, like the rainy season, depends on the oscillation of the intertropical front. About 15° north the northeast winds persist throughout the greater part of the year, and consequently there is a long period of little or no precipitation. Toward the south the dry season becomes shorter until finally along the Guinea coast the northeast winds, or harmattans, blow for only short intervals of time. The dust storms of the Sahara do not extend to the Guinea coast. Also, the extreme dryness of the air is tempered by the vegetation and moisture sources south of 10° north. Thus the rainfall regime depends on the progress and recession of the southwest monsoon.

TABLE 73.—PRECIPITATION REGIME OF WEST AFRICA FROM 0° TO 17°N
(Precipitation amounts in inches)

Station	Lat.	Long.	Primary maximum	Secondary maximum	Rainfall minimum	Location
St.-Louis	16°N	17°W	Aug., 7.4	None	Dec.–Apr., 0.0	Coastal
Timbuktu	17°N	3°W	July, 3.5	None	Nov.–Apr., 0.0	Inland
Bathurst	13°N	17°W	Aug., 20.4	None	Dec.–Apr., 0.0	Coastal
Niamey	14°N	2°E	Aug., 6.9	None	Nov.-Mar., 0.1	Inland
Conakry	9°N	14°W	July, 55.8	None	Jan.–Feb., 0.0	Coastal
Sokodé	9°N	1°E	Aug., Sept., 9.8	May, 6.9	Dec.–Feb., 0.1	Inland
Misahöhe	7°N	1°E	June, 10.0	Sept., 7.7	Jan., 0.9	Inland
Grand Bassam	5°N	4°W	June, 27.9	Nov., 7.2	Aug., 0.8	Coastal
Lomé	6°N	1°E	June, 6.7	Oct., 2.4	Aug., 0.1	Coastal
Lagos	6°N	3°E	June, 19.2	Oct., 8.6	Dec.–Jan., 1.0	Coastal
Akassa	4°N	6°E	Oct., 24.7	June, 18.6	Jan., 2.6	Coastal

The average cloudiness varies in a similar fashion to the rainfall frequency. Slight cloudiness, especially in the north, is experienced during the dry season and almost overcast skies during the rainy season, especially about the western coast line. Even though the Bight of Benin experiences relatively light rainfall during August, the mean cloudiness remains high because of coastal fog and low stratus clouds.

Since the region has a monsoon regime and is situated at low latitudes, the highest temperatures are observed just before the beginning of the true rainy season. For example, around Kissidougou (32, Table 75) three seasons may be noted:

1. Cool dry season from November to February
2. Hot season from March to May
3. Warm rainy season from May to November

Farther north toward Timbuktu (36), the dry season is accompanied by lower winter temperatures (average for December and January, 71°F); higher hot-season temperatures (average for May and June, 94°F); and, because of the decreased precipitation, higher rainy-season temperatures (average for July and

August, 88°F). On the other hand, in the rainy south, for example, at Lagos (31), the mean temperature range is slight. High temperatures prevail prior to the season of very heavy rainfall.

The combination of the foregoing rainfall and temperature distributions results in a gradual transition from true tropical rain-forest climate along the Guinea coast to tropical savanna climate farther north, in turn to a steppe climate, and finally to the true desert climate of the western Sahara.

East of 10° east the climate belts are similar to those in the west. However, the continental influence results in important differences. Because of the remoteness of the source region of *mT* air and as a result of the continental heating, respectively, the rainy periods are less intense and the mean temperatures during the summer months are higher than those near the west coast. As mentioned previously, the desert region extends from almost the Mediterranean coast to about 18° north. From 18° north to the equator, the climate is again controlled by the meridional displacement of the doldrum belt. Table 74 serves to illustrate the similarity of the rainfall regime to that farther west, *viz.*, a single maximum in the north, double maxima farther south, and finally rain throughout the year near the equator. A comparison with Table 73 verifies the diminished rainfall in the east as compared with the very heavy precipitation near the west coast.

TABLE 74.—PRECIPITATION REGIME OF CENTRAL AFRICA FROM 0° TO 16°N
(Precipitation amounts in inches)

Station	Lat., °N	Long., °E	Primary maximum	Secondary maximum	Rainfall minimum
Khartoum.	16	33	Aug., 2.2	None	Nov.–Apr., 0.0
El Obeid..	13	30	Aug., 4.4	None	Nov.–Apr., 0.0
Kodok....	10	32	Aug., 7.6	June, 5.2	Dec.–Mar., 0.0
Mongalla .	5	32	Aug., 5.9	June, 4.6; Oct., 5.4	Dec.–Jan., 0.1
Wadelai...	3	31	Oct., 6.5	May, 4.7	Dec.–Jan., 1.1

Except during the rainy season, the mean monthly cloudiness is low. During the dry season, especially in the northern Anglo-Egyptian Sudan, severe dust storms accompany the northeast winds. These storms, called "haboobs," are more frequent with the freshening of the wind in connection with shallow depressions along the intertropical front.

d. Cameroons to South-West Africa. The climate of western Africa, south of the equator, in a broad sense is the mirror image of northwestern Africa, *viz.*, a gradual trend from equatorial rains to desert conditions about 20° south. However, because of the shape of the continent, there are two important differences, which may be summarized as follows:

1. Throughout the year, along the west coast, the tropical low-pressure area remains north of the equator. Consequently, there is no strong onshore flow toward the interior of the continent in the Southern Hemisphere; *i.e.*, unlike the Guinea lands, the coastal areas of southwestern Africa are not flooded with $mTKu$ air masses.

2. The west coast of South Africa is washed by the cool Benguela Current almost to the equator, while the Canaries Current is replaced by warmer water at about 15° north latitude. These two factors account for less precipitation along the Southern Hemisphere coastal areas than along those of the Northern Hemisphere.

In the northern Cameroons the rainfall regime is very similar to that of Nigeria, with the heaviest rainfall from May to October. About 2° north or 3° north the rainfall distribution changes so that at Libreville (42, Table 75), on the equator, the rainy period is from September to May, while July and August are dry months. Farther south, along the coast, there are a gradual lengthening of the dry winter period and a gradual diminution of the yearly precipitation total, as shown by Libreville (42), Banana (45), Luanda (47), and Swakopmund (49) in Table 75. By 20° south the yearly total is less than 2 in. Inland, the north to south trend in the rainfall distribution is similar to that along the coast line. The upper Congo has a Northern Hemisphere rainfall regime, with a maximum in July and August; immediately north of the equator, appreciable rainfall is observed in all months; finally, the southern Congo basin [see Luluabourg (46, Table 74)] has a pronounced dry season from May to September.

Throughout the Congo the rainfall is of the thundershower type and, as in the Guinea lands, westward-moving tornadoes are not uncommon. The total rainfall is less than in the Amazon basin. This fact can be explained by the unfavorable topography to the east; the high mountains shield the Congo basin from direct invasions of *mT* air. At the same time, the cool western coast line and unfavorable wind circulation prohibit eastward invasions of *mT* air. The pronounced inland winter dryness, even near the equator, is also the result of the prevailing air motion from the east or southeast.

As in the Northern Hemisphere, the rainfall decreases steadily poleward, and finally desert conditions are reached in southwestern Africa. Except along the coast line the mean cloudiness follows the mean precipitation. Consequently, relatively clear skies are common inland during the winter months. Along the coast the cool Benguela Current favors fog and low stratus similar to that along the Peruvian coast of South America. The influence of the coastal fog on the mean cloudiness is illustrated by the values for Swakopmund (49) and Windhoek (50) in Table 75. Whereas the inland region, south of 10° south, has clear winter skies, there may be appreciable dry-season cloudiness in the Congo basin. The abundance of surface moisture and tropical vegetation aids in the production of surface fog or low-lying clouds.

The data in Table 75 demonstrate that the temperature regime resembles that of northwestern Africa, *viz.*, uniformly high equatorial temperatures changing gradually to high summer and low winter temperatures over the southern desert. The higher topography, south of 10° south, favors lower mean temperatures than those in similar sections of northwestern Africa. As in the case of the Atlantic coast of northern Africa, the persistent sea breezes and cool coastal water have a cooling influence on the coastal temperatures.

e. South Africa. The climate of South Africa varies from place to place, mainly because of the seasonal variation of the prevailing wind systems and the topographical features. In winter, when the subtropical belt of high pressure is displaced farthest north, the Cape of Good Hope receives rainfall with the passage of fronts that are connected to cyclones centered over the southern ocean. The frontal precipitation is augmented by orographical precipitation in the rear of the cold fronts. This type of rainfall varies from traces about Lüderitz Bay to almost 4 in. a month around Cape Town. With a wedge of high pressure located about 30° south, the east coast receives only slight precipitation. Similarly, anticyclonic circulation accounts for the very slight winter rainfall over the interior.

During the summer months, the subtropical high-pressure wedge is located south of the Cape of Good Hope, while a thermal low pressure controls the circulation over the land. Hence from Cape Agulhas north, there is an inflow of tropical maritime air across the east coast. Because of the warm Mozambique Current this air mass is very warm and moist. Furthermore, it is unstable over the land partly because of the convergence toward the low-pressure area and partly because of the continental heating. The net result is summer precipitation over a vast area comprising the east coast and the interior. This occurs principally in the form of showers and thundershowers. The proximity of the moisture source region and orographical lifting results in greater rainfall along the coast than over the interior [compare Durban (55) and Kimberley (56) in Table 75]. Along the west coast, except for sea breezes, the prevailing air flow is offshore. Because of the downslope motion, foehn winds, locally called *berg* winds, produce periods of warm, clear weather. However, sea breezes usually predominate directly along the coast. Since the upwelling of cold water produces cool sea-surface temperatures, cool summer weather generally prevails along the west coast, as shown by the mean temperatures at Port Nolloth (52). The spring rainfall, which is fairly uniformly distributed, occurs in the form of convective showers and thundershowers. On the other hand, the autumn rainfall is restricted mainly to the east coast and is connected with the southward-moving tropical cyclones.

The temperature distribution is affected to a considerable extent by the topography. Since the highest land is to the north, the latitudinal variation of temperature is slight. Longitudinal temperature variations are more pronounced; for example, the lowlands are appreciably warmer than the high plateaus. A comparison of Port Nolloth (52) and Durban (55) clearly illustrates the important influence of the ocean currents; the Benguela Current and cold-water upwelling result in a mean January temperature that is 15°F lower than the mean January temperature along the east coast, where the warm Mozambique

Current causes hot, humid summer weather. In winter, radiational cooling results in low temperatures over the plateaus, with frequent occurrences of frost.

Considering the country as a whole, the mean cloudiness is low. Relatively clear skies predominate along the east coast and over the interior during the winter months. Since the summer rainfall is convective in nature, prolonged periods of overcast skies are very rare. Coastal fog and stratus clouds increase the mean cloudiness of the west, especially during summer.

The climate of South Africa may be summarized as follows: Mediterranean type of climate, with winter rains and summer dryness about Cape Town; steppe and desert climate along the west coast, with slight winter rainfall; a *C* type of climate along the east coast and east-central regions, with summer rainfall; a steppe climate with slight summer rainfall over the west-central area. Only the inland valleys of the Zambezi and Limpopo rivers experience the warm, humid conditions so frequently observed in low latitudes.

f. Eastern Africa. In this section the climate of the east coast and the east-central portion of the continent, from about 20° south to Ethiopia, will be considered.

Since Portuguese East Africa is south of 10° south, the precipitation shows a decided summer maximum. Here the rainfall arises from the general instability in the *mTu* air mass, which streams inland toward the thermal low-pressure area. During the winter season the air mass is relatively stable, and in the absence of continental heating only slight precipitation is observed. During late summer some heavy rain falls in connection with the southwestward motion of tropical cyclones from the Indian Ocean. The inland areas of Nyasaland and Northern Rhodesia experience a similar rainfall regime, as shown by the various stations in Table 75. As in the case of western Africa, the land areas in this latitude have a single precipitation maximum about midsummer, when the doldrum belt is in its southernmost position. The precipitation is convective in nature, and consequently thunderstorms are common. In general, the entire land area south of 10° south has a rainy season from November to April and a dry season from May to September. The dry season is more pronounced inland.

Table 75.—Meteorological Means at Selected Stations
(Temperature in degrees Fahrenheit, precipitation in inches, cloudiness in tenths)

Station	Lat.	Long.	Temperature				Precipitation				Cloudiness				Elevation ft
			Jan.	Apr.	July	Oct.	Jan.	Apr.	July	Oct.	Jan.	Apr.	July	Oct.	
North															
13. Santa Cruz	28°N	16°W	61	65	73	71	2.7	1.0	0.0	1.5	4	4	1	4	130
14. Cape Yubi	28°N	13°W	61	64	68	67	0.5	0.2	0.0	0.5	4	4	6	5	0
15. Mogador	31°N	10°W	57	63	68	67	2.2	0.7	0.0	1.3	3	3	1	2	30
16. Marrakech	32°N	8°W	52	67	82	70	1.3	1.1	0.2	0.5	5	4	2	3	1540
17. Tangier	36°N	6°W	55	61	75	67	4.7	3.6	0.1	3.6	..	..	..	..	250
18. Algiers	37°N	3°E	53	61	77	69	4.2	2.3	0.1	3.1	5	5	3	4	70
19. Geryville	34°N	1°E	39	52	78	56	0.9	1.7	0.2	1.5	4	4	3	4	4280
20. Biskra	35°N	6°E	51	67	89	70	0.5	1.2	0.2	0.8	2	2	1	2	410
21. Bizerte	37°N	10°E	51	60	77	69	4.4	1.9	0.2	2.1	..	..	..	..	20
22. Gabès	34°N	10°E	51	64	81	72	0.9	0.7	0.0	1.0	..	..	..	..	30
23. Bengasi	32°N	20°E	55	66	78	74	3.1	0.1	0.0	0.5	2	1	1	2	30
24. Alexandria	32°N	30°E	57	65	78	74	2.3	0.1	0.0	0.3	5	3	1	3	100
Sahara															
25. Ghardaïa	33°N	4°E	47	67	93	69	0.9	0.3	0.0	0.2	2	1	0	2	1770
26. In-Salah	27°N	3°E	54	76	98	80	....	....	....	....	..	..	..	..	980
Sudan-Guinea Coast															
27. St.-Louis	16°N	17°W	68	68	81	81	0.2	0.0	2.6	0.7	..	..	..	..	20
28. Bathurst	13°N	17°W	74	76	80	81	0.0	0.0	12.5	4.4	..	..	..	..	10
29. Sierra Leone	8°N	13°W	79	81	76	77	0.6	5.4	36.8	15.2	..	..	..	..	390
30. Cape Gold Coast	5°N	1°W	80	83	77	80	0.6	3.5	2.2	2.4	..	..	..	..	70
31. Lagos	6°N	3°E	80	82	76	78	1.1	6.3	10.2	8.6	..	..	..	..	0
32. Kissidougou	9°N	10°W	76	81	78	78	0.9	5.3	8.4	11.9	..	..	..	..	1560
33. Fort-Lamy	12°N	15°E	75	92	81	84	0.0	0.1	5.1	1.5	0	1	5	3	890
34. Kayes	14°N	12°W	77	94	84	84	0.0	0.0	8.3	1.9	..	..	..	..	200
35. Niamey	14°N	2°E	71	93	81	85	0.4	0.2	5.8	0.6	1	1	6	3	1640

TABLE 75.—METEOROLOGICAL MEANS AT SELECTED STATIONS.—(*Continued*)

Station	Lat.	Long.	Temperature				Precipitation				Cloudiness				Elevation ft
			Jan.	Apr.	July	Oct.	Jan.	Apr.	July	Oct.	Jan.	Apr.	July	Oct.	
36. Timbuktu	17°N	3°W	71	92	89	89	0.0	0.0	3.5	0.4	..	..	..	..	820
37. Khartoum	16°N	33°E	71	87	88	88	0.0	0.0	1.7	0.4	..	..	..	..	1260
38. Mongalla	5°N	32°E	81	82	77	79	0.1	3.9	4.0	5.4	..	..	..	..	1440
39. Suakin	19°N	37°E	72	77	93	83	1.0	0.0	0.1	1.5	..	..	..	..	20
Cameroons															
40. Douala	4°N	10°E	79	79	75	76	1.2	8.9	28.1	16.6	..	..	..	..	40
41. Yaoundé	4°N	12°E	74	72	70	71	1.6	9.1	2.6	8.9	..	..	..	..	2461
42. Libreville	0°	10°E	78	78	73	76	8.9	12.8	0.1	14.1	..	..	..	..	70
Congo															
13. Mobaye	5°N	21°E	80	80	77	77	0.2	5.7	4.7	8.3	..	..	..	..	1310
44. Bolobo	2°S	16°E	78	78	77	77	5.0	7.2	0.0	6.5	..	..	..	..	1080
45. Banana	6°S	12°E	80	80	72	79	2.1	6.1	0.0	1.6	..	..	..	..	10
46. Luluabourg	6°S	23°E	76	77	76	76	7.2	6.1	0.1	6.6	6	6	4	8	2030
Angola															
47. Luanda	9°S	13°E	77	79	67	74	1.0	4.3	0.0	0.2	..	..	..	..	190
48. Caconda	14°S	15°E	71	69	65	70	8.1	8.3	0.0	4.6	..	..	..	..	5510
South-West Africa															
49. Swakopmund	23°S	15°E	63	60	56	58	0.0	0.0	0.0	0.1	7	6	5	6	20
50. Windhoek	23°S	17°E	74	68	57	71	3.9	1.7	0.1	0.4	4	3	1	3	5460
51. Kubub	27°S	16°E	70	59	47	59	0.6	0.4	0.0	0.1	3	2	2	2	5020
South Africa															
52. Port Nolloth	29°S	17°E	60	58	55	58	0.0	0.2	0.2	0.0	..	..	..	..	20
53. Cape Town	34°S	18°E	70	63	55	61	0.7	1.9	3.7	1.6	..	..	..	..	40
54. Port Elizabeth	34°S	25°E	69	65	58	62	1.2	2.0	1.9	2.6	..	..	..	..	180
55. Durban	30°S	31°E	76	72	64	70	4.6	3.4	1.2	5.1	..	..	..	..	260
56. Kimberley	29°S	25°E	76	64	51	67	2.5	1.4	0.3	1.0	..	..	..	..	4040
57. Pretoria	26°S	28°E	72	63	52	68	5.5	1.1	0.1	1.8	..	..	..	..	4470
58. Bulawayo	20°S	29°E	72	66	57	72	5.9	0.7	0.0	0.9	..	..	..	..	4470

TABLE 75.—METEOROLOGICAL MEANS AT SELECTED STATIONS.—(*Continued*)

Station	Lat.	Long.	Temperature				Precipitation				Cloudiness				Elevation ft
			Jan.	Apr.	July	Oct.	Jan.	Apr.	July	Oct.	Jan.	Apr.	July	Oct.	
59. Juliasdale	18°S	33°E	63	58	50	64	10.6	2.2	0.8	1.3	..	..	..	..	6070
60. Salisbury	18°S	31°E	68	64	55	68	7.5	1.1	0.0	0.7	..	..	..	..	4880
Mozambique															
61. Mopeia	18°S	36°E	83	78	70	82	7.4	3.3	0.8	1.1	..	..	..	..	80
62. Inhambane	24°S	36°E	79	76	69	74	5.4	3.1	1.5	2.4	..	..	..	..	50
63. Beira	20°S	35°E	80	77	69	76	13.7	4.5	1.4	1.9	..	..	..	..	20
64. Mossuril	15°S	41°E	81	79	72	78	6.1	3.0	0.7	0.3	..	..	..	..	50
65. Muchena	16°S	34°E	83	74	67	80	5.4	1.5	0.1	0.7	..	..	..	..	1640
East Africa															
66. Mumbwa	15°S	27°E	73	69	61	78	8.5	0.8	0.0	0.9					
67. Mongu	15°S	23°E	78	75	66	82	9.4	1.4	0.0	2.2	..	..	..	..	3320
68. Lundazi	12°S	33°E	69	66	56	71	7.4	1.1	0.0	0.0	..	..	..	..	3360
69. Fort Rosebery	11°S	29°E	73	71	62	76	8.5	1.4	0.0	1.5	..	..	..	..	3830
70. Abercorn	8°S	31°E	69	68	63	73	7.7	3.7	0.0	1.0	..	..	..	..	5110
71. Dar es Salaam	7°S	39°E	82	78	74	77	3.3	11.9	1.7	1.2	..	..	..	..	40
72. Tabora	5°S	33°E	71	72	71	78	5.7	5.2	0.0	0.5	..	..	..	..	3980
73. Mombasa	4°S	40°E	80	81	76	79	0.8	8.0	3.5	3.3	..	..	..	..	70
74. Nairobi	1°S	37°E	64	64	58	65	1.9	8.9	0.9	2.3	..	..	..	..	5450
75. Kisumu	0°	35°E	77	73	72	75	2.6	6.6	1.8	3.0	..	..	..	..	3800
76. Entebbe	0°	32°E	71	70	69	70	3.0	10.0	3.0	3.6	..	..	..	..	3860
77. Gondokoro	5°N	32°E	82	84	78	78	0.2	3.1	5.9	5.4	..	..	..	..	1440
78. Harar	10°N	42°E	66	69	66	68	0.8	4.0	5.2	0.6	..	..	..	..	6090
79. Addis Ababa	9°N	38°E	60	64	62	62	0.4	3.3	12.0	0.6	..	..	..	..	8010
80. Gondar	13°N	37°E	67	73	61	65	0.0	0.0	11.4	1.8	..	..	..	..	6230
81. Addi Ugri	15°N	39°E	64	70	65	67	0.0	0.9	6.0	0.4	..	..	..	..	6630
82. Massaua	16°N	39°E	79	85	95	87	1.7	0.2	0.1	0.4	..	..	..	..	30
Madagascar															
83. Tamatave	18°S	50°E	79	76	68	74	15.0	12.0	13.2	5.3	..	..	..	..	20

In view of the irregular topography the rainfall distribution over land is somewhat erratic; this is particularly true of Kenya colony, Uganda, and Tanganyika. The seasons of maximum precipitation, in general, coincide with the occurrence of light and variable winds within the doldrum belt. For example, southern Tanganyika has one prolonged rainy season like Nyasaland, whereas, farther north, there is a double maximum corresponding to the twice-yearly passage of the low-pressure belt, *i.e.*, March to May and November to December. Relatively slight precipitation occurs during the periods of the well-established northerly and southerly winds, except through orographical lifting over the high mountain ranges. In these mountainous countries the total rainfall is considerably less than in the Congo basin. Because the air masses are more unstable to the west, heavier precipitation occurs west of Lake Victoria than east of it. Farther north, the diminished precipitation near the east coast is probably due to the lower moisture content and greater stability of the air masses from the North Indian Ocean; for example, during the Northern Hemisphere winter the air that reaches northeastern Africa has had only a short trajectory over the ocean after leaving the Asiatic mainland.

Ethiopia, Eritrea, and Somaliland have a varied rainfall distribution. The precipitation occurs during the Northern Hemisphere summer. Since the air masses from the east, and especially the northeast, are dry during the Northern Hemisphere winter, very slight winter precipitation is to be expected. During the summer months the prevailing air flow is from the southwest. This air has crossed a vast land area since leaving the Atlantic ocean and therefore arrives as a modified form of *mTKu* air. The net result is less precipitation than is observed farther west. Moreover, with the sheltering effect of the high mountains to the west, the east coast of northeastern Africa remains dry even during the summer. The immediate coast line of the Red Sea experiences slight winter precipitation.

As in the case of western Africa, the highest temperatures generally occur at the end of the dry season. In regions of slight precipitation, for example, in northeastern Africa, very high summer temperatures are common. The lowest temperatures occur during the winter dry season. Throughout all eastern

Africa, the great variability of topography causes wide geographical temperature variations. The high mountains experience cool weather throughout the year, whereas low-lying valleys and the eastern coast line remain warm, even during the winter months. The warmth of the east coast, south of the equator, as compared with the west coast is mainly the result of the warm Equatorial Current.

In general, the mean monthly cloudiness corresponds closely to the mean rainfall distribution. Relatively clear skies predominate in regions of slight rainfall, for example, in eastern Somaliland, while cloudy skies predominate during periods of heavy rainfall, for example, in western Ethiopia in the summertime. One exception is worthy of note. Even though the Gulf of Aden has slight precipitation, the upwelling of cool water during the summer months favors fog and low stratus clouds. Fog is frequent during the summer half year.

From the foregoing discussion it is apparent that no portion of eastern Africa experiences a tropical rain-forest climate. Even in equatorial latitudes, there is sufficient winter dryness to justify a tropical savanna classification. Because of the high elevation, large sections of the interior have a mild temperate climate, with summer rain and winter dryness. Finally, the slight precipitation of northern Kenya and eastern Somaliland results in a steppe type of climate.

g. Madagascar. Since the mountain range is normal to the prevailing air motion, orographical precipitation is to be expected over the eastern portion of the island in all seasons of the year. Through adiabatic descent the west coast has little or no precipitation during seasons of straight east-west motion. Consequently, in winter there is moderate to heavy precipitation on the eastern half of the island with very slight precipitation on the western half. In midsummer, the doldrum belt of convergence (intertropical front) crosses the northern portion of the island. Heavy rainfall is therefore observed in the northern section of the island as well as over the eastern portion. Furthermore, moderate summer rainfall occurs over the generally dry southwestern area. This precipitation results from the general instability of the *mT* air masses. The late summer and autumn rainfall is augmented by heavy downpours, which accompany the passage of tropical cyclones close to the coast line. Madagascar is situated about the latitude of recurvature of these tropical disturbances.

Besides the variation with altitude, the most important feature of the temperature distribution is the occurrence of high summer temperatures over the southern portion of the island. The diminished cloudiness of this region favors strong thermal heating.

The heavy rainfall of the east coast produces a tropical rain-forest climate as compared with the tropical savanna climate of the west coast. The highlands of the interior have a typical *C* type of climate.

CHAPTER XIV

AUSTRALIA AND NEW ZEALAND

Since Australia and New Zealand are within reasonable proximity, they will be discussed together. By this grouping it is not suggested that the two countries have similar climatic types. Actually, there is a vast difference between the factors that control the weather of the respective regions.

Because continental Australia extends from 10° south to 40° south, tropical types of climate predominate. Furthermore, in view of the large land mass, continental influences are very important. The subtropical belt of high pressure, which is characterized by clear skies, dominates the weather of a large section of the country and consequently arid conditions prevail over western and central Australia. This vast region of desert-type climate is the second largest in the world, being exceeded in size only by the Sahara. Another striking feature is the complete absence of a true tropical rain-forest type of climate. Even along the north and east coasts there is only a relatively small area of tropical savanna climate. Finally, another outstanding feature is the absence of high mountain chains to provide effective climatic barriers, like the Andes of South America, the Rocky Mountains of North America, or the Himalaya of Asia. However, even though the highlands of the east coast are comparatively low, they have a noticeable effect on the climate.

Unlike Australia, Tasmania and New Zealand come under the direct influence of the prevailing westerlies, south of 35° south. The narrow extent of these regions favors a marine type of climate. In contrast to the low average rainfall of Australia, New Zealand and Tasmania enjoy copious precipitation during all seasons of the year. Consequently, except for the high mountain chains, they experience a moderate marine-forest climate (*Cf*) with mild winters.

76. Physical Features. The continent of Australia is characteristically devoid of high mountain chains. However, there is some variation in topography; the principal features may be summarized as follows:

1. Extending from the southeast to the far north, a more or less continuous chain of highlands is located near the east coast. In the southeast, there are few peaks above 5000 ft, while across Queensland the highest ridges are generally less than 2000 ft in elevation. Since the highlands are near the coast, the ground slopes off rapidly to the South Pacific Ocean.

2. The major part of western Australia has an elevation of between 1000 and 2000 ft, with isolated plateaus about 3000 or 4000 ft above sea level.

3. Except for the small Flinders Range, central-eastern Australia is below 500 ft.

4. The prominent indentations of the Australian continent are the Gulf of Carpentaria and the Great Australian Bight.

In contrast to the absence of mountains throughout Australia, Tasmania and New Zealand are relatively mountainous. For example, central Tasmania attains an elevation of 5000 ft. One general mountain system may be identified in New Zealand. The Southern Alps, with numerous minor ranges, extend to 12,000 ft and form the backbone of South Island; across Cook Strait there is a similar mountain system from Wellington to East Cape. Whereas the highlands are nearer the west coast in South Island, the North Island chain is nearer the east coast. However, tall, isolated peaks are located in the central and western sections of North Island.

The effect of the varying relief on the broad-scale features of the climate is illustrated by the following:

1. Since, north of 30° south, the prevailing air flow is from east to west, orographical lifting results in considerably heavier precipitation along the east coast than over the central or western sections of Australia.

2. South of 30° south, the southeastern highlands also favor higher rainfall over Victoria and New South Wales than across the low-lying lands around the Great Australian Bight.

3. Besides favoring orographical precipitation, the eastern highlands shelter the interior from direct invasions of maritime air. Consequently, the dry air masses of central Australia result in very high temperatures, especially during the summer months.

4. The highland of Tasmania aids in the production of moderate to heavy precipitation, especially across the windward western section of the island.

5. The high mountains across South Island, New Zealand, provide a very effective barrier to the prevailing westerlies; therefore, heavy orographical precipitation is observed along the west coast. The North Island mountains aid in the production of precipitation to a smaller extent because they are a less formidable barrier.

6. Adiabatic heating results in higher temperatures over the east coast than over the west coast of South Island.

These general considerations demonstrate the importance of topography on the distribution of precipitation and temperature.

77. Mean Distribution of Pressure and Wind. The following features of the mean July circulation are worthy of note:

1. Since Australia lies in low latitudes, the air flow is dominated by the circulation around the subtropical high-pressure belt. Wintertime continental cooling results in a separate high-pressure cell across Australia; the mean center is located about 30° south and 145° east. With a low-pressure area north of the equator, the prevailing air flow north of 25° south is southeast, as demonstrated by the stations in Table 76.

2. Across the interior of the continent, from 25° south to 30° south the mean velocities are light, consistent with the high-pressure ridge. However, along the coast, the increased gradient causes moderate winds. The coastal stations, for example, Geraldton (7) and Port Macquarie (9), have mean wind directions consistent with the presence of high pressure over the land.

3. With high pressure to the north and low pressure to the south, general westerly or northwesterly winds prevail south of 30° south. As shown by the New Zealand stations, the westerly flow around the mean subpolar low dominates the winter air motion.

The principal difference between the summer and winter surface circulation is the southward displacement of the subtropical high-pressure and the equatorial low-pressure areas during summer. Whereas, in July, the mean latitude of the anticyclonic center is about 31° south, the January position of the high-pressure ridge is approximately 37° south. The meridional displacement of the doldrum belt is more pronounced. As a result of the intense continental heating, the low-pressure trough extends across northern Australia. For example, the prevailing northwest winds at Darwin (3) and Thursday Island (2) substantiate the presence of a mean low-pressure area to the south or southwest. South of the subtropical anticyclone, westerly winds around the deep cyclone to the far south dominate the summer circulation. The principal difference between the two seasons is the higher wind velocity of the winter season.

From Table 76 it is evident that in winter the most persistent winds occur mainly north of 25° south. Even though the mean-pressure maps show a high-pressure system centered over the continent in winter, it should be noted that this is the integrated effect of many pressure oscillations. Actually, the anticyclones migrate across Australia along a general west to east trajectory. In contrast to South America and the other continents, the flat Australian relief does not offer any obstacle to the west-east circulation. The migratory anticyclonic centers rarely penetrate farther north than 25° south, and consequently the winter southeast trades and the cyclonic circulation in summer are reasonably constant across northern Australia. On the other hand, the circulation around the subpolar cyclone is continually interrupted by moving cyclones and anticyclonic wedges. Therefore, there is a considerable variation of wind direction south of 35° south.

Lack of mean upper-air charts prohibits a detailed discussion of the free air flow. Nevertheless, the following general features may be deduced from the known surface distribution of temperature and pressure. During winter, the high-pressure belt about 31° south is displaced northward with elevation. For example, at Sydney and Melbourne, west or northwest winds are observed at 3 km about 75 per cent of the time. Across New Zealand they become west or southwest. The latter direction arises from the flow in the rear of the mean upper-level trough located farther east. Furthermore, the decrease of temperature southward produces an increase in the wind velocities with elevation.

In summer, the westerly winds again dominate the upper-air flow except in low latitudes. Again because of the north-south temperature gradient, the high-pressure belt is displaced northward with elevation. Unlike conditions in the Northern Hemisphere, the mean temperature gradient between 30° south and

50° south differs slightly from winter to summer. Therefore, the increase in the west winds with elevation is approximately as great in summer as in winter.

TABLE 76.—PERCENTAGE FREQUENCY OF SURFACE-WIND DIRECTION

Station	Lat.	Long.	January								July							
			N	NE	E	SE	S	SW	W	NW	N	NE	E	SE	S	SW	W	NW
North of 25°S																		
1. Townsville.....	20°S	147°E	13	11	*27*	25	11	3	3	7	2	1	16	35	*39*	4	2	1
2. Thursday Island	11°S	142°E	5	7	11	26	2	2	3	*44*	0	0	1	*98*	0	0	0	1
3. Darwin........	12°S	131°E	2	6	7	18	7	10	24	*26*	1	5	20	*63*	4	4	2	1
4. Mundiwindi....	24°S	120°E	5	19	*29*	24	4	11	3	5	7	18	*34*	18	6	10	2	5
5. Hall's Creek....	18°S	128°E	10	8	19	12	6	12	*23*	10	1	12	*59*	19	2	2	4	1
6. Alice Springs...	24°S	134°E	13	7	26	*38*	3	3	1	9	15	4	12	*42*	12	6	1	8
South of 25°S																		
7. Geraldton......	29°S	115°E	1	2	12	19	*40*	16	9	1	3	15	21	*32*	5	7	15	2
8. Thargomindah .	28°S	144°E	20	*23*	18	19	5	3	4	8	19	*22*	19	12	8	7	8	5
9. Port Macquarie	31°S	153°E	11	*27*	2	9	21	15	7	8	2	1	0	0	1	16	*70*	10
10. Perth..........	32°S	116°E	2	7	18	16	*28*	22	5	2	17	*18*	8	8	10	10	17	12
11. Eucla..........	32°S	129°E	8	3	21	*46*	9	11	1	1	29	6	2	3	2	3	20	*35*
12. Melbourne.....	38°S	145°E	8	4	11	19	*28*	18	5	7	*32*	13	14	11	7	6	4	13
13. Sydney........	34°S	151°E	4	*24*	22	12	21	5	8	4	6	5	4	3	9	7	*49*	17
14. Hobart........	43°S	147°E	*28*	6	4	13	9	8	4	*28*	35	3	1	1	3	7	6	*44*
New Zealand																		
15. Auckland......	37°S	175°E	7	22	8	4	10	*31*	10	8	5	11	6	10	16	*28*	13	11
16. Wellington.....	41°S	175°E	19	5	2	16	13	3	2	*40*	16	8	2	21	18	5	2	*28*
17. Chatham Islands..........	44°S	176°W	11	6	5	9	4	*33*	10	22	9	6	10	14	7	*37*	8	9

78. Ocean Currents. The principal currents that control the oceanic circulation around Australia and New Zealand may be described briefly as follows:

1. South of the land areas the oceanic circulation is from west to east—the West Wind Drift. Because of the high latitude, this is essentially a cool current, and consequently the land areas near this water, for example, southern New Zealand and the adjacent islands, remain cool even during the summer months. However, the true cold water of this drift is entirely south of New Zealand; consequently, its principal influence is to control the temperature of the maritime air masses that move northward from high latitudes.

2. Branching off from the West Wind Drift, the West Australian Current transports cool water northward along the coastline of western Australia. However, this current is not so

pronounced as either the Peru or the Benguela Current; for example, at 25° south the August sea-surface temperatures off South America and South Africa are about 15°C as compared with 21°C off the Australian coast. The difference probably arises from the less permanent and weaker wind circulation of the eastern South Indian Ocean as compared with the stronger circulations of the South Pacific and Atlantic oceans. Nevertheless, the West Australian Current results in relatively cool surface temperatures along the coast line. Furthermore, the cool water favors fog and low stratus clouds. However, these hydrometeors are considerably less frequent and less significant than along the west coasts of Africa and America. Consequently, the *Bn* type of climate does not occur.

3. Like the east coasts of all other continents, eastern Australia is washed by a warm current, the East Australian Current. The water flows southward and then branches eastward so that western and northern New Zealand are washed by relatively warm water. The warm water favors higher coastal temperatures in eastern Australia and western New Zealand than along the opposite coasts. For example, at 35° south the February sea-surface temperatures along the Australian west and east coasts are 19°C and 22°C, respectively. Of even greater significance is the influence of the warm current on the production of precipitation. Its importance lies in the availability of heat and moisture for the formation of warm, humid air masses, which frequently contribute to the production of moderate or heavy precipitation.

4. Other minor currents are the warm Equatorial Current along northern Australia and the narrow warm current that flows westward along the south coast of Australia. The latter is an extension of the East Australian Current.

The entire system of oceanic currents is less pronounced than those around the other continents. The flow off the east coast of Australia may even become reversed. This indefiniteness of the water circulation is undoubtedly the result of a change in the wind circulation from summer to winter, as well as frequent day-to-day changes. A strong and persistent oceanic circulation requires a persistent wind circulation; therefore, in regions of changing or weak wind circulations the ocean currents are weak and variable.

79. Air Masses. The three air masses that influence Australia are tropical continental, tropical maritime, and polar maritime air. The properties of these air masses may be described briefly as follows:

a. Tropical Continental Air (*cT*). The source region of *cT* air is the vast arid region of central and western Australia. In summer, with the southward displacement of the subtropical high-pressure belt, the driest *cT* air is located over the southern section of the continent, west of 135° east. As a result of the low moisture content, the continental heating results in only slight cloudiness and rainfall. Farther north, the air mass is moister because of an inflow of moisture from the east, and consequently this modified *cT* air is accompanied by convective precipitation.

In winter, north of 30° south, *cT* air dominates the weather. Besides the dryness, continental cooling and anticyclonic subsidence favor a stable air mass. Consequently, wintertime *cT* air is accompanied by clear skies. Even though the air originally comes from the ocean, average morning relative humidities are about 50 per cent in central and northern Australia.

b. Tropical Maritime Air (*mT*). Whereas Australia may be invaded by tropical maritime air from either the Pacific or the Indian Ocean, the Pacific Ocean is the usual source of *mT* air for New Zealand. In winter, the air is stable on the eastern edge of the Indian anticyclone and consequently the *mT* air that invades western Australia is not only relatively cool for the latitude but also convectively stable. In contrast, the *mT* air that invades eastern Australia or New Zealand is warm, moist, and convectively unstable like the *mT* air mass that affects the weather of the southeastern United States. When this air mass is subject to frontal or orographical lifting, moderate or heavy precipitation results.

During the summer season, convergent flow toward the low-pressure belt across northern Australia produces an unstable *mTu* air mass. Furthermore, the continental heating over the northern continent contributes to the instability. This *mTKu* air mass is associated with the heavy summer convective precipitation of northern Australia. Between 30° south and 40° south the air is more stable as a result of the high-pressure circulation. Nevertheless, heating over the land, for example, across northern New Zealand, produces convective precipitation.

c. Polar Maritime Air (*mP*). The extensive oceanic region south of 40° south forms the source region for *mP* air. On some occasions the *mP* air originates as a *cP* air mass from the Antarctic Continent, which subsequently gains heat and moisture from below before reaching the land areas of Australia and New Zealand. On other occasions, the polar maritime air masses are returning air masses that have had a very long trajectory over the water. In either case, *mP* air is characteristically cool and moist. The stability of the air mass depends on the direction of the flow and the nature of the circulation. When the air flows northward in a strong cyclonic flow, the *mPKu* mass is accompanied by showers and instability-type clouds. However, when the air flows southward around a pronounced anticyclone, the *mPWs* mass is associated with stable-type clouds and slight precipitation. During winter, *mP* invasions, especially *mPKu* air masses, account for a large portion of the precipitation of southern Australia and New Zealand.

Because of the low latitude and low elevation, true *cP* air masses are never formed over Australia. Continental cooling does result in the formation of a cool air mass across the highlands of the southeast during winter. However, a mean July temperature between 40°F and 50°F cannot be considered characteristic of a *cP* source region.

80. Fronts, Cyclones, and Anticyclones. *a. Anticyclones.* Even though Australia is located in the subtropical high-pressure belt, there is a lesser tendency toward quasi-stationary anticyclones than elsewhere within these latitudes. This characteristic is in contrast to the behavior of subtropical highs near South and North America, where the steady motion of a subtropical anticyclone across the continent is uncommon. Since the anticyclones determine the mean frontal positions, it is desirable, in the case of Australia, to discuss the movement of the anticyclones before treating the principal frontal regions.

The warm anticyclones generally move from west to east with a mean velocity of about 20 to 25 mph. Consistent with the meridional displacement of the mean high-pressure belt, the paths of the highs vary from winter to summer as shown in Table 77. Since this table is based on only 1 year's data, the absolute values may not represent true means. The approximate frequency of the anticyclones is about five a month.

b. Frontal Zones. It is not very significant to speak of a mean position of the subtropical anticyclone in this region. One can expect the polar front, which separates the *mT* air from *mP* air, to vary its position considerably. Nevertheless, in winter, with a tendency toward high pressure over the continent, the average position of the polar front probably stretches from western Australia toward the southeast. The northern section of the front lies in the trough between two migratory highs. Because of the movement of the anticyclones, pronounced cold fronts occur throughout the entire region from western Australia to eastern New Zealand. Another winter frontal zone of importance is a second polar front, which extends southeastward from South Africa. Again, this front separates *mP* air from *mT* air and resembles in mean position the polar front off the east coast of South America.

TABLE 77.—MOTION OF ANTICYCLONIC CENTERS
(After Taylor*)

Months	Mean latitude, °S	Mean velocity, mph
June–Aug	32	18
Sept.–Nov	33.5	27
Dec.–Feb	37	23
Mar.–May	35	18

* TAYLOR, G., "Australian Meteorology," Clarendon Press, Oxford, 1920.

In summer, the polar front off South Africa is displaced southward in accordance with the meridional displacement of the subtropical anticyclone. As with respect to winter, it is somewhat misleading to discuss the mean position of the polar front in the Australian area. Nevertheless, with the tendency for a low-pressure trough to extend from central Australia across New South Wales, this may be considered as a mean frontal zone whose southern portion extends southeastward. The other important frontal zone exists off the north coast of Australia—the intertropical front. This zone of converging *mTu* and modified *cT* air masses may be sharply defined, but more frequently it is a wide zone of convective activity.

c. Cyclones. The cyclones that affect the weather have been discussed by Palmer[1] and may be summarized as follows:

[1] PALMER, C. E., *New Zealand Met. Office, Prof. Note* No. 1, Wellington, 1942.

1. Wave cyclones develop on the polar front off South Africa during all seasons of the year. They usually deepen and move southeastward so that the center of the cyclone generally passes south of New Zealand. However, the cold front, or occluded front, in the rear of the low center may produce widespread precipitation across southern Australia and New Zealand. Besides prefrontal and frontal precipitation, instability showers within the *mPKu* air mass strongly contribute to the winter rainfall of southern Australia. Also, orographical lifting greatly intensifies the precipitation over western Tasmania and western New Zealand. The cold front may have traversed the entire Indian Ocean and then crossed Australia or may have been subjected to frontogenesis off western Australia.

In the rear of the deepening cyclone, an *mP* anticyclone generally moves northeastward and finally reaches subtropical latitudes, where it becomes a new subtropical high-pressure cell replacing the one that moved eastward across southern Australia in advance of the cold front. In contrast to the unsettled frontal weather, the passage of an anticyclonic center is accompanied by relatively clear skies.

2. When a cold front becomes stationary over Queensland, a cyclone may develop and move southeastward, either north of New Zealand or across North Island. Because of the moisture content of western Pacific *mT* air, these cyclones are associated with widespread precipitation. Sometimes the cyclones develop farther south, near New South Wales, and subsequently cross southern New Zealand as deep lows. They may even develop along the south coast of Australia.

These developments off eastern Australia result in frontal precipitation not only over New Zealand but also over the east coast of Australia.

3. During the summer months, tropical cyclones that develop over the western Pacific may recurve and cross New Zealand as very deep extratropical cyclones. Hurricanes which recurve far to the east do not affect New Zealand, whereas those which recurve near Australia generally cross or pass near New Zealand. Heavy downpours along the Queensland coast sometimes result from these storms.

4. A secondary family of summer cyclones are those which develop on the secondary front across southeastern Australia.

They form in the north, where the high moisture content and instability of the air masses produce heavy precipitation. The shallow lows may move across New South Wales and gradually develop into deep southeastward-moving cyclones.

The cyclones of types 1 to 4 in Fig. 38 are preceded by warm highs or high-pressure wedges and are followed generally by cold northeastward- or eastward-moving anticyclones. As mentioned previously, the central portions of the anticyclones are accompanied by clear weather.

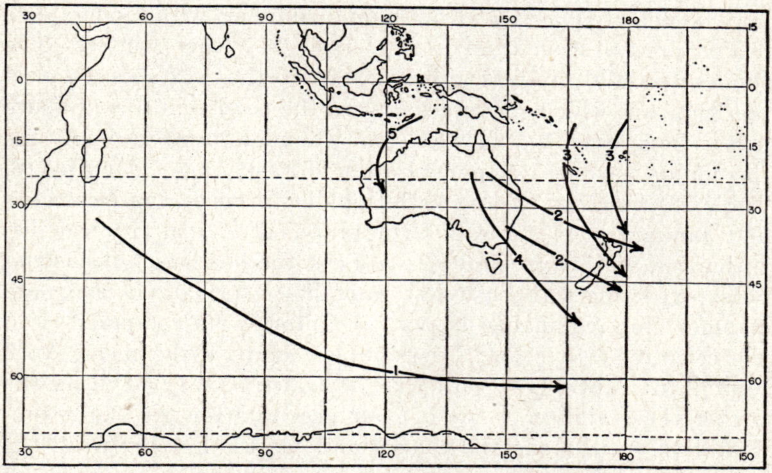

FIG. 38.—Cyclone tracks of Australia and New Zealand.

5. A final group of cyclones are the tropical disturbances that develop north of Australia and move in a parabolic path on to the northwest coast of the continent. Like all tropical disturbances, they are accompanied by heavy rainfall and strong winds. They occur in summer and autumn.

81. Climatic Regions. Plates I and II demonstrate that the mean temperature distribution over Australia depends much more on the latitude than is the case over other continents. Because of the absence of high mountains, topographical influences are small. Likewise, the absence of pronounced cold or warm currents results in a fairly uniform longitudinal temperature distribution. Continental influences, however, cause somewhat cooler inland temperatures in winter and warmer temperatures in summer than along the coast lines. Except for this continental factor, the temperature distribution favors a regular variation of climate from north to south. On the other hand, the rainfall distribution is asymmetrical, and consequently it is the important factor that accounts for longitudinal climatic variations. Therefore, the division of Australia into various climatic regions is based primarily on the rainfall regime.

a. Northern Coastal Region. This region includes northern Queensland and the land areas that border upon the Gulf of Carpentaria and the Timor Sea. Here two distinct seasons may be detected, *viz.*, a warm, humid summer, with moderate to heavy precipitation, and a dry, relatively cool winter. The summer rainfall results primarily from the unstable tropical maritime air, which converges toward the persistent low-pressure belt. Consequently, showers and thunderstorms account for the major part of the precipitation; for example, Darwin (18, Table 78) averages 32 days with thunderstorms during December and January. As a result of orographical lifting, tropical disturbances, and its proximity to *mT* air, the northern Queensland coast experiences the heaviest rainfall. Especially during late summer this region has heavy precipitation as compared with the central north coast [for example, compare the mean April rainfall at Cairns (19) with that at Darwin (18)]. During the winter the stability of the *mT* air accounts for the slight precipitation, even along the Queensland coast. Farther west, the prevailing offshore flow produces no precipitation. As shown by the mean values for Darwin (18), the cloudiness distribution parallels the rainfall distribution.

Relatively high temperatures prevail even during the winter months. Especially toward the interior, very high summer maxima are common; for example, 111°F has been recorded at Wyndham. As in other tropical regions, the hottest weather is experienced during the early summer months. As a result of the temperature and precipitation regimes, the northern coast experiences a tropical savanna type of climate, with a winter dry season.

b. Eastern Queensland. North of about 25° south, eastern Queensland has a rainfall regime similar to that of the northern coast. The causes are identical with those operating in the latter region, *viz.*, summer instability showers and occasional tropical cyclones. Orographical lifting produces heavy coastal rainfall and comparative dryness over the interior. Again, because of the air-mass stability, winter rainfall is slight. Unlike conditions in northern Queensland, some of the precipitation occurs in connection with cyclogenesis and frontal activity, as described in Sec. 80.

The summer mean temperatures are almost identical with those of northern Queensland. Very high maxima are commonly experienced. Because of the higher latitude, however, eastern Queensland experiences lower winter temperatures, and hence the climate is a *C* type, with summer rain and winter dryness.

c. Southeast Australia. Southeastern Queensland, New South Wales, and Victoria have a reasonably uniform distribution of precipitation, as shown by the stations in Table 78. In the north, for example, Brisbane (23), the precipitation regime resembles that of northern Queensland, with heavy summer rainfall; farther south winter rainfall becomes more important. Southeastern Australia experiences frontal rainfall in all seasons of the year as well as summer showers. The latter predominate north of the subtropical high-pressure belt, *i.e.*, north of 33° south. However, summer cyclonic activity (type 4 cyclones) also contributes to the thermally produced precipitation. In the extreme south, frontal and orographical precipitation favor a slight winter precipitation maximum.

Since the east coast is nearer to the direct source of *mT* air, this region experiences heavier precipitation than the interior, especially the portion west of the eastern highlands. As a result of the less intense heating and the more stable *mT* air, southeastern Australia has fewer thunderstorms than northern Australia. Nevertheless, thermal and frontal thundershowers are not uncommon; for example, Newcastle (33° south, 153° east) averages 15 a year. Consistent with the relatively low latitude, the average cloudiness is not very high. This region is free of the prolonged periods of low cloudiness that are commonly observed about 50° latitude. During the winter months, radiational cooling produces some fog over the inland areas of Victoria and New South Wales.

Because of the higher latitude, milder temperatures are observed in summer than in northern Australia. The cool winters produce a *C* type of climate for this general region.

d. Southern Coastal Region. With the passage of cold fronts in the rear of type 1 cyclones, the southwest and extreme southern coast of Australia experience slight to moderate winter rainfall. Anticyclonic circulation predominates north of this region, and consequently the cyclonic activity is confined to the immediate coast line. A further factor contributing to the absence of winter rainfall over the interior is the dryness of the continental air masses. As shown by the stations in Table 78, the summer rainfall is also very slight. The summer dryness arises from the prevailing easterly winds. The air is somewhat desiccated along the east coast, and consequently the air masses over southwestern and southern Australia are relatively dry. Even though the summer heating is intense, the low humidities account for the slight precipitation.

The general decrease of rainfall along the Great Australian Bight, as compared with the southwest or the Spencer Gulf, is accounted for by the sheltered location. The climate of this region closely resembles that of the interior of the continent.

The mean-cloudiness pattern is very similar to the precipitation pattern, *viz.*, a winter cloudiness maximum with a decided summer minimum. Partly as a result of the low summer humidities and partly as a result of the winter air-mass stability, heavy thundershowers are infrequent.

The cool winter temperatures help to produce a Mediterranean climate except north of the Great Australian Bight. Here the low rainfall, in all seasons, results in a steppe type of climate.

e. Interior and West. Along the west coast from 28° south to 18° south and over the vast interior to about 140° east or 145° east, the continent is almost devoid of precipitation. During the winter, the dry air masses connected with the continental anticyclonic circulation favor clear skies and very slight precipitation. Similarly, with the east-west summer circulation the interior and the west coast are again under the dominance of dry *cT* air. In general, the intense summer heating accounts for the principal precipitation—mainly in the form of light convective showers. In the south and west this region borders upon area *d*, which experiences winter rainfall. However, the lower latitude and the dry air masses of the interior again favor only minor precipitation.

During the summer season, except along the immediate coast line, the intense heating results in very high daytime temperatures. On the other hand, radiational cooling accounts for the cool winters.

From the temperature and precipitation regimes, it is apparent that this large portion of Australia has a desert type of climate.

f. Tasmania. Since Tasmania is always located in the prevailing westerlies, it experiences moderate rainfall. This rainfall, in all seasons of the year, results from frontal activity and orographical lifting. Furthermore, convective showers contribute to the summer precipitation. Because of the west-east circulation, western Tasmania, through orographical lifting, experiences more rainfall and is generally cloudier than the eastern portion of the island [for example, compare Cape Sorell (36) and Hobart (37) in Table 78].

Because of the higher latitude, with the more frequent cyclonic activity, the average cloudiness over Tasmania is greater than over the Australian mainland. Even Hobart (37), which is sheltered from the prevailing westerlies, averages 200 hr less sunshine than Melbourne (27) or Sydney (25).

Maritime influences produce relatively mild summer and winter temperatures. Consequently, Tasmania enjoys a moderate marine-forest climate (*Cf*).

g. New Zealand. Except near Auckland, New Zealand is continually within the belt of prevailing westerly winds. Consequently, cyclonic activity is an important contributing factor to the general moderate precipitation. As shown in Sec. 80, cyclones cross New Zealand either from the west or the northwest. With the general southward displacement of cyclonic activity in summer, the extreme north is primarily influenced by winter cyclones, whereas South Island experiences migratory low-pressure systems during all seasons of the year. Besides the frontal precipitation, the high mountain ranges produce local variations of the precipitation regime. For example, the westerly flow across South Island favors heavy rainfall over the west coast and considerably diminished precipitation over the east coast and the southeastern interior.

As a result of the frontal and orographical influences, together with summer convective showers, no section of New Zealand is arid. On an average, the annual rainfall is about 40 in. Except

TABLE 78.—METEOROLOGICAL MEANS AT SELECTED STATIONS
(Temperature in degrees Fahrenheit, precipitation in inches, cloudiness in tenths)

Station	Lat., °S	Long., °E	Temperature				Precipitation				Cloudiness				Elevation, ft
			Jan.	Apr.	July	Oct.	Jan.	Apr.	July	Oct.	Jan.	Apr.	July	Oct.	
Northern Coast															
18. Darwin	12	131	84	84	77	85	15.3	4.1	0.1	2.0	7	4	1	3	100
19. Cairns	17	146	82	78	70	77	15.5	11.9	1.5	1.8	..	..	..	..	20
20. Georgetown	18	144	..	..	..	..	8.3	1.3	0.3	0.7	..	..	..	..	990
East Queensland															
21. Mackay	21	149	80	74	62	73	15.0	7.3	2.3	2.4	..	..	..	..	40
22. Sandy Cape	24	153	80	71	60	70	7.6	5.0	3.2	1.9	..	..	..	..	330
Southeast															
23. Brisbane	27	153	77	70	58	70	6.3	3.7	2.2	2.5	6	4	4	4	140
24. Goondiwindi	29	150	80	68	53	69	3.0	1.5	1.7	1.6	..	..	..	..	720
25. Sydney	34	151	72	65	53	64	3.5	5.5	4.6	2.8	6	5	4	5	140
26. Albury	36	147	..	..	..	..	1.5	2.0	2.8	2.5	..	..	..	..	540
27. Melbourne	38	145	67	59	49	58	1.9	2.2	1.8	2.6	5	6	6	6	110
South															
28. Perth	32	116	74	67	55	61	0.4	1.6	6.5	2.1	3	4	6	5	200
29. Katanning	34	118	71	61	50	57	0.4	1.2	3.0	1.6	..	..	..	..	1020
30. Nullarbor	31	131	..	..	..	..	0.4	0.6	0.9	0.5	..	..	..	..	0
31. Adelaide	35	139	74	64	52	61	0.7	1.7	2.6	1.7	4	5	6	5	140
Interior and West															
32. Peak Hill	26	119	87	74	56	72	0.4	1.0	0.6	0.2	..	..	..	..	1930
33. Alice Spring	24	134	83	73	53	74	1.7	0.7	0.4	0.7	4	3	1	3	910
34. Farina	30	138	82	67	51	69	0.5	0.4	0.4	0.5	..	..	..	..	310
35. Windora	25	143	87	73	56	76	1.5	1.0	0.6	0.6	..	..	..	..	390
Tasmania															
36. Cape Sorell	42	145	..	..	..	..	3.0	4.4	5.7	4.3	..	..	..	..	80
37. Hobart	43	147	62	55	46	54	1.8	1.9	2.1	2.2	6	6	6	6	160
New Zealand															
38. Auckland	37	175	66	60	52	57	2.6	3.5	5.0	3.6	6	6	6	6	150
39. New Plymouth	39	174	63	58	50	55	4.4	4.8	6.7	5.9	5	5	6	6	60
40. Taihape	40	176	64	57	48	55	3.4	3.0	3.1	4.0	..	..	..	..	2160
41. Napier	39	177	65	58	49	57	2.9	2.8	3.6	2.2	4	4	5	4	10
42. Wellington	41	175	62	57	48	54	2.6	3.2	4.5	3.3	6	6	6	6	10
43. Hokitika	43	171	59	54	45	52	9.5	9.5	8.6	11.6	6	6	6	7	10
44. Christchurch	44	173	61	53	43	53	2.2	2.0	2.7	1.7	6	6	6	6	30
45. Ophir	45	170	63	52	38	52	2.0	1.5	0.7	1.4	..	..	..	..	980
46. Dunedin	46	171	59	52	43	52	3.3	2.9	3.1	3.2	6	6	5	6	240
47. Invercargill	46	168	57	51	42	51	3.8	4.3	3.1	4.4	..	..	...	..	10

over Auckland and the west coast, thunderstorms are infrequent; for example, Wellington (42, Table 78) averages 6 days a year with thunder as compared with 15 days at Auckland (38). These convective showers may be caused either by diurnal heating or by frontal lifting.

The mean cloudiness resembles that of Tasmania and, as shown by Table 78, is nowhere so high as that of western Europe. Even though the country enjoys moderate precipitation, there are no prolonged periods of stagnant cyclonic activity. Instead, the migratory pressure systems favor short periods of rain followed by relatively clear periods. Consistent with the heavy rainfall, the west coast of South Island has the highest mean cloudiness.

Because of the maritime influence New Zealand has mild summer and winter temperatures. Because of the temperature and precipitation regimes, the entire country has a moderate marine-forest climate.

CHAPTER XV

ARCTIC AND ANTARCTIC

In this chapter a brief discussion will be given of the polar land and sea areas beyond the Arctic and Antarctic circles. Unfortunately, as a result of incomplete meteorological data, it is possible to describe only the broad-scale features of the climate. Except at a few stations, the mean values of the various elements are based on short periods of observation.

As a result of the high latitude the thermal heating, even during the long summer days, is slight, whereas the winter radiational cooling is very pronounced. Consequently, throughout the year, the Arctic and Antarctic experience cold temperatures, typical of a glacial climate. The principal climatic variations arise from the irregular topography and the difference between water and land. The high land masses of Antarctica and Greenland favor lower surface temperatures than over the neighboring low-lying regions, such as the Arctic Ocean. Hence, the South Pole is considerably colder than the North Pole. Likewise, even though Greenland is principally south of 80° north, northern Greenland is colder than the North Pole.

The outstanding feature of the precipitation regime is the relatively light annual rainfall. Besides the predominance of anticyclonic conditions, the slight precipitation is caused by the low specific humidities of all cold air masses.

82. Physical Features. The main topographical features may be summarized briefly as follows:

1. The entire continent of Antarctica consists of a very high plateau with numerous mountain ranges. Many of the high ridges and isolated peaks exceed 10,000 feet in elevation. The two principal indentations are the Ross and Weddell seas. Because of the elevation, Antarctica is a source region for very cold *cP* air and is much colder than the Arctic. Owing to the irregular topography, strong drainage winds are common and may account for the blizzard conditions that are frequently

observed. For example, the strong winds experienced at the base of the plateau in the Ross Sea region may be essentially katabatic.

2. Greenland consists almost entirely of a high plateau with occasional peaks extending above 10,000 ft. As in the case of Antarctica, the elevation favors the presence of cold, dry air masses. Again, strong downslope winds are observed along the periphery of the continent similar to those of the southern continent. The high plateau also prohibits the passage of pronounced frontal systems across Greenland; consequently, frontal precipitation is slight over the interior.

3. North of Canada there exists a series of islands from Banks Island in the west to Baffin Island in the east. Unlike Greenland, these islands rarely extend above 800 ft.

4. Except for isolated islands, the remainder of the Arctic comprises the Arctic Ocean. During the winter months the continuous ice cover favors the formation of cold *cP* air masses. However, during summer the breaking up of the ice results in relatively mild temperatures. Of more importance to the general weather is the presence of a moisture source region. Consequently, the summer air masses are moist as compared with the dry winter air masses.

83. Mean Distribution of Pressure and Wind. In view of the paucity of meteorological data, only very general statements may be made regarding the mean circulation.

a. January. The principal features of the January pattern may be summarized in the following manner:

1. As a result of the asymmetrical distribution of the vast land masses, the North Pole definitely does not coincide with the mean position of the polar anticyclone. Radiational cooling over North America and Siberia contributes to the production of a high-pressure belt across the Pacific side of the polar sea, while the Atlantic side is under the influence of the low-pressure trough that extends northeastward from the Icelandic Low. Consequently, the region around the North Pole is in a zone of prevailing easterlies, whereas a calm area exists north of Bering Strait. As shown by the "Fram" (1) and "Maud" (2) observations in Table 79 the easterly winds, which generally predominate in high latitudes, are frequently interrupted by winds from other directions.

2. Strong continental cooling produces high pressure over Greenland. This anticyclonic wedge separates the westward and eastward extensions of the Icelandic low-pressure trough. However, as a result of the topography, the surface coastal winds are frequently quite different from the geostrophic winds; for example, strong downslope motion is commonly observed. Because of the elevation, the air motion above the Greenland plateau is probably controlled by the mean pressure distribution at about 3 km.

3. Since they lie between the high-pressure ridge to the west and the Icelandic Low to the southeast, the islands north of Canada, in general, experience northwesterly winds.

4. Spitsbergen and Fridtjof Nansen Land (Franz Josef Land) are in the zone of the northeasterly winds on the northern edge of the Icelandic Low.

5. Unlike the North Pole, the South Pole approximately coincides with the center of the polar anticyclone, which persists over the very cold Antarctic Continent. However, the indentation of the Ross Sea causes a displacement of the high pressure toward the Indian Ocean side of the pole. Similarly, the thermal influence of the Weddell Sea is to produce a trough between the northern extremities of the anticyclonic ridges. Because of the high pressure to the south and a low-pressure trough about 65° south, the continent comes within the regime of surface easterlies. However, like Greenland, topographical factors cause decided local variations in wind force and direction. Furthermore, above the friction layer of the continental plateau the air flow is probably controlled by the mean circulation at about 3 km. The stations in Table 79 confirm the general conclusions regarding the low-level stations; for example, the easterly regime at the continental stations may be compared with the prevailing westerlies of the South Orkney Islands (9). The latter station lies north of the circumpolar cyclone.

b. July. The mean-pressure chart for July shows that the north polar anticyclone is weakly developed. Furthermore, the Icelandic and Aleutian cyclones are less pronounced in summer than in winter. As a result of the weak pressure gradients, light and variable winds prevail within the north polar regions. On the other hand, intense radiational cooling during the winter months favors very high sea-level pressures about the South

Pole. Again, the deep cyclones about the Weddell and Ross seas cause an asymmetrical distribution of the anticyclone about the South Pole. The coastal stations in Table 79 demonstrate the high frequencies of east winds. However, as in summer, local topographical features cause many variations from the sea-level geostrophic winds; for example, strong southerly drainage winds are quite common. Furthermore, over the high antarctic plateau, the air flow parallels the upper-level pressure pattern rather than the sea-level isobars.

TABLE 79.—PERCENTAGE FREQUENCY OF SURFACE-WIND DIRECTION

Station	Lat.	Long.	January								July							
			N	NE	E	SE	S	SW	W	NW	N	NE	E	SE	S	SW	W	NW
Northern Hemisphere																		
1. "Fram".......	83°N	90°E	8	14	16	*28*	18	7	6	3	7	6	7	10	12	*26*	23	9
2. "Maud".......	75°N	164°E	7	21	*28*	10	5	15	9	5	13	13	12	12	10	10	14	*16*
3. Banks-Strait...	74°N	115°W	22	7	8	5	6	*23*	8	21	20	14	3	6	9	13	13	*22*
4. Barrow-Strait..	74°N	92°W	16	5	9	17	7	4	5	*37*	9	8	13	18	9	9	14	*20*
Southern Hemisphere																		
5. Gauss Station..	66°S	90°E	0	4	*55*	18	6	2	12	3	0	0	*79*	6	1	2	8	4
6. Adélie Land....	67°S	142°E	2	0	2	*86*	6	1	2	1	0	0	0	16	*84*	0	0	0
7. Framheim......	79°S	164°W	1	4	*42*	11	14	24	4	0	0	3	*49*	10	14	19	4	1
8. Snow Hill......	64°S	57°W	8	28	5	2	18	*30*	8	1	4	10	4	1	21	*55*	5	0
9. South Orkney..	61°S	45°W	7	5	6	10	16	20	*25*	11	6	3	2	4	12	27	*30*	16

c. Upper-air Circulation. Since the polar anticyclones are extremely cold, the pressure distribution changes from high pressure at the surface to low pressure aloft. Consequently, the prevailing easterlies of high latitudes reverse in direction and become west winds aloft. The latter increase in velocity with increasing elevation. In the northern polar regions, the level at which there is a reversal of wind direction depends on the location of the region with respect to the upper-level Icelandic and Aleutian cyclones. For example, in January, north of the upper-level cyclones centered over northern Hudson Bay and northeastern Siberia east winds prevail to at least 4 km, whereas general westerly winds are observed over the remainder of the Arctic regions.

Even though the south polar sea-level anticyclone is very pronounced, the strong north-south temperature gradient probably results in a polar cyclone at low altitudes. The east winds gradually become westerly aloft.

84. Air Masses. Polar continental and polar maritime air masses predominate in high latitudes. Their principal properties will be discussed briefly.

a. Polar Continental Air (*cP*). During the arctic winters, *cP* air is characterized by cold surface temperatures, low specific humidities, and, in regions of anticyclonic flow, stable lapse rates aloft. Unlike the *cP* air of Siberia and northern Canada, the ground inversion is frequently absent over the frozen polar sea. This arises partly from the conduction of heat through the ice and partly from the moderate wind velocities, which produce strong mixing in the lowest layer of the atmosphere. A subsidence inversion is generally present over this shallow surface layer. In the rear of the deep upper-level cyclones, steep lapse rates prevail, and therefore the coldest temperatures aloft occur across northeastern Siberia and Canada. Over Greenland, *cP* air masses resemble those of America and Asia.

Over the land areas continental heating results in a summer air mass with relatively steep lapse rates in the lower layers. The partial breaking up of the arctic ice provides a moisture source region, and consequently the air over the polar sea is considerably modified by the underlying surface. The air mass is therefore classified as a maritime air mass. It differs from the air over the continental regions mainly because of the increased low-level humidities and the stable lapse rate near the surface. The latter is due to the advection of mild air over the melting ice, whose temperature is near the freezing point.

The air masses of the continental Antarctic resemble the *cP* air of northeastern Asia. The principal differences are the high frequency of steep lapse rates near the stirred surface layer and the very cold temperatures at all levels over the southern continent. During summer, antarctic *cP* air is still very cold and, except for a gradual rise in temperature, appears to resemble the winter air mass. Throughout the year, low specific humidities are characteristic of the southern *cP* air masses.

b. Polar Maritime Air (*mP*). The *mP* air masses that invade high latitudes have usually originated in the intense low-pressure regions. In winter, as a result of the strong cyclonic circulation, the lapse rate is practically moist adiabatic. Furthermore, in contrast to the clear skies of *cP* air, overcast skies are commonly observed within *mP* air. During the winter, this air mass affects the surface weather conditions of southern Greenland, Spitsbergen, and the Bering Strait region and supplies the upper-level moisture for the occasional occluded cyclones that invade the Arctic. With the approach of summer, *mP* air masses are observed farther north because of the melting of the arctic ice. At the same time, *mP* air becomes more stable through the decreased cyclonic activity and the stabilizing effect of cold water temperatures. While towering instability clouds are common in winter near the deep cyclones, the summer hydrometeors are principally surface fog and stratus and stratocumulus clouds. When *mP* air invades the relatively warm lands of high latitudes, flat cumulus clouds arise from the mild convection.

Because of the high topography, *mP* air masses affect only the periphery of Antarctica, where the cyclonic circulation is reasonably intense in summer and winter. Consequently, the summer air mass probably lacks the stability of the Northern Hemisphere air mass. Therefore, the lapse rate is likely to be approximately moist adiabatic throughout the year. Besides increasing the temperature and cloudiness of the coastal area, *mP* air must also supply the moisture for the light precipitation of the interior plateau.

85. Fronts, Cyclones, and Anticyclones. Incomplete synoptic information prohibits a detailed discussion of the cyclones and anticyclones that traverse the Arctic and Antarctic.

a. Arctic and Greenland. During the winter half year, numerous cyclones move toward and away from the semipermanent Aleutian and Icelandic lows. Cyclones that develop over the western Atlantic usually pass southeast of Greenland and frequently deepen in the vicinity of the arctic front, west of Iceland. In most cases, these lows fill as they move eastward and rarely move north of 75° north, because of the tendency toward anticyclonic conditions over the cold ice. This cyclone type results in frontal and air-mass precipitation over southern and southeastern Greenland and the southern Arctic. Sometimes the cyclones from the American continent move west of

Greenland and gradually fill as they traverse the Baffin Bay region. They produce appreciable precipitation along the south and west shores of Greenland. Occluded fronts apparently cross the Greenland plateau; but, because of the upslope motion to the west, they produce very little frontal cloudiness and precipitation. Furthermore, through the downslope heating, eastward-moving fronts are associated with relatively clear skies over eastern Greenland.

Cyclonic activity in the Aleutian Islands sometimes causes cyclones to move eastward along the arctic front. Apparently these lows move across the col between the Siberian and American anticyclones. Since there is no immediate moisture source region, the cyclones that traverse the Arctic produce only slight precipitation.

In summer, as in lower latitudes, cyclonic activity is less intense. However, over the Arctic the disturbances appear to be more numerous in summer than in winter. Besides producing slight precipitation, these cyclones also affect the weather of coastal regions, through causing a change in wind direction. With the high frequency of fog over the ocean, onshore winds produce fog over the land, while offshore winds are accompanied by good visibility. For example, the fiords of southern Greenland are fog bound when the air flow is toward the continent.

The principal anticyclonic movements conform with the deep cyclonic activity. For example, fresh outbreaks of cold air with Greenland anticyclones occur in the rear of deep eastward-moving Icelandic cyclones. Likewise, highs move southward in the rear of Aleutian cyclones.

b. Antarctic. Throughout the year the cyclones of middle latitudes move into the region of the polar cyclone. This cyclone is located about 60° to 65° south in summer and slightly farther north in winter. Unlike the two in the Northern Hemisphere, the polar cyclone in the Southern Hemisphere is appreciably deeper in summer than in winter. However, the cyclones rarely move south, except toward the Weddell Sea and Ross Sea indentations, and therefore cyclonic precipitation is slight over the high plateau. As in the Northern Hemisphere, cold *cP* highs move toward lower latitudes in the rear of the deep cyclones that are centered about 60° south.

86. Climatic Regions. Almost the entire region under discussion experiences the tundra or frost type of climate. Hence, according to the broad-scale features of Köppen's classification, there are only minor climatic variations. However, these variations are worthy of attention, and a brief description of the climate will be presented. In connection with Table 80, it should be borne in mind that the meteorological means are based upon scanty data.

a. Polar Sea and the Canadian Archipelago. Precipitation observations are difficult owing to the frequent occurrence of blowing snow. However, it appears that the annual snowfall north of 80° north is about 4 in. Farther south, with the more frequent cyclonic activity and higher specific humidities, the annual precipitation increases considerably; for example, see Cumberland Sound (13) in Table 80. Because of the anticyclonic control, the winters are very dry beyond 80° north, and the major part of the precipitation is due to weak summer cyclonic activity. Farther south, the deep cyclones produce a winter maximum, east of Greenland, as shown by Green Harbor (14). However, north of Alaska, there is a summer precipitation maximum.

The annual cloudiness variation is very marked. Whereas the winter half year experiences relatively clear skies, the summers are very cloudy, principally because of the stratus clouds that form over the cold broken ice of the polar sea. On the other hand, except with onshore winds, moderately clear skies are observed over the Canadian archipelago in summer. The air-mass cloudiness of the Arctic is considerably more important than occasional frontal clouds. Owing to the cold underlying surface (melting ice), fog is very common over the Arctic during the summer months. In summer, it is probably one of the foggiest regions of the world.

As shown by the data in Table 80, there is a moderately large annual temperature variation. The winter temperatures are higher than in Siberia, partly because of the heat conduction through the ice and partly because of the stirred surface conditions, which prevent large temperature inversions. On the other hand, the summer temperatures, except over land areas, remain close to the freezing point because of the melting ice. The heating over the land results in appreciably higher summer temperatures.

b. Greenland. Since the precipitation is principally associated with the cyclones from eastern America, the heaviest rainfall occurs over southern Greenland and decreases rapidly northward. Because some cyclones move along the west coast and others along the east coast, there is no great difference between the total precipitation of each coast. Like the Arctic, northern Greenland has a summer precipitation maximum, while southern Greenland has an autumn and early winter maximum because of the intense cyclonic activity. The cloudiness regime parallels the precipitation distribution except for the increased summer cloudiness, which is due to stratus clouds. During the summer months, fog is very frequent over the cold water; with a favorable onshore wind, it may move up the coastal fiords.

Over the high Greenland plateau the precipitation and cloudiness are slight owing to the dryness of the *cP* air. Fogs are not uncommon. Unlike the advection fogs of the coastal areas, they are formed through radiational cooling.

The data in Table 80 indicate that there is a consistent temperature decrease with increasing latitude. The low winter temperatures along the coast line, north of 65° north, are made possible by the freezing of the adjacent water areas. In summer, the cold coastal currents aid in maintaining low mean temperatures. Over the high plateau very cold weather persists throughout the year.

One striking feature of the coastal areas are the strong katabatic winds that blow through the fiords on both coasts. In the rear of a cyclone these winds are frequently very strong along the east coast. This tendency toward drainage winds partly accounts for the moderate cloudiness of the coastal regions. Whereas fog and summer stratus may exist over the sea, the interior of the fiord enjoys clear skies as a result of the downslope motion. Since the air originates over the very cold plateau, adiabatic heating does not produce very high foehn wind temperatures.

c. Antarctic. In the absence of reliable statistics only general remarks may be made regarding the precipitation of the Antarctic. As a result of the shift of the maximum cyclonic activity to higher latitudes, more precipitation would be expected in summer than in winter. Furthermore, in Table 80, the precipitation decrease from Point Charcot (28) and Laurie Island (29)

TABLE 80.—METEOROLOGICAL MEANS AT SELECTED STATIONS
(Temperature in degrees Fahrenheit, precipitation in inches, cloudiness in tenths)

Station	Lat.	Long.	Temperature				Precipitation				Cloudiness				Elevation, ft
			Jan.	Apr.	July	Oct.	Jan.	Apr.	July	Oct.	Jan.	Apr.	July	Oct.	
Arctic excluding Greenland															
10. "Maud"	75°N	164°E	−26	−7	32	7	0.9	0.1	0.7	0.2	4	5	9	8	
11. Lady Franklin Bay	82°N	63°W	−34	−11	37	−9	0.4	0.2	0.7	0.2	3	4	7	4	
12. Simpson-Strasse	69°N	97°W	−29	−3	44	10	0.2	0.2	0.4	0.4	3	4	6	7	
13. Cumberland Sound	67°N	67°W	−20	8	42	21	0.3	0.5	3.4	0.6	5	6	8	7	
14. Green Harbor	78°N	14°E	3	8	42	22	1.4	0.9	0.6	1.1	6	6	7	7	10
15. Lake Harbor	63°N	70°W	−15	8	45	26	1.1	1.5	2.6	1.5	..	..	..	..	70
16. Ponds Inlet	75°N	78°W	−30	−8	43	18									
Greenland															
17. North Star Bay	77°N	69°W	−21	1	40	14	0.1	0.1	0.7	0.1	..	..	..	..	20
18. Jacobshaven	69°N	51°W	0	15	46	25	0.4	0.5	1.3	0.9	4	4	5	5	100
19. Godthaab	64°N	52°W	14	25	44	31	1.4	1.2	2.2	2.5	7	7	7	7	70
20. Ivigtut	61°N	48°W	19	31	50	34	3.3	2.5	3.1	5.7	6	6	6	6	100
21. Angmagssalik	66°N	38°W	18	25	45	30	3.3	2.4	1.9	5.7	7	6	6	7	100
22. Scoresby Sound	70°N	22°W	1	10	40	19	1.8	1.5	0.6	1.1	6	7	5	5	60
23. Mygg-Bukta	74°N	22°W	−5	2	39	15	0.2	0.2	0.5	0.2	4	4	6	5	10
Antarctic															
24. Gauss Station	66°S	90°E	30	4	−1	9	..	..	..	..	7	6	8	7	
25. Adélie Land	67°S	143°E	30	2	−3	5	..	..	..	..	7	7	4	5	20
26. Victoria Land	71°S	170°E	32	9	−12	−1	..	..	..	..	7	8	6	6	20
27. McMurdo Sound	78°S	166°E	24	−9	−15	−2	..	..	..	..	6	6	5	7	
28. Point Charcot	65°S	64°W	33	23	−3	19	1.8	0.9	0.4	0.9	9	7	6	7	30
29. Laurie Island	61°S	45°W	32	26	12	24	1.3	1.7	1.3	1.1	9	9	8	9	20
30. Deutschland-Trift	69°S	38°W	28	3	−15	14	0.6	0.4	0.3	0.1					

to Deutschland-Trift (30) suggests that the annual snowfall south of 80° south must be very slight. This is also indicated by the very strong anticyclonic circulation. The scanty data indicate about 15 in. of precipitation at 65° south, decreasing to less than 4 in. south of 75° south. The coastal stations in Table 80 show a relatively high degree of cloudiness. This cloudiness arises from their oceanic exposure; it is to be noted that the maximum cloudiness corresponds to the season of strongest cyclonic activity. On the other hand, the sheltered station of McMurdo Sound (27) shows considerably diminished cloudiness. In view of this trend as well as the anticyclonic control, it is probable that the skies of the interior are relatively cloudless.

Because of the elevation, Antarctica is very cold; in the interior the warmest month probably has a mean temperature well below 32°F. Even at coastal stations very low winter temperatures are common. They arise partly from the offshore winds and partly from the freezing of the ocean. Since the ice only partly breaks in summer, the coastal temperatures are also quite low during this season.

Like Greenland, the plateau experiences severe blizzards, when the visibility is reduced to almost zero by blowing snow. Furthermore, the coastal regions are subject to strong katabatic winds, which reach gale force on occasion, like the similar down-slope winds of eastern Greenland.

It is evident, therefore, that practically all Antarctica, an area considerably larger than Australia, experiences a true frost climate.

CHAPTER XVI

OCEANIC REGIONS

Since the factors that control the weather are similar from one ocean to another, the climate of the oceanic regions will be discussed in one chapter. At the same time, the broad-scale climatic features of the numerous island groups will be included. In this respect it should be remembered that local factors cause differences in the weather between the islands and neighboring water areas. For example, daytime heating produces isolated convective showers over tropical islands.

Except for topographical influences, the factors which control the climate of the oceans are identical with those which influence the continental climates. Consequently, large longitudinal and latitudinal variations may be expected in the climate of the oceanic regions, as discussed in the first part of the book.

87. Mean Distribution of Pressure and Wind. *a. Sea Level.* The predominant features of the mean sea-level circulation may be briefly summarized as follows:

1. The semipermanent anticyclones of low latitudes vary in position and intensity from summer to winter, as illustrated in Table 81. The values are based on more detailed pressure charts than those reproduced in Plates IV and V. It will be noted that there is a distinct difference between the two hemispheres; in the Northern Hemisphere the central pressures are higher in summer, while in the Southern Hemisphere the higher pressures occur during winter. However, in both hemispheres the anticyclones are nearer the equator in winter than in summer. The strong monsoon circulation over central Asia accounts for the absence of a semipermanent high-pressure center over the North Indian Ocean. Finally, it is to be noted that the strongest anticyclonic circulation occurs over the eastern portions of the oceans.

These pressure systems have a profound influence on the climate from the equator to 40°. For example, the circulation controls the ocean currents; the strong subsidence over the eastern half of the oceans produces clear skies, while the absence of subsidence in the doldrum belt favors convective activity; the asymmetry of the anticyclones favors convectively unstable air over the western oceans and convectively stable air over the eastern oceans (see page 70).

TABLE 81.—MEAN STATE OF SUBTROPICAL ANTICYCLONES
(Latitude of center and central pressure)

Ocean	January		July	
	Lat.	Pressure, mb	Lat.	Pressure, mb
North Atlantic	34°N	1022	36°N	1025
North Pacific	28°N	1022	39°N	1027
North Indian	None		None	
South Atlantic	28°S	1021	23°S	1024
South Pacific	32°S	1021	26°S	1024
South Indian	36°S	1021	30°S	1026

2. At about 60° latitude are located the quasi-permanent cyclones whose seasonal variations have a distinct effect upon the climate of the oceans north of the subtropical latitudes. The variations are summarized in Table 82. In contrast to the Aleutian and Icelandic lows, it is to be noted that the polar cyclone of the Southern Hemisphere is deeper in summer than winter. Over the Southern Hemisphere oceans, as shown by the average pressure difference between the subtropical anticyclone and the polar cyclone, the circulation in mid-latitudes is slightly stronger in summer than in winter.

Since numerous cyclones traverse the regions surrounding the semipermanent lows, these are areas of frequent cloudiness, precipitation, and strong winds. In view of the intensity of the polar cyclones, the Northern Hemisphere experiences stronger cyclonic activity in winter than in summer. On the other hand, in the high latitudes of the Southern Hemisphere there is probably little seasonal variation.

3. Between the subtropical anticyclones, the equatorial low-pressure belt is generally a region of variable winds. However, the wind regime is different from one section of the world to

another. For example, true doldrum conditions, with light or calm winds, exist over the western Pacific, whereas steady winds are observed across the eastern Pacific. Consequently, it is incorrect to consider that a continuous belt of calms encircles the globe.

TABLE 82.—MEAN STATE OF POLAR CYCLONES
(Δp represents the average pressure difference between subtropical anticyclone and polar cyclone per degree of latitude)

Ocean	January			July		
	Lat.	Pressure, mb	Δp, mb/deg lat.	Lat.	Pressure, mb	Δp, mb/deg lat.
North Atlantic	64°N	995	0.9	64°N	1008	0.6
North Pacific	52°N	998	1.0	No definite center		0.6
South Atlantic	63°S	987	1.0	58°S	994	0.9
South Pacific	63°S	987	1.1	59°S	994	0.9
South Indian	62°S	990	1.2	56°S	996	1.2

b. Upper Air. The mean temperature, as shown by Plates I and II, decreases in all seasons of the year from the equator to the pole. From the temperature field and the mean surface-pressure charts it is evident that:

1. The axes of the subtropical anticyclones lean toward the equator during all seasons of the year; for example, whereas the Northern Hemisphere surface anticyclones are located about 30° north in January, the high-pressure centers at 4 km are situated about 18° north. At the same time the axes tilt westward toward the warmer temperatures. Because of this upper-level pressure pattern, it is apparent that the easterly trade winds on the equatorward side of the anticyclones decrease with elevation (neglecting the friction layer) and finally become westerly aloft. Approximately, the depth of the easterly zone varies indirectly with distance from the heat equator.

2. In the Northern Hemisphere, the axes of the deep polar cyclones tilt northwestward toward the cold air so that at 4 km the low-pressure centers are found over Hudson Bay and northeastern Siberia. The Southern Hemisphere polar cyclone tilts toward the South Pole. As a result of the strong temperature

gradient, the general westerly winds of mid-latitudes increase in speed with elevation.

Numerous cyclones in middle and high latitudes produce many day-to-day changes in the wind direction and velocity. These changes are more pronounced near the surface than at higher levels. On the other hand, within the trade-wind zone, the relative absence of migratory pressure systems favors very persistent and steady east winds. Observations from tropical islands do not always verify this persistency; here local factors cause variations from the general easterlies. Furthermore, near large continental areas like South America and South Africa, the easterly flow is interrupted. In the extreme eastern Pacific and Atlantic the low-latitude winds tend to be north or south rather than northeast or southeast.

88. Ocean Currents. The ocean currents are maintained by the stresses that the winds exert on the sea surface. Consequently their general distribution conforms closely with the prevailing wind systems as discussed in Sec. 15. The principal ocean currents may be summarized as follows:

1. Over the Atlantic and Pacific oceans the persistent trade winds produce the North and South Equatorial currents, which flow from east to west. The current is weakly defined in the Indian Ocean because of the marked seasonal wind variations.

2. With the exception of the North Indian Ocean, warm currents flow poleward in the western portion of all oceans. This flow persists to about 40° latitude, under the influence of the wind circulation around the subtropical anticyclones. Beyond this latitude the current flows east-northeast (Northern Hemisphere) or east-southeast (Southern Hemisphere) as a result of the predominant westerly winds on the equatorward side of the semipermanent high-latitude cyclones. Examples of these warm currents are the Gulf Stream system in the North Atlantic, the Kuroshio Current in the North Pacific, the Brazil Current in the South Atlantic, the Agulhas Current in the South Indian Ocean, and the East Australian Current in the South Pacific Ocean. These vast ocean currents have a profound influence on the climate since the properties of the underlying surface largely determine the properties of the various air masses.

From tropical latitudes to about 40°, as a result of the warm water, the surface air temperatures are higher over the western sections of the oceans than over the eastern sections. Likewise, in the Northern Hemisphere, the warm currents result in relatively high air temperatures over the northeast oceanic regions. On the other hand, south of 40° south, there are only minor temperature differences because of the absence of continents in high latitudes that can deflect the warm water southeastward.

3. The equatorward flow on the western edges of the subtropical anticyclones accounts for the cool ocean currents between latitude 40° and the equator, for example, the Canaries (North Atlantic), Benguela (South Atlantic), California (North Pacific), Peru (South Pacific), and West Australian (South Indian) currents. The coldness of the coastal surface water in subtropical latitudes is due partly to the cold ocean currents and partly to the upwelling of cold water near the coast lines, for example, off the western coast lines of North and South Africa and North and South America and, to a considerably lesser extent, Australia. Besides producing relatively cool air temperatures the cold water is a favored region for low stratus clouds. In the South Pacific and South Atlantic oceans this cold water is observed some distance from the coast line as shown by the mean surface temperatures south of the Galápagos Islands.

4. The monsoon winds, north of the equator, produce a complicated system of ocean currents in the North Indian Ocean, Arabian Sea, and Bay of Bengal. With the complete wind reversal from summer to winter, the ocean circulation is variable.

5. In the Northern Hemisphere, the easterly winds on the north side of the Icelandic and Aleutian cyclones produce a westward flow of cold water, the Iceland–East Greenland system and the similar current about the Bering Sea. These currents finally flow southward along the northeastern coasts of America and Asia in the rear of the semipermanent cyclones, as the extremely cold Labrador and Oyashio currents. The presence of this water, to almost 40° north, accounts for the high frequency of fog and low clouds off the northeastern continental coast lines.

6. South of 40° south there is a circumpolar current—West Wind Drift. This is warm water as compared with the melting ice north of Antarctica. The drift moves from west to east as a result of the air circulation south of the semipermanent cyclone.

In so far as this discussion is confined to the major currents, the change from winter to summer is relatively unimportant. Consequently, no description of the seasonal oceanic circulation will be given.

89. Air Masses.—The properties of the various air masses depend principally on the temperature of the underlying surface and on the type of wind circulation.

a. Tropical Maritime Air (mT). At the earth's surface, *mT* air is characteristically warm and moist during all seasons of the year. The principal variations within the air mass depend on the location of the air with respect to the subtropical anticyclones. In the doldrum belt, between the subtropical highs, surface heating together with convergence toward the low-pressure trough produces a warm, unstable air mass. The lapse rate of temperature is practically moist adiabatic. On the western sides of the highs, in both hemispheres, the poleward flow favors warmer *mT* air than over the eastern edges of the anticyclones, where the flow is equatorward. Furthermore, the asymmetry of the anticyclones produces very stable *mTs* air in the east and relatively unstable *mTu* air in the west, especially adjacent to the polar front.

The *mT* air that leaves subtropical latitudes for higher latitudes usually flows poleward in the warm sector of a deep cyclone. The increased wind velocities and convergence generally produce an air mass with a moist-adiabatic lapse rate. Through surface cooling, stratus clouds and the drizzle type of precipitation frequently accompany this poleward flow.

As a result of the larger areas covered by the anticyclones in summer, there is a difference between the summer and winter air masses over the Northern Hemisphere. Increased subsidence produces a more stable type of *mT* air in summer; for example, very strong subsidence inversions occur near the western coasts of Africa and North America. Except for the intensification of the anticyclones off the west coasts in summer, there is little change in the circulation of the Southern Hemisphere from winter to summer. Consequently, only a slight annual variation is to be expected in the air-mass properties of the Southern Hemisphere. However, in both hemispheres, *mT* air near the western shores of the continents is made more stable in summer through cooling over the cool coastal water.

b. Polar Maritime Air (mP). Polar maritime air masses are found within the circulation around the high-latitude cyclones. The temperature of the air mass depends on the sea-surface temperature and the direction of flow. Air that flows equatorward is slightly colder than the sea surface, while air on the eastern side of the cyclone, which flows poleward, is slightly warmer than the sea surface. The stability and moisture distribution vary considerably. Off the east coasts of the continents in the Northern Hemisphere, the short trajectory over the ocean accounts for a low moisture content at high levels. On the other hand, near the west coasts of the continents the long oceanic trajectory produces a deep layer of moist air. During the winter, in the Northern Hemisphere, the *mP* air off the east coasts is decidedly unstable because of the surface heating of *cP* air and the strong cyclonic circulation. In contrast, the *mP* air off the western coasts may be stable or unstable depending on whether it is being cooled from below and under anticyclonic control or whether it is being heated from below as fresh *mP* air in the rear of a deep cyclone. During the summer months the relatively cool ocean and diminished cyclonic activity favor a more stable type of air mass in the Northern Hemisphere. Fog is very common where the warm continental or warmer maritime air flows over cold-water regions.

In the Southern Hemisphere the sole source of *cP* air is Antarctica; consequently, fresh *mP* air or modified *cP* air is restricted to the oceanic area around the polar continent. Hence, the most unstable type of *mP* air is found about the circumpolar cyclone. North of 60° south, the vast expanse of ocean accounts for an air mass with a high moisture content to great elevations. In contrast to conditions in the Northern Hemisphere, the air-mass properties vary only slightly from summer to winter. This uniformity arises from the slight annual variation of cyclonic activity as well as from the absence, south of 45° south, of large continents where cold air masses may be produced in winter and relatively warm air masses may be produced in summer. It is these continental air masses which contribute to the annual variations within the *mP* air of the Northern Hemisphere.

Besides *mT* and *mP* air masses, the oceanic regions near the continents come under the influence of *cP* and *cT* air masses. However, the latter are in the transition stage from a continental to a maritime type. They may be considered as fresh maritime air masses or modified continental masses. Their properties, in any region, may be deduced from the descriptions in the preceding chapters. One important guiding principle is that relatively cold continental air is made unstable, while relatively warm continental air is made stable by flowing over the ocean. In each case, moisture is absorbed from the sea surface.

90. Fronts, Cyclones, and Anticyclones. *a. Frontal Zones.* Plates XVI and XVII show the principal frontal regions. They may be subdivided into three classes, as follows:

1. The intertropical front is located in the low-pressure belt between the large anticyclones of both hemispheres. Actually, the usually negligible temperature contrast and the smallness of the Coriolis force at low latitudes cause this front to differ from those of middle and high latitudes. The convergence toward the low-pressure belt is the prime cause of the heavy precipitation that is common about the intertropical front. The annual oscillation of this zone of convergence depends on the meridional displacement of the subtropical anticyclones as well as on the movement of the thermal low-pressure belt. For example, the strong January heating over Australia and South Africa and the similar July heating over southeastern Asia and North Africa account for the pronounced oscillation of the front across the Indian and western Pacific oceans. In contrast, the slight meridional variation of the front over the eastern Pacific and Atlantic oceans results from the relative permanency of the Southern Hemisphere subtropical anticyclones. In general, the front is everywhere farthest north in July and August and farthest south in January and February.

Convective activity within this zone results in persistent cloudiness and frequent precipitation. Cumuliform clouds predominate and are attended by thunder and lightning.

2. The polar fronts are primarily located off the east coasts of the continents. In the Northern Hemisphere, in winter, they separate *mT* air from *cP* or fresh *mP* air masses. The frontogenetical wind field is the field of deformation between the maritime and continental anticyclones. In the Southern Hemisphere, in summer, the fronts lie between *mT* and *mP* air masses. Because, on daily weather maps, there are usually two anticyclones over the North Pacific in winter, a second

Pacific polar front is frequently located in the mid-Pacific. The frontogenetical temperature field is due to the difference between *mT* air to the east and *mP* air to the west.

In summer, the Northern Hemisphere polar fronts are located much farther north than in winter because of the seasonal variation of the subtropical anticyclones. The diminished frontogenetical wind and temperature fields account for the relatively weak summer fronts. In the Southern Hemisphere the fronts do not differ very considerably from season to season.

As described in previous chapters, the fronts oscillate from day to day. This oscillation is more evident in the South Indian and South Pacific oceans, as a result of the migration of the low-latitude anticyclones. Probably a mean position of the polar front across Australia and New Zealand is less significant than mean locations in other regions.

3. Arctic and antarctic fronts lie in the troughs that extend from the deep cyclones of high latitudes. They separate *mP* air from *cP* or fresh *mP* air. In the Northern Hemisphere the seasonal oscillation is slight. However, the intensity of the front decreases from winter to summer because of the weaker frontogenetical fields. In the Southern Hemisphere the mean position of the antarctic front is doubtful.

When the fronts are sharp, a widespread band of precipitation may be expected on the poleward side. This precipitation is frequently in the form of snow, and consequently the mean frontal zones are also mean zones of poor visibility and low ceilings.

b. Cyclones and Anticyclones. The migratory cyclones that traverse the oceans may be generally subdivided as follows:

1. Tropical Cyclones. Small, intense tropical disturbances develop beyond 5° north and 5° south. They generally form within the convective zone called the "intertropical front." Widespread thunder activity and light winds are commonly observed prior to cyclone formation. From the description of the oscillation of the intertropical front, it is to be expected that these lows will develop in the summer hemisphere, will be rare in the South Atlantic and eastern South Pacific, and will have a different seasonal frequency in the North Indian Ocean from elsewhere. The cyclones of the Bay of Bengal and the Arabian Sea generally occur between the monsoon seasons, when the

intertropical front is located in low latitudes in the Northern Hemisphere. Other regions of tropical cyclones are the Philippine Islands and China Sea area, the Solomon Islands and New Hebrides, the middle of the South Pacific (Society Islands), the region off the west coast of Central and North America, the southwest portion of the North Atlantic (Caribbean Sea), the area about Madagascar, and the region off the northwest coast of Australia. In general, the paths of these disturbances are controlled by the circulation around the large-scale pressure systems, *viz.*, the subtropical anticyclones. Hence, the typical path of a cyclone is first westward in low latitudes and then toward the northeast (Northern Hemisphere) or southeast (Southern Hemisphere) in higher latitudes. In the Arabian Sea and Bay of Bengal, and sometimes in other regions, the indefiniteness of the pressure pattern accounts for rather erratic paths.

The cyclones may vary in intensity from weak, flat pressure troughs on the intertropical front to very intense low-pressure centers. They usually move slowly before recurvature and are almost invariably accompanied by heavy rainfall. After recurving they frequently increase in horizontal extent and move poleward like extratropical cyclones. As mentioned in previous chapters, tropical cyclones may move on land before recurving. Such a path is commonly observed in the Caribbean Sea and China Sea regions.

2. Extratropical Cyclones. The cyclones of middle and high latitudes usually develop as wave disturbances on the polar front. The region of formation varies with the wide oscillation of the front; for example, a cyclone that crosses the North Atlantic may develop over Oklahoma or off the east coast of Florida. However, most cyclones follow the same general path. That is, in the Northern Hemisphere, they move northeastward toward the Aleutian or Icelandic lows and, in the Southern Hemisphere, southeastward toward the circumpolar low. Their path, velocity, and intensity vary within a wide range. In the Northern Hemisphere there is a marked seasonal variation. Summer cyclones usually form farther north, move more slowly, and are shallower than the winter cyclones.

Besides frontal precipitation, widespread cloudiness and precipitation may occur because of convergence within the maritime air masses in the vicinity of a cyclone.

The principal developments on the arctic front are secondary lows, which form in connection with primary cyclones that have moved into the semipermanent area of low pressure. For example, a primary may move west of Greenland, and a secondary cyclone may develop upon the arctic front east of Greenland. The rapid deepening of some cyclones in the region of the arctic front may be caused by the movement of the cyclone into this region of marked temperature contrast. This phenomenon is not uncommon southeast of Greenland.

Associated with the various cyclones already described are the migratory anticyclones. In contrast to the former, the high-pressure areas are usually regions of relatively fair weather. However, the high moisture content of all maritime air masses produces more air-mass cloudiness over the ocean than over land. Therefore, except in regions of pronounced subsidence and light winds, at least stratocumulus type clouds may be expected.

91. Climatic Regions. A number of island stations are included in Table 83 in order to illustrate the broad-scale climatic features. In this connection, it should be remembered that, especially in low latitudes, local conditions may cause appreciable differences between the weather on land and on sea. For example, since land air masses are more unstable during the day and ocean air masses are more unstable at night, different precipitation and cloudiness regimes may be expected. The ocean temperature means in Table 83 represent the average of shipboard observations.

The coldest and warmest ocean months are February and August. However, in order to make comparisons with data in previous chapters, it has been decided to give the meteorological means for January, April, July, and October. In most cases, therefore, Table 83 does not indicate the yearly amplitude of temperature.

a. Equatorial Regions. In general, the equatorial regions experience moderate to heavy precipitation during every month of the year. However, since the rainfall is due to convergence within the doldrum zone, there is a seasonal variation that depends primarily on the meridional displacement of the intertropical front. In the vicinity of the intertropical frontal zone, heavy convective showers are quite common. Across the Atlantic and eastern Pacific oceans the front is usually north of the equator throughout the year; consequently, as shown by Plate IX, the belt of heavy precipitation lies entirely north of the equator. Over the western Pacific, west of 180°, the doldrum belt oscillates considerably. Therefore, heavy precipitation is observed over a wide area, with the maximum at any place occurring when, in the mean, the doldrum zone is over that place. For example, Saipan (39), Guam (40), and Yap (41) in Table 83 experience very heavy rainfall from June to September, while Apia (60) and Vila (59) have a maximum between December and March. Because stations near the equator are never very distant from the intertropical front, they have no real wet and dry seasons but rather heavy rain throughout the year; for example, in Table 83 compare Palau (42), Apaiang (43), Fanning Island (47), and Nauru (57) with stations farther north or south. Owing to the extreme oscillation of the doldrum belt, the Indian Ocean generally experiences wet and dry seasons. For example, the Seychelles (79), only 5° south, have a moderately dry season during the southwest monsoon of the Northern Hemisphere.

The annual rainfall chart (Plate IX) shows that there are longitudinal variations within the equatorial zone. These variations arise partly from different stability and humidity conditions within the tropical maritime air masses and partly from the extent of the latitudinal oscillation of the doldrum zone.

The mean cloudiness is not very high in spite of the heavy rain, for this is frequently in the form of heavy showers, which may be followed by a temporary clearing. In general, the land areas have a cloudiness maximum in the afternoon, while the ocean areas show an early-morning maximum. Tropical cyclones, waterspouts, and frequent thundershowers provide the most violent weather phenomena.

The temperature is uniformly high, and therefore the equatorial regions experience an *A* type of climate.

b. Tropical and Subtropical Regions. The rainfall distribution from equatorial regions to about 30° or 35° latitude depends primarily on the seasonal variation of the subtropical anticyclones. Subsidence within the center and the eastern sector of the high produces relatively clear skies. The increased air-mass instability around the periphery and the western sector of the high, the tropical and extratropical cyclones along the western sector of the high, and the activity along the equatorward edge of the polar front together produce appreciable rainfall. Consequently, the rainfall regime shows the following characteristics:

1. There are wet and dry seasons on the equatorward side of the anticyclone, as shown by the North Atlantic islands between 10° north and 18° north and at St. Helena (27), Suva (64), Niue (65), Papeete (67), Rodriguez (80), and Mauritius (81 in Table 83). The wet summer season (February or August) corresponds to the poleward advance of the doldrum belt. The dry season occurs in the winter hemisphere. At this time, relatively stable air masses predominate.

2. An area of light precipitation coincides with the eastern-central portion of the subtropical high. The driest zone is generally on the equatorward side where no rainfall results from frontal activity, for example, at Saint Vincent (2) in the North Atlantic, in the regions off the coast of Peru in the South Pacific, off South-West Africa in the South Atlantic, and off western Australia in the Indian Ocean.

3. An area of light to moderate precipitation is found on the poleward side of area 2. This precipitation is due to the frontal activity at low temperate latitudes. Especially in the Northern Hemisphere, the rainfall occurs mainly during the winter half year, for example, at Santa Cruz (3), Funchal (4), and Ponta Delgada (5).

4. Over the western parts of the oceans, cyclones, frontal action, and convective showers produce a large area of moderate to heavy precipitation, for example, in the West Indies and the islands of the western Pacific. The maximum generally occurs during the late summer and autumn. This is the period of frequent tropical cyclones and unstable mT air masses.

The cloudiness regime closely follows the precipitation regime, except off the west coasts of the continents. Along the west coasts, through the upwelling of cold water, low stratus clouds are very common, for example, the "California stratus." These are regions of slight precipitation yet high air-mass cloudiness. During the Southern Hemisphere winter this cloudiness extends westward off Peru and South Africa because of the cold water west of the continents.

Owing to the high temperatures, the A type of climate prevails to about 25° or 30° latitude, except in the following areas:

1. Off the coast of China the cold water and cold winters produce a *C* type of climate to 20° north.

2. The regions off California, Peru, western Africa, and western Australia receive little rainfall and hence experience a *BS* or *BW* climate.

While the latitudinal temperature control is very pronounced, there is also a striking difference in temperature between the eastern and western sides of the ocean when the subtropical anticyclone is well developed.

c. Temperate and Subpolar Regions. Beyond 30°, precipitation occurs mainly in connection with extratropical cyclones and frontal lifting. Consequently, the zone of heaviest precipitation coincides approximately with the mean position of the polar and arctic fronts and with the mean cyclone path. Over the North Atlantic this zone extends from the southeastern United States to the Icelandic Low. Cyclonic activity west and north of the British Isles also produces moderate precipitation. A similar southwest to northeast orientation of the precipitation zone exists over the North Pacific. However, two general zones may be detected, corresponding to the two mean positions of the double Pacific polar front. The frontal lifting and cyclones of the Australia and New Zealand sector as well as similar activity northwestward from Cape Horn produce two zones of moderate precipitation over the South Pacific. Because Africa and Australia are mainly north of 30° south, the cyclones of middle latitudes produce a more zonal distribution of precipitation over the Indian Ocean, with a maximum about 45° south to 55° south. Nevertheless, in lower latitudes (30° south to 40° south), the frontal activity off southeastern South America and South Africa causes more rainfall over the western portions than over the eastern portions of the oceans.

Beyond 55° the generally colder temperatures favor a poleward decrease in precipitation. Especially in late winter, when the ice cover extends to lower latitudes, cyclones are less frequent and therefore precipitation is light.

The seasonal distribution of precipitation is illustrated by the stations in Table 83. It is evident that, in temperate latitudes, the maximum generally occurs in autumn or early winter, for example, at Thorshavn (7), Vestmannaeyjar (8), South Georgia (29), Dutch Harbor (37), and Chatham Islands (71). Toward the South Pole, the southward movement of the circumpolar low produces a summer precipitation maximum over the extreme southern oceans.

The mean cloudiness corresponds to the annual precipitation distribution, except over those areas where advection fog is common, *viz.*, off the east coasts of North America and Asia from 35° north to the polar ice. The advection of warm air over the cold ocean currents readily produces fog. Even though fog occurs throughout the year, summer is the season of maximum frequency. In this season, the wind circulation is more favorable, and fog may arise from the cooling of either maritime or continental air masses. Frontal fog also occurs over the oceans but is quite infrequent as compared with the air-mass type.

Besides a poleward decrease in temperature, there is a longitudinal variation that arises from the ocean currents. In the North and South Atlantic and North Pacific, the ocean currents in high latitudes produce warm temperatures over the eastern oceanic regions. As a result, the northeast portions have a *C* type of climate, whereas the extreme northwest portions experience a *D* type. In the latter areas the cold air flow off the continents produces unusually cold winter temperatures. Beyond 40° south, longitudinal variations of temperature over the southern oceans are slight because of the tendency toward zonal ocean currents. Hence, in the Southern Hemisphere, there is a more regular change of climate with latitude, with a *C* type of climate centered about 40° south, changing to an *ET* type of climate about 50° to 55° south. For example, the warmest months at Macquarie (72), Kerguelen (82), Cape Pembroke (28), and South Georgia (29) have mean temperatures of less than 50°F.

The climate of the ocean areas within the Arctic and Antarctic circles has been discussed in the previous chapter.

TABLE 83.—METEOROLOGICAL MEANS AT SELECTED STATIONS
(Temperature in degrees Fahrenheit, precipitation in inches, cloudiness in tenths)

Station	Lat.	Long.	Temperature				Precipitation				Cloudiness				Elevation, ft
			Jan.	Apr.	July	Oct.	Jan.	Apr.	July	Oct.	Jan.	Apr.	July	Oct.	
North Atlantic															
1. Saint Thomas	0°	7°E	71	72	67	68	7.6	10.2	0.9	20.1	8	8	7	8	2260
2. Saint Vincent	17°N	25°W	71	71	76	79	0.2	0.0	0.3	1.5	4	4	5	5	40
3. Santa Cruz	28°N	16°W	61	65	73	71	2.7	1.0	0.0	1.5	4	4	1	4	130
4. Funchal	33°N	17°W	59	61	70	68	3.3	1.9	0.1	3.9	6	6	5	6	80
5. Ponta Delgada	38°N	26°W	58	59	70	66	2.8	2.0	0.8	3.7	6	6	5	6	70
6. Horta	39°N	29°W	59	60	71	66	4.5	3.0	2.0	4.3	..	..	..	..	210
7. Thorshavn	62°N	7°W	38	41	51	44	6.6	3.7	3.1	6.0	8	7	8	8	30
8. Vestmannaeyjar	63°N	20°W	35	39	51	42	6.0	3.8	2.9	5.5	6	6	6	6	30
9. Grimsey	67°N	18°W	28	30	45	37	1.0	0.7	1.1	2.0	8	8	8	8	70
10. Hamilton	32°N	65°W	64	66	80	75	4.5	4.1	4.5	6.0	7	6	5	6	150
11. Nassau	25°N	77°W	72	75	83	80	1.8	2.1	5.7	6.5	5	5	6	6	10
12. Kingston	18°N	77°W	77	79	82	80	0.7	1.3	1.2	6.3	4	5	5	6	20
13. Grand Turk	21°N	71°W	76	78	82	81	2.6	1.3	1.9	4.4	5	5	5	6	10
14. Santo Domingo	18°N	70°W	76	77	80	80	2.5	3.7	7.1	7.0	4	5	4	4	60
15. San Juan	18°N	66°W	75	77	80	80	4.1	4.4	5.8	5.7	4	4	5	4	100
16. Roseau	15°N	61°W	78	79	81	82	4.8	2.9	10.2	7.6	5	4	6	5	30
17. St. George	13°N	62°W	77	79	79	80	4.4	1.8	9.4	8.1	3	4	4	4	510
18. Trinidad	10°N	61°W	75	78	78	78	2.6	1.9	8.6	6.6	..	..	..	..	130
19. 10°N 25°W	10°N	25°W	78	78	80	82	....	....	....	....	5	5	5	5	
20. 30°N 25°W	30°N	25°W	67	67	75	75	....	....	....	....	6	5	4	5	
21. 50°N 25°W	50°N	25°W	50	52	61	57	....	....	....	....	7	6	7	7	
22. 10°N 50°W	10°N	50°W	79	80	81	82	....	....	....	....	5	5	5	4	
23. 30°N 50°W	30°N	50°W	69	69	79	77	....	....	....	....	5	5	4	5	
24. 50°N 50°W	50°N	50°W	36	40	51	47	....	....	....	....	7	7	7	7	
South Atlantic															
25. Fernando de Noronha	4°S	32°W	78	79	76	77	2.5	9.9	2.5	0.2	5	6	4	4	310
26. Ascension	8°S	14°W	78	81	77	74	....	....	....	....	..	..	..	..	50

27. St. Helena	16°S	6°W	64	66	58	58	1.9	3.6	3.6	1.7	5	6	8	7	1770
28. Cape Pembroke	52°S	58°W	49	44	37	42	2.8	2.4	2.0	1.7	7	7	7	7	70
29. South Georgia	54°S	37°W	40	35	29	34	3.6	5.9	5.5	2.2	7	7	7	7	10
30. Laurie Island	61°S	45°W	32	26	12	24	1.3	1.7	1.3	1.1	9	9	8	9	20
31. 20°S 20°W	20°S	20°W	..	..	..	..	....	....	....	....	5	5	5	5	
32. 40°S 20°W	40°S	20°W	..	..	..	..	....	....	....	....	6	6	6	7	
33. 20°S 0°	20°S	0°	..	..	..	..	....	....	....	....	6	5	6	6	
34. 40°S 0°	40°S	0°	..	..	..	..	....	....	....	....	6	6	6	6	
35. 40°S 40°W	40°S	40°W	..	..	..	..	....	....	....	....	6	6	6	6	
North Pacific															
36. St. Paul	57°N	170°W	24	27	46	39	2.2	1.5	2.8	3.5	..	..	..	..	20
37. Dutch Harbor	54°N	167°W	32	35	51	39	5.5	3.8	2.2	8.4	..	..	..	..	10
38. Bonin	27°N	142°E	62	69	80	77	2.5	1.3	4.6	6.8	..	..	..	..	10
39. Saipan	15°N	146°E	76	78	78	78	4.1	3.7	15.8	14.2	7	6	8	7	680
40. Guam	13°N	145°E	78	80	80	79	2.4	2.2	14.1	12.4	7	7	8	8	60
41. Yap	10°N	138°E	80	81	80	80	7.0	5.2	16.5	11.7	7	7	8	8	110
42. Palau	7°N	134°E	80	81	80	81	15.7	8.0	24.4	15.3	8	7	9	8	100
43. Apaiang	2°N	173°E	82	82	82	83	13.6	11.1	6.8	10.2	7	7	6	6	10
44. Midway	28°N	177°W	65	68	77	75	4.3	4.4	4.1	2.5	..	..	..	..	20
45. Hilo	20°N	155°W	70	71	74	74	13.1	13.1	9.7	11.5	..	..	..	..	40
46. Honolulu	21°N	158°W	70	72	77	76	3.4	2.1	1.0	1.8	4	5	4	5	60
47. Fanning Island	4°N	159°W	80	81	82	82	10.1	13.7	8.4	3.6	6	6	5	5	10
48. 0° 140°W	0°	140°W	78	80	80	78	....	....	....	....	4	4	4	4	
49. 20°N 140°W	20°N	140°W	71	71	73	74	....	....	....	....	5	6	6	6	
50. 40°N 140°W	40°N	140°W	53	53	62	62	....	....	....	....	7	7	7	6	
51. 0° 180°	0°	180°	81	81	81	82	....	....	....	....	4	4	4	4	
52. 20°N 180°	20°N	180°	75	75	80	80	....	....	....	....	4	4	4	4	
53. 40°N 180°	40°N	180°	49	50	63	59	....	....	....	....	6	7	7	6	
54. 0° 160°E	0°	160°E	82	82	83	84	....	....	....	....	4	5	4	5	
55. 20°N 160°E	20°N	160°E	76	77	81	81	....	....	....	....	4	4	4	4	
56. 40°N 160°E	40°N	160°E	42	46	62	61	....	....	....	....	7	7	8	7	
South Pacific															
57. Nauru	0°	167°E	81	82	82	82	11.9	5.6	9.7	5.2	5	4	4	4	20
58. Malden	4°S	155°W	82	82	83	82	....	....	....	....	..	..	..	..	30
59. Vila	18°S	169°E	80	79	71	75	11.8	9.8	2.3	3.9					
60. Apia	14°S	172°W	79	79	77	79	17.0	10.2	3.0	6.6	7	6	4	5	10
61. Pukapuka	11°S	166°W	80	80	80	80	....	....	....	....	6	6	5	4	10

TABLE 83.—METEOROLOGICAL MEANS AT SELECTED STATIONS.—(*Continued*)

Station	Lat.	Long.	Temperature				Precipitation				Cloudiness				Elevation, ft
			Jan.	Apr.	July	Oct.	Jan.	Apr.	July	Oct.	Jan.	Apr.	July	Oct.	
62. Atuana	10°S	139°W	79	80	78	79	....	....	....	....	4	4	5	4	30
63. Nouméa	22°S	166°E	79	76	69	72	3.7	5.2	3.6	2.0	4	4	4	4	30
64. Suva	18°S	178°E	79	78	73	75	11.3	12.8	5.2	8.5	7	7	7	7	20
65. Niue	19°S	170°W	79	77	73	75	10.5	7.5	3.3	4.6	7	6	6	7	70
66. Rarotonga	21°S	159°W	77	75	70	72	9.8	7.3	4.6	5.1	6	6	6	6	20
67. Papeete	18°S	150°W	80	80	76	78	8.3	4.5	1.6	3.8	6	5	4	5	20
68. Norfolk Island	29°S	168°E	72	69	62	65	3.5	4.3	6.3	4.0	6	6	6	6	60
69. Mangareva	23°S	135°W	80	79	71	74	5.2	4.3	4.8	12.1	..	..	..	..	10
70. Easter Island	27°S	109°W	73	70	63	65	5.2	5.1	3.2	2.2	5	6	6	6	100
71. Chatham Islands	44°S	176°W	58	53	46	50	2.3	2.7	3.7	2.2	..	..	..	..	150
72. Macquarie	54°S	159°E	43	41	37	38	5.1	4.5	3.9	4..	..	..	..	..	10
73. 20°S 180°	20°S	180°	..	..	..	..	....	....	....	....	5	5	5	5	
74. 40°S 180°	40°S	180°	..	..	..	..	....	....	....	....	5	6	6	6	
75. 50°S 180°	50°S	180°	..	..	..	..	....	....	....	....	6	7	6	7	
76. 20°S 100°W	20°S	100°W	..	..	..	..	....	....	....	....	6	5	5	6	
77. 40°S 100°W	40°S	100°W	..	..	..	..	....	....	....	....	5	6	6	6	
78. 50°S 100°W	50°S	100°W	..	..	..	..	....	....	....	....	6	7	6	6	
Indian Ocean															
79. Seychelles	5°S	55°E	78	80	77	79	15.3	8.1	2.8	6.2	6	5	6	6	10
80. Rodriguez	20°S	63°E	79	77	70	74	6.4	4.9	2.8	1.3	..	..	..	..	10
81. Mauritius	20°S	58°E	79	75	67	72	7.2	5.8	2.3	1.7	6	6	5	6	180
82. Kerguelen	49°S	70°E	44	39	33	34	3.0	1.5	2.5	2.8	7	6	6	8	50
83. 0° 60°E	0°	60°E	..	..	..	..	....	....	....	....	5	4	5	5	
84. 20°S 60°E	20°S	60°E	..	..	..	..	....	....	....	....	5	5	5	5	
85. 40°S 60°E	40°S	60°E	..	..	..	..	....	....	....	....	5	6	6	6	
86. 0° 100°E	0°	100°E	..	..	..	..	....	....	....	....	6	6	6	6	
87. 20°S 100°E	20°S	100°E	..	..	..	..	....	....	....	....	5	6	6	5	
88. 40°S 100°E	40°S	100°E	..	..	..	..	....	....	....	....	6	6	6	6	

INDEX

Page references for the continents refer only to the chapters discussing these continents, but not to the many cross references in other chapters.

R

S